The Mathematics in Our Hands

Christina M. Krause

The Mathematics in Our Hands

How Gestures Contribute to Constructing Mathematical Knowledge

With a foreword by Angelika Bikner-Ahsbahs and Ferdinando Arzarello

Christina M. Krause
University of Duisburg-Essen
Germany

Dissertation University of Bremen, 2015
Date of thesis defense: April 21st, 2015
Assessors:
Prof. Dr. Angelika Bikner-Ahsbahs (Doctoral supervisor) (University of Bremen, Germany)
Prof. Dr. Ferdinando Arzarello (University of Turin, Italy)

ISBN 978-3-658-11947-8   ISBN 978-3-658-11948-5 (eBook)
DOI 10.1007/978-3-658-11948-5

Library of Congress Control Number: 2015956041

Springer Spektrum
© Springer Fachmedien Wiesbaden 2016
This work is subject to copyright. All rights are reserved by the Publisher, whether the whole or part of the material is concerned, specifically the rights of translation, reprinting, reuse of illustrations, recitation, broadcasting, reproduction on microfilms or in any other physical way, and transmission or information storage and retrieval, electronic adaptation, computer software, or by similar or dissimilar methodology now known or hereafter developed.
The use of general descriptive names, registered names, trademarks, service marks, etc. in this publication does not imply, even in the absence of a specific statement, that such names are exempt from the relevant protective laws and regulations and therefore free for general use.
The publisher, the authors and the editors are safe to assume that the advice and information in this book are believed to be true and accurate at the date of publication. Neither the publisher nor the authors or the editors give a warranty, express or implied, with respect to the material contained herein or for any errors or omissions that may have been made.

Printed on acid-free paper

Springer Spektrum is a brand of Springer Fachmedien Wiesbaden
Springer Fachmedien Wiesbaden is part of Springer Science+Business Media
(www.springer.com)

Foreword

"anyway the exponential function would ‚wouldn't go up again' because one but it would flatten further that would than haven u-h ‚as its asymptote" is Tim's commentary as he watches the computer screen in front of him. On the computer screen we find a coordinate system. Starting from the first quadrant, a parabola is swept out, crossing the y-axis in the curve's lowest point. However, Tim does not directly refer to the parabola but to the graph of an exponential function, that is not visible on the screen. In his commentary, Tim wants to say that the visual piece of the curve on the screen cannot represent an exponential function, "… because one but it would flatten further that would than haven u-h ‚as its asymptote". The last part of his utterance is accompanied by a gesture: directly in front of the computer screen, Tim's finger draws the graph of an exponential function in the air, passing from the right top downwards, a bit up again before going to the left and approximating the x-axis.

Such situations often happen in mathematical learning situations. Even when unobserved, students may use gestures to approach mathematical ideas or clarify aspects, specifically when solving problems. This does not only happen with graphical or geometrical topics. Learners may represent all kinds of things by gestures, for instance if they indicate the substitution of an algebraic expression by a "grasping hand" as is described in this book. Such representational gestures are relevant tools for constructing mathematical knowledge, as has been known for quite some time. However, what has not been known is how gestures enable students to advance their epistemic processes. In her thesis, Christina Krause investigates this research problem. Her main approach is to merge two models, an epistemic action model, representing the three epistemic actions of gathering and connecting mathematical meaning as a basis for structure seeing, and the semiotic bundle model, connecting semiotic means, such as gestures, language and representations. Empirically based, she identifies 'specifying gestures', specifying the *Where*, the *What* and the *How* of mathematical ideas and also *Relationships* between mathematical aspects. In the example above, Tim not only specifies what he refers to verbally, the exponential function, but also its graphical style, hence the *How* and, in terms of direction, also the *Where*. Through comparison, Tim also constitutes a relationship between two graphical representations: one represented on the screen, the other by tracing the style of its shape with a gesture in front of the screen in the air. In the investigated data more than 90% of the gestures referring to mathematical objects were specifying gestures, and on average 40% of them specified several things at the same time.

For her empirical investigation, Christina Krause uses given data about pairs of students solving mathematical problems. These data, already transcribed, were re-worked in detail and enriched by pictures of gestures and inscriptions. In her fine-grained analysis, she focused on so called "epistemic-dense episodes", the most relevant parts for structure seeing. For that she clarified the epistemic role of gestures: two planes of analyses are distinguished, one in which gestures represent mathematical aspects related to speech and inscriptions, the second in which gestures relate to epistemic actions. For the latter, Christina Krause is able to reconstruct two different types of epistemic functions affecting the epistemic processes: *performing functions* and *forming function*. This frame of concepts linked to results from gesture studies allows her to reconstruct processes of coming to know mathematics as they take place in epistemic-dense situations. The results show that gestures may prepare, shape, substitute or directly advance epistemic actions. In the example above, Tim's gesture sources parts of the argumentation out from speech to visualize them by gesturing. By this, gesturing simplifies verbal argumentation.

Gestures do not only advance epistemic insights, they may even assist to create them and to elaborate them. For example, as the abovementioned "grasping hand" was repeated, the idea of substitution afforded by a task on continued fractions evolved towards new insights. Hence, gestures may indicate on which idea a student is working, thus providing diagnostic potential for teachers' noticing. For this fresh area of research, Christina Krause has made a significant contribution by showing how gestures are used spontaneously to complement speech and advance mathematical insights.

She succeeds not only in presenting her research results in an exciting manner, her book is also shaped in an attractive way. This shaping is not just an end in itself: it makes her fine-grained gesture analyses accessible and transparent to the reader. Her analyses do not only affect the reader by consistent argumentation but they often happen to be much more directly convincing, e.g. when one realizes that while reading the meaning of the "grasping hand" is grasped by gestural imitation.

At first sight the data base seems rather small, but the added value of this research lies in the theoretical and methodological deepness having led to a framework for studying a new topic in a fresh field of research. Only now it will be possible to explore the epistemic function of gestures systematically with qualitative as well as quantitative methods. Therefore, we wish that the national and international community of researchers interested in gesture research and beyond will adopt the results of this dissertation and will be inspired by this book and its exciting contribution to our understanding of mathematics education practices and research.

July 2015,   *Angelika Bikner-Ahsbahs (University of Bremen),*
Ferdinando Arzarello (University of Torino)

Acknowledgments

There are several people and institutions that made it possible for me to carry out this study and to write this book by supporting me in many different ways:

First, I want to thank my supervisor, *Angelika Bikner-Ahsbahs*, for the great opportunity she offered when accepting me as her PhD-student. I appreciate all the trust she set in me and thank her for all the creative ideas, the support, and the encouraging words right when I needed them. But especially, I thank her for the contagious curiosity and enthusiasm she herself showed in my research topic, and in the results that finally arose.

Second, I want to thank *Ferdinando Arzarello* as my second advisor and cooperation partner of the project that made my studies possible in the first place. Despite his numerous engagements and the spatial distance between Bremen and Turin, he found the time to sit down with me, providing helpful advices as an expert in the field of semiotics and for the study of gestures in mathematics education.

I thank the Central Research Development Fund of the University of Bremen for making this study possible by funding the project "Epistemic processes from a semiotic point of view – On the role of signs in the construction of mathematical knowledge" as collaboration project with the University of Turin. Additional thanks are provided to the principle investigators of the project "Effective knowledge construction in interest-dense situations", *Angelika Bikner-Ahsbahs*, *Tommy Dreyfus*, and *Ivy Kidron*, for their consent to use data of the project. Their project was funded by the German-Israeli-Foundation for Scientific Research and Development (GIF); therefore this research was also supported by the GIF. Also, I thank the German team of the GIF-project, *Angelika Bikner-Ahsbahs*, *Julia Cramer*, *Thomas Janßen*, and *Jakob Priwitzer* for data collection and preparation, and the entire project group (*Angelika Bikner-Ahsbahs*, *Tommy Dreyfus*, *Ivy Kidron*, *Miriam Adi*, *Julia Cramer*, *Nava Gilboa*, and *Raz Harel*) for their task designs.

I also thank the entire mathematics education working group at the University of Turin for their collaboration in the project and the hospitality during my three stays at their department. Thank you for providing many helpful advices and for introducing me into a multimodal analysis in general and the study of gestures in particular. Especially, I need to send out a very warm 'grazie' to *Cristina Sabena* for being a great professional support and such a kind and friendly person I appreciate to have gotten to know through this project.

The mathematics education working groups of the University of Bremen always provided a very supportive and collaborative environment for me. Special thanks are given to my PhD-student colleagues *Julia Cramer, Thomas Janßen, Jonathan von Ostrowski, Jenny Cramer, Stephanie Lachky,* and *Daniela Behrens* (in order of appearance) for all the talks we had and for their support. Furthermore, I want to acknowledge my current working group and collaborators at the University of Essen for backing me in the very last stage.

Benjamin Rott, Johannes Kautzsch, and *Daniela Behrens,* thank you for reading (more or less mature) parts of the writing during the last couple of years. *Raja Herold,* thanks a lot for finding my file after my laptop's 'hiccup'. Thanks to my friends for being patient in those periods I hided somehow from social live. Especially to mention *Johannes* once more: thank you for cake and cookies, coffee, and words in long office hours.

I also thank *my family* for their steady support and for always encouraging me. I found a lot of words in this book, but I do not find proper words to thank my parents as much as they deserve it. Therefore, just a simple "Danke für alles!". To my brother thanks also for the technical advice during the last couple of years, and double thanks for two wonderful nephews that always make me smile.

Last, but quite the opposite of least: Thank you, *Jan.* Thanks for your patience, your hugs, your words, for simply being there and supporting me with the best timing possible.

July 2015

Abstract

When learning mathematics, students do not only use language to communicate, but also gestural and written modes of expression. Gestures seem to play an important role in a pre-explicit stadium of an idea, when it is not yet explicitly accessible to the students. This is especially the case when learners construct new mathematical knowledge in social interactions. While previous studies mostly examined these processes of social con-struction of knowledge on the basis of verbal utterances, the present work addresses the question: *How do gestures contribute to processes of constructing mathematical knowledge in social interactions?*

To answer this question, I aimed to enrich a pre-existing theory of epistemic processes with a semiotic perspective based on empirical findings. Thereby, a more extensive comprehension of social processes of constructing mathematical knowledge shall be attained. Three pairs of high performing students were filmed while solving three tasks that differ both in mathematical topic, as well as in the variety of representations provided for solving the task. Following an interpretive approach, the traditional, speech-based reconstruction of the epistemic processes by means of an epistemic action model was extended by an analysis of gestures in their synergy with language and inscriptions. The contribution of gestures to the epistemic processes is captured by two functions of gestures: The *representational function* of gestures concerns the ways in which gestures can refer to a mathematical object in processes of knowledge construction. I systematically expose how gestures can enrich the verbal utterance and how speech, gesture, and inscription act together to shape the mathematical object. The *epistemic function* of gestures concerns the ways in which gestures can contribute to the accomplishment of the three epistemic actions 'gathering', 'connecting', and 'structure-seeing' (Bikner-Ahsbahs, 2005). I found that gestures not only support, but can also prepare, and even realize these epistemic actions. More specifically, I differentiate between four *forming* and six *performing* epistemic functions of gestures: The forming functions are related to the ways in which gestures can provide visual access to the mathematical object of an epistemic action; the performing functions of gestures concern how gestures can actively affect the accomplishment of epistemic actions. A comparison of the different data sets reveals that gestures influence the epistemic process more significantly than previously thought, and gives an idea about what this influence may depend on.

Zusammenfassung

Beim Lernen von Mathematik nutzen Schüler nicht nur Sprache, um sich mitzuteilen, sondern auch gestische und schriftliche Ausdrucksformen. Gesten scheinen eine besondere Rolle zu spielen, wenn Ideen für die Lernenden noch nicht ‚greifbar', noch nicht explizit zugänglich sind. Dies ist vor allem dann der Fall, wenn Lernende neues mathematisches Wissen in sozialer Interaktion konstruieren. Während bisherige Studien diese Prozesse sozialer Wissenskonstruktion zumeist auf Basis der verbalen Äußerungen untersuchten, stellt die vorliegenden Arbeit die Frage: *Wie tragen Gesten zu Prozessen der Konstruktion mathematischen Wissens in sozialer Interaktion bei?*

Zur Beantwortung dieser Frage wurde an eine bestehende Theorie epistemischer Prozesse angeknüpft, um diese auf empirischer Grundlage durch eine semiotische Perspektive anzureichern und so ein umfassenderes Verständnis sozialer Prozesse der Konstruktion mathematischen Wissens zu erlangen. Hierzu wurden leistungsstarke Schülerpaare bei der Bearbeitung dreier Aufgaben gefilmt, die sich sowohl in der mathematischen Thematik, wie auch in der Vielfalt und Art der zur Verfügung stehenden Repräsentationen unterscheiden. Einem interpretativen Ansatz folgend wurde die traditionelle, sprachbasierte Rekonstruktion der sozialen epistemischen Prozesse durch ein epistemisches Handlungsmodell um eine Analyse der Gesten im Zusammenspiel mit Sprache und Inskriptionen erweitert. Der Beitrag von Gesten zu den Erkenntnisprozessen wird anhand zweier Funktionen der Gesten gefasst: Die *Repräsentationsfunktion* von Gesten betrifft die Arten und Weisen, in denen Gesten in Prozessen der Wissenskonstruktion dazu beitragen können, auf ein mathematisches Objekt zu verweisen. So wird systematisch herausgestellt, wie Gesten die sprachliche Äußerung anreichern können und wie Sprache, Geste und Inskription das mathematische Objekt hierdurch gemeinsam formen. Die *Erkenntnisfunktion* von Gesten bezieht sich darauf, wie Gesten zu der Ausführung der drei epistemischen Handlungen Sammeln, Verknüpfen und Struktursehen (Bikner-Ahsbahs, 2005) beitragen können. Es hat sich gezeigt, dass Gesten diese epistemischen Handlungen nicht nur unterstützen, sondern sie auch vorbereiten und sogar selbst realisieren können. Hierbei wurden vier *ausformende* und sechs *ausführende* epistemische Funktionen von Gesten unterschieden: Die ausformenden Funktionen beziehen sich auf die Arten und Weisen, in denen Gesten dazu beitragen mathematische Objekte epistemischer Funktionen visuell zu gestalten; die ausführenden Funktionen darauf, wie Gesten aktiv auf die *Ausführung* epistemischer Handlungen wirken können. Ein Vergleich der Analysen verschiedener Datensätze führt zu der Vermutung, dass Gesten mehr Einfluss auf den epistemischen Prozess haben als bisher angenommen und gibt Hinweise darauf, wovon dieser Einfluss abhängen kann.

Table of Contents

Abstract ix
Table of Contents xi
Table of Abbreviations xvii
Mathematical Notations xix

1 Introduction 1
   1.1   Why Look at Signs in the Learning of Mathematics: Becoming Aware of a Chicken-Egg Problem and the Importance to Address it 1
   1.2   The Project 2
   1.3   On the Structure of this Book 3

Part I: State of the Art and Theoretical Background   7

2 State of the Art 9
   2.1   A Psychologist's Approach to Gestures in the Learning of Mathematics 9
   2.2   Research in Mathematics Education on the Role of Gestures in Learning Processes 10
   2.3   The Epistemic Role of Gestures Against the Background of Different Theoretical Frameworks 12
   2.4   The Epistemic Role of Gestures in this Project 14

3 Shaping the Semiotic Approach 19
   3.1   Signs: The Peircean Semiotics 19
        3.1.1   Definition of Signs and the Process of Semiosis 20
        3.1.2   The Trichotomy of the Object: Iconic, Indexical, and Symbolic .. 22
        3.1.3   The Representational and the Epistemic Function of Signs 25
   3.2   Gestures: A Framework for this Study 26
        3.2.1   Understanding of Gestures in this Study 26
        3.2.2   Gesture and Speech: An Integrative System 28
        3.2.3   Gesture Classification 31

4   Perspectives on Learning 35
  4.1   Learning as a Social Process: An Epistemic Action Model 35
  4.2   Embodied Thinking and Learning: How our Body Shapes our Mind 37
      4.2.1   The Role of Metaphors in the Conceptualization of Mathematics 38
      4.2.2   Multimodality and Perceptuo-Motor Experiences in Learning Mathematics 42
  4.3   Learning as a Multimodal Process 44
      4.3.1   The Basic Idea of the Semiotic Bundle 44
      4.3.2   The Analysis Of and Within the Semiotic Bundle 46
      4.3.3   Implications for the Mathematics Classroom 48
  4.4   Summarizing the Theoretical Frame for this Study 49

Part II: The Study   51

5   Methodological Assumptions 53
  5.1   The Goal of this Study: Focus, Aims, and Questions against the Theoretical Background 53
  5.2   The Interpretative Approach 55
      5.2.1   Handling the Data within the Interpretative Approach 55
      5.2.2   Transcript versus Video Data: Dealing with Gestures 56
      5.2.3   Analyzing Interaction by Tracing the Development of Signs 58
          5.2.3.1   A *Re*constructivist Approach, Based on a Peircean Understanding of Signs 58
          5.2.3.2   Considering a Multimodal Approach by Using the Model of the Semiotic Bundle 67
          5.2.3.3   Combining Both Approaches 76
  5.3   Theoretical Sampling as a Focusing Strategy 79
  5.4   Forming the Buidling Blocks for a Theory on Gestures' Contribution to Social Epistemic Processes 79
      5.4.1   Identifying the Gestures to be Analyzed 79
      5.4.2   Elaborating Functions of Gestures as Empirically Based Categories 80
      5.4.3   The Scope of the Theoretical Building Blocks 81
  5.5   Quality Criteria and Methodological Concerns from Gesture Studies 81

| | | 5.5.1 | Quality Criteria Considered in Qualitative Research .... 81 |
| | | 5.5.2 | Criteria Considered in Gesture Studies .... 85 |

6 Methods 89

- 6.1 Setting 89
  - 6.1.1 General Aspects of the Study and on Gathering the Data 89
  - 6.1.2 The Students 90
  - 6.1.3 The Tasks 91
    - 6.1.3.1 Geometric-Algebraic: The Parabola as Geometrical Locus (PA) 91
    - 6.1.3.2 Arithmetic-Analytic: Continued Fraction (CF) 94
    - 6.1.3.3 Logical Reasoning: Induction (IN) 95
- 6.2 Methods of the Analysis 96
  - 6.2.1 Data Preparation 96
  - 6.2.2 Choice of Data 97
  - 6.2.3 Analysis of the Data 98
    - 6.2.3.1 Steps 1-3: Preparatory Steps 99
    - 6.2.3.2 Steps 4 and 5: Main Steps 104
    - 6.2.3.3 Step 6: Comparative Step 108
- 6.3 Presentation of the Results 108

Part III: Results   111

7 Representing *Within* the Multimodal Sign 113

- 7.1 Within-Functions of Gestures: A Tool to Integrate How Gestures May Shape Mathematical Meaning 113
  - 7.1.1 Specifying the Immediate Object: Adding What is not Revealed in Speech 114
    - 7.1.1.1 Clarifying Imprecise Terms: Gesture can Complement the Immediate Object Presented in Speech by Filling a Semantic Gap 114
    - 7.1.1.2 From Clarifying to Specifying: Providing a Real Surplus of Meaning 119
    - 7.1.1.3 Specifying-Gestures in *Mismatch*-Situations: Gestures Take the Lead 121

		7.1.1.4	Summarizing Overview of the Specifying-Function of Gestures on Speech 125
	7.1.2	Specification and Non-Specification: Further Details 126
		7.1.2.1	Coding Guideline for Illustrators 127
		7.1.2.2	Non-Specifying Illustrators 128
	7.1.3	Gestures can Refer to Mathematical Objects on Three Spatial Levels 134
		7.1.3.1	Presentation of the Three Levels of Gestural Reference 134
		7.1.3.2	Summarizing Overview about the Referential Levels of Gestures 138
	7.1.4	Closing Remarks on the Within-functions of Gestures 139
7.2	Reconstructing Meaning in Social Interaction: The Development of Associated Signs and Information Bundles 140
	7.2.1	Detailed Reconstruction of Two Illustrative Examples 141
	7.2.2	Evidence on the Use of Associated Signs 150
		7.2.2.1	Evidence through Irritation when the Use of a Gesture does Not Fit the Context 151
		7.2.2.2	Evidence through Students' Joint Packing and Shared Use of the Associated Sign 155
	7.2.3	Closing Remarks on Associated Signs 157
7.3	Representing Within the Multimodal Sign: Within-Functions as a Methodological Tool 158
7.4	Summary 159

8 The Epistemic Process in Progress 161

8.1 Epistemic Functions of Gestures: Detailed Descriptions of the Categories as Reconstructed within the Data 162

    8.1.1 Forming-Functions: How Gestures can Provide Visual Access to a Mathematical Object that is Involved in an Epistemic Action 162

        8.1.1.1 Sourcing Out 162

        8.1.1.2 Depicting 168

        8.1.1.3 Extracting 174

        8.1.1.4 Illustrating Generality 180

        8.1.1.5 Summarizing Overview on the Forming-Functions of Gestures 181

        8.1.2   Performing-Functions: How Gestures can Take Part in the Accomplishment of an Epistemic Action 181

              8.1.2.1   Focusing 182

              8.1.2.2   Exemplifying 190

              8.1.2.3   Contrasting 193

              8.1.2.4   Making More Precise 197

              8.1.2.5   Gluing 202

              8.1.2.6   Structuring Verbal Discourse 205

              8.1.2.7   Summarizing Overview on the Performing Functions . 206

   8.2   Extensive Reconstruction of a Larger Scene: Merging the Possible Variety of Gestures' Contribution to the Epistemic Process 208

        8.2.1   Semiotic Analysis: Epistemic Functions of Gestures 212

        8.2.2   Reflection on the Benefit of Using Gestures with Respect to the Epistemic Process 222

   8.3   Summary 224

9   Towards the Role of Gestures: A Comparison **227**

   9.1   Horizontal Comparison: One Task, Different Pairs of Students (PA5, PA7, PA6) 227

   9.2   Vertical Comparison: One Pair of Students, Different Tasks (PA7, CF7, IN7) 231

   9.3   Summarizing Conclusions Resulting from the Horizontal and the Vertical Comparison 242

10 Reflection, Synopsis, Perspectives **245**

11 References **251**

12 Glossary **265**

   12.1   What is Meant by… 265

   12.2   Glossary of Important Gestures Referred to in This Study 272

Part IV: Appendix   275

A   The Tasks **277**

   a)   Parabola 277

	b)	Continued Fraction	282
	c)	Induction	286
B	Analysis of the Tasks		**287**
	a)	Parabola	287
	b)	Continued Fraction	297
	c)	Induction	305
C	Condensed Process Diagrams of the Epistemic-Dense Episodes as Used for Comparisons Between the Data Sets		**309**
D	The Inscriptions Produced by the Students		**313**
	a)	PA5	313
	b)	PA7	320
	c)	PA6	326
	d)	CF7	330
	e)	IN7	336
E	Example of a Multimodal SRA (PA5e1.1)		**339**

Table of Figures 341

Table Directory 349

Table of Abbreviations

CF Abbreviation for the task dealing with a **c**ontinued **f**raction (see section 6.1.3.2)

DGS Short for **D**ynamic **G**eometry **S**ystem (GeoGebra in this study)

EDE Short for **e**pistemic-**d**ense **e**pisodes (see chapter 6.2.3.1, step 2); EDEs are named by a code consisting of the name of the data set, the number of the episode within this data set, and sometimes also the number of the scene within the episode: As an example, PA5e1.1 is the name of the first scene in the first EDE of the data set PA5 ('e' stands for 'episode')

GCSt Name of the epistemic action model, embracing the three epistemic actions **G**athering, **C**onnecting, and **S**tructure-seeing (first mentioned in chapter 1.2, described in detail in chapter 4.1)

GP Short for **G**rowth **P**oint (see chapter 3.2.2)

IN Abbreviation for the task on logical reasoning by using the concept of mathematical **in**duction (see chapter 6.1.3.3)

IPH Short for **I**nformation **P**ackaging **H**ypothesis (see chapter 3.2.2)

IS Short for **i**nterpretation **s**pace, the space of possible interpretations of a sign (e.g. a verbal utterance as a sign, or a gesture as a sign) (see section 5.2.3.1)

PA Abbreviation for the task dealing with the **parabola** as geometrical locus (see section 6.1.3.1)

SRA Short for **s**emiotic **r**econstructive **a**nalysis, the first step of Bikner-Ahsbahs' semiotic sequence analysis (see section 5.2.3.1)

Mathematical Notations

$(f(x))_{x \in \mathbb{N}_0}$	Sequence with elements $f(x)$ with $x \in \mathbb{N}_0$
$[AB]$	Segment with end points A and B
$\overline{AB}$	Length of the segment $[AB]$
$\triangle BCO$	Triangle with vertices B, C, and O
$\angle ACG$	Angle at the vertex C enclosed by the segments $[AC]$ and $[CG]$, i.e. the anticlockwise angle in point C from A to G
$\lvert \angle ACG \rvert$	The value of the angle $\angle ACG$
$\mathbb{N}_0$	Set of natural numbers including zero
$\mathbb{R}_{>0}$	Set of real numbers larger than zero

1 Introduction

From Signs to Gestures: The General Semiotic Approach as a Starting Point

1.1 Why Look at Signs in the Learning of Mathematics: Becoming Aware of a Chicken-Egg Problem and the Importance to Address It

Being a science of structures, mathematics deals with relations and general rules. A direct access to the mental objects of mathematics is impossible (Hoffmann, 2003, pp. 5-6; Steelker-Weithofer, 2003, p. 2571); they only become tangible through using signs that represent these objects. This fact causes a variety of obstacles students have to cope with, starting with the ambiguous meaning of signs that represent a mathematical idea and not ending with the fact that this ambiguity is not considered by everybody teaching those ideas. The meaning of a sign is not given per se but needs to be learned together with the knowledge of the mathematics potentially represented by it. At the same time, the representation of a mathematical object must not be confused with the object itself, as Duval points out in his '*cognitive paradox* of access to mathematical knowledge objects': "*How can [students] distinguish the represented object from the semiotic representation used if they cannot get access to the mathematical object apart from the semiotic representation?*" (Duval, 2006, p. 107, italics in the original). Concluding from a study investigating this issue, he states that the coordination and variation of different representations of a mathematical object is essential to the learning of mathematics. Here, he includes the transfer of signs within the same sign system and translating them into another sign system (Duval, 2006, pp. 126-127). This importance of variation and coordination of different representations can be concretized by means of an example concerning mathematical functions: A function can be represented by its graph, its function equation, by a table, but also by a verbal description of the course of the curve. Every representation provides advantages and disadvantages concerning the exploration of the mathematical object - the function in this case - and the investigation of its properties. Changing between representations of mathematical objects thus needs to be learned together with the knowledge of the object.[1]

Learning here is understood as a social event of mutually constructing knowledge within social interaction. What we understand as a certain mathematical object can however not per se be considered identical for two distinct individuals since the abstract knowledge needs to be constructed individually. That is, mathematical objects are seen as mental individual constructions that become established in social interaction. In the mathematics classroom, this social interaction takes place between students and the teacher, but also among students. This highlights the role of signs even more, since using signs is the only way to communicate observations and insights and to organize collaborative work (Seeger, 2006, p. 267).

1.2 The Project

The aforementioned considerations led to this project which aims to apply a semiotic perspective to knowledge construction in order to get a better understanding of epistemic processes by looking at the signs used by the students. This research project has been funded by the Central Research Development Fund[2] of the University of Bremen as a cooperative project with the mathematics education research group of the University of Turin, headed by Prof. Dr. Arzarello. This collaboration led to four exchange stays in which I was introduced to the application of their *Semiotic Bundle model* for analyzing the interplay of signs. In return, I presented the *semiotic sequence analysis* and the *GCSt-model*, developed by Bikner-Ahsbahs (2005, 2006, see also chapter 4) to reconstruct and describe epistemic processes in social interaction based on speech acts.

A Semiotic Perspective on Social Epistemic Processes: The Role of (Hand) Gestures as the Focus for This Study
This thesis aims to combine the Turinese model, describing how *signs* and the relations between them develop in social learning processes, with the epistemic action model by Bikner-Ahsbahs, describing how *knowledge* develops in these processes. According to Bikner-Ahsbahs, epistemic processes in social interaction can be described by the three epistemic actions of *gathering* minimal mathematical meaning units, the *connecting* of such units, and the *seeing* of *structures*, hence regularities (Bikner-Ahsbahs, 2006).[3] As mentioned above, the reconstruction of these actions is primarily based on a speech act

---

[1] Arzarello and Sabena (2011) refer to the ability of "[choosing] a suitable semiotic representation for solving a task (e.g. an algebraic formula vs a Cartesian graph)" (Arzarello & Sabena, 2011, p. 191) as *semiotic control*.
[2] "Epistemic processes from a semiotic point of view – On the role of signs in the construction of mathematical knowledge", grant #03/112/08
[3] A more detailed description of this model will be given in chapter 4.1.

analysis, following Austin (1962). He distinguishes three levels of analysis of a speech act: In addition to a level of semantic content (locutionary level: *saying something in a commonly used sense*), it can also be acted in saying something (illocutionary: *Doing something in saying something, for example asking a question, issuing a warning, or uttering a threat.*) to prompt an effect (perlocutionary: *doing something by saying something; upsetting somebody, changing someone's mind,...*). Assuming speech to be a "socially objectivated system of meanings" (Jungwirth, 2003, p. 191, my own translation), the analysis of speech acts allows reconstructing the epistemic processes within social interactions. This approach however reduces the analysis to linguistic signs. Adding to this a more comprehensive semiotic perspective, I intend to get further insights on the performance of epistemic actions, and on conditions that foster or hinder the epistemic process. For this reason, the semiotic bundles developing within the epistemic processes have been taken into consideration; that is, the relations between different sets of signs as playing together at the same time, as well as developing during the process.[4] First, tentative analyses confirmed an already suggested, significant influence of gestures on social processes of knowledge construction (see also Krause, 2012; Krause & Bikner-Ahsbahs, 2012). While these analyses still treated sign use more generally, including the production and interpretation of inscriptions, the results led to a focus on the use of gestures and the leading question:

> How can gestures contribute to social processes of constructing mathematical knowledge?

This question will be approached from two angles: First, considering gestures as a specific kind of sign in a Peircean sense[5] allows them to be integrated in the reconstruction of the epistemic process as a means of expression shaping social interaction. Second, respecting the development of relations within evolving semiotic bundles embraces the relationship of gestures to other signs, such as speech and inscription. We will see how answering the question from these two angles will not only enrich understanding of the epistemic processes, but will also reflect back to provide methodological implications on the models used.

1.3 On the Structure of this Book

The structure of this book is comprised of three parts:

The **first part** (chapters 2-4) presents the **theoretical background** that frames this study. First, the study is placed within the current state of research concerning gestures' role in learning mathematics (chapter 2) as approached from psychology as well as from

---

[4] See chapter 4.3 for a description of the semiotic bundle model and the analysis of signs within this model.
[5] See chapter 3.1 for the Peircean definition of signs.

mathematics education. Chapter 3 is dedicated to the semiotic background theories that direct the semiotic perspective in this study. This first comprises the Peircean concept of signs as a suitable semiotic tradition considered for this study and second, the understanding of gestures as adopted in it. Furthermore, an introduction to the theory of gesture studies illuminates the importance of gestures in thinking and speaking. In chapter 4, learning mathematics is approached from three perspectives: as a *social event* (4.1) in which mutual knowledge is constructed based on common *bodily experiences in the physical world* (4.2), developing *in*, and *by* the use of various signs, that is in a *multimodal way* (4.3).

The **second part** (chapters 5 and 6) concerns the **methods and methodology** of the study. Chapter 5 starts with a résumé of the research hypotheses that underlie this study to pinpoint the research focus, the research questions, and the goal against the theoretical landscape that has just been drawn. Following this, the basic methodological concerns are disclosed: First a mental model is presented as locally combining speech act analysis and gesture analysis to reconstruct epistemic processes in social interaction by adopting an interpretative approach. Second, the research logic underlying the study embraces the choice of data as well as the generation of hypotheses. Third, criteria considered to measure the quality of the analyses and of the research products are explicated, embracing also a discussion of criteria considered important for research in the field of gesture studies. The methods chapter (chapter 6) draws a detailed picture of the how and why of the data collection and preparation as well as of the analytical steps carried out to investigate and answer the research questions.

Part three (chapters 7-10) is the main part of this book, in which the **results** are presented. Chapter 7 and 8 provide a detailed analysis of gestures' influence on the epistemic processes, emphasized differently in both chapters: In chapter 7, the question of how mathematical meaning is shaped by the use of gestures is answered by presenting a methodological tool describing how gestures bridge between speech and inscription. Besides providing the possibility to draw conclusions about information that is only provided non-verbally, this tool also helps to reconstruct how knowledge about an object and about its representations are mutually developed in social interaction. Chapter 8 concerns functions of gestures that have been identified to benefit the social process of knowledge construction by supporting the accomplishment of epistemic actions. The different epistemic functions will be presented as sorted according to two categories, depending on whether the benefit consists of giving visual access to epistemic action (forming function) or in actively enhancing their accomplishment (performing function). In chapter 9, the analyses are summarized in order to give a comprehensive answer to the leading question of this study, pinpointing the insights on gestures' contribution to epistemic processes. In chapter 10, a review is given of the study, its results and potential implications for theory and teaching practice. Based on this, possible further research will be proposed.

1.3 On the Structure of this Book

This study presents results that might be insightful not only from the perspective of mathematics education, but also for the field of gesture studies. On the one hand, they shed light on a rarely researched approach to mathematical learning *processes* rather than existing knowledge. On the other hand, they describe aspects of gestures' interplay with speech and inscription in this specific context. Having this in mind, an overview of the central notions, concepts, and ideas as referred to in this study are provided in the form of a **glossary**, attached at the end as chapter 12, following the list of references (chapter 11). It is hoped that this will facilitate the reading of this book as bridging between the 'language' of both disciplines, mathematics education and gesture studies.

The book makes use of footnotes, all having in common that they provide additional information that may be of interest to the reader. Their function is manifold, such as recalling references within this book, giving illustrative examples to theoretical concepts, providing further information by means of reference to other texts, or offering a more comprehensive explanation.

Part I
State of the Art and Theoretical Background

In the first part of this book, I will provide the basic concepts and notions this study is based on. Here, the study's position as a piece of the larger mosaic of research, related to the role of gestures in the learning of mathematics, is clarified. Furthermore, I expound on the theoretical approach undertaken to answer the research question, taking into account the multiple factors that need to be considered as being involved.

2 State of the Art
Towards a Gesture Perspective on Epistemic Processes

The starting point for this study stems from an interest in how gestures contribute to learning processes in social interaction. By investigating this issue, I aim to provide a theoretical basis from which can be hypothesized on the potential of gestures as a didactical means. This chapter seeks to embed this into recent and current research, singling out how the study aims to take part in investigating the epistemic role of gestures. However, the study of gestures in mathematics teaching and learning is a still young research field. Having just started to gain attention during the last two decades, this interest can be considered being initiated by the rise of the *embodied cognition theory* that embraces the body's role in shaping the mind (Varela, Thompson, & Rosch, 1991).

2.1 A Psychologist's Approach to Gestures in the Learning of Mathematics

Based on this and on the close relationship between gesturing, thinking and speaking as proposed by McNeill (1992; 2005) and Kendon (1988; 2004), researchers especially from the field of psychology started to investigate gestures in relation to students' mathematical thinking. Their studies shed light on how students' use of gestures reveals mathematical conceptualization and learning strategies (Alibali & Goldin-Meadow, 1993; Alibali, 1999; Goldin-Meadow, Nusbaum, Garber, & Breckinridge Church, 1993) and on how it influences their mathematical skills (Alibali & Goldin-Meadow, 1993; Alibali & diRusso, 1999; Goldin-Meadow, 2003; Broaders, Cook, Mitchell, & Goldin-Meadow, 2007; Goldin-Meadow, Cook, & Mitchell, 2009; Goldin-Meadow, 2010). The insights gained from these studies can be considered as groundbreaking from a psychological perspective to investigate students' mathematical knowledge, providing a diagnostic potential the teacher can react to[6].

---

[6] See also chapter 3.2.2 for their results on the concept of gesture-speech mismatches and its role in teaching and learning as described in these studies.

Nevertheless, it does not cover two main issues that are sought to be addressed in this study:

First, learning mathematics is seen as a social process that takes place in social interaction, as for example in the mathematics classroom. The aforementioned experimental studies did not consider the acquisition of knowledge in collaboration with peers and the teacher but observed gestures related to the students' current state of mathematical conceptualization. Second, the mathematics within these studies concerns elementary topics such as 'learning to count' (e.g. Alibali & diRusso, 1999) and mathematical equivalence problems.[7] The studies do not make any statements about students' gestures related to mathematical ideas on a higher abstract level, comprising generalization and reasoning.

2.2 Research in Mathematics Education on the Role of Gestures in Learning Processes

Recent research on gestures in mathematics education points out the crucial role gestures fill in communication in mathematical talk (Edwards, 2009) and in collaborative learning processes (Arzarello, Robutti, Paola, & Sabena, 2009; Bjuland, Cestari, & Borgersen, 2008; Reynolds & Reeve, 2002; Yoon, Thomas, & Dreyfus, 2011), giving a special remark about students' reasoning and argumentation (Bjuland et al., 2008; Chen & Herbst, 2013; Marghetis, Edwards, & Núñez, 2014; Rasmussen, Stephan, & Allen, 2004). Furthermore, studies also highlight the potential of gestures for exploration and experimentation while working out mathematical ideas (Chen & Herbst, 2013; Sabena, 2007; Yoon et al., 2011), and for coordinating the work with given inscriptions, for example through focusing by pointing on different representations of the same situations, and by sliding between them (Bjuland et al. 2008).

Reynolds and Reeve (2002) conducted a problem solving situation to find out how students use gestures to communicate their interpretations of unknown graphs. The students worked on different tasks, each concerning the graphical representation of speed as a rate. In both cases, graphs were given to be investigated in dyadic collaboration: First, by assigning specific situations to given graphs; second, by inventing a description of a bus trip that fits a speed-time graph provided by the task. Reynolds and Reeves observed that gestures are used to direct the others' attention to certain aspects of the graphs, and also as a means to express an understanding of a concept that has not been fully elaborated at that point in time (Reynolds & Reeve, 2002, p. 457). They show how missing or insecure mathematical terms can be overcome by providing additional reference to a

---

[7] That is, problems that concern filling in a missing number in equations of the kind 2+7+8 = _+8 (Alibali & Goldin-Meadow, 1993; Goldin-Meadow, 2003, pp. 41-47).

2.2 Research in Mathematics Education on the Role of Gestures in Learning Mathematics

mathematical entity. Gesture thus can support mathematical communication by providing a joint visual reference for the participants.

This has also been identified as benefiting reasoning in collaborative problem solving in a study conducted by Bjuland, Cestari, and Borgersen (2008). Concluding from results of their empirical study, they determined that gestures and discourse "constitute mathematical reasoning" (ibid., p. 289). In this study, they asked 12-year-old students to transfer a real world representation of a situation to its corresponding representation in a coordinate system: Four people were displayed, all differing in age and height. The coordinate system displayed the age on the y-axis and the height on the x-axis. The students were asked to assign one person from the real world representation to each point within the coordinate system. In order to convince their classmates, the students reasoned their hypotheses first and foremost helping themselves by pointing gestures to complement their verbal utterances. In this regard, gesture is considered a *necessary* complementation to verbal discourse. Furthermore, this study shows how gesture helps coordinating and relating the different representations between which information needs to be transferred.

Chen and Herbst (2013) set a special focus on students' use of diagrams in conjecturing and reasoning processes in geometry class. According to them, the amount of information given in a diagram influences the configuration of gestures: They argue that "when limited information is given in a diagram, students make use of gestural and verbal expressions to compensate for those limitations as they engage in making conjectures" (ibid., p. 285). This has been concluded from observations made about students' work on similar diagrams in two different geometry classrooms. They compared students' use of the diagram in an 'intact lesson' in which a complete, "intact" diagram was handed in, with the use of the diagram in an 'intervention lesson' in another class. In the intact lesson, marks and labels in the given diagram provided a lot of information, whereas the students in the intervention lesson worked with a similar diagram that contained less information. According to Chen and Herbst, gestures have been used differently: While the students in the intact lesson exclusively used pointing gestures to refer to facts given in the diagram (ibid., p. 291), the students in the intervention lesson also made use of iconic gestures to refer to the mathematical object. They "were extensively employed to represent the object that had not been represented in the diagrams, and so to represent the students' conception of figures" (ibid., p. 293).

These three examples make apparent that gesture, speech, and inscription crucially work together when knowledge is constructed in social interaction. To investigate this interplay more profoundly, Arzarello developed a model to analyze the relationships between semiotic resources of different kinds as they are used in learning processes (Arzarello, Ferrara, Robutti, Paola, & Sabena, 2005; Arzarello, 2006, pp. 280-288). He adopts a multimodal approach to learning and proposes to always observe signs within the intertwined relationships among gesture, speech, and inscription, constituting *semiotic bundles*.[8] These semiotic bundles are dynamic structures, the relations between them changing over time. Because of that, both a synchronic analysis of mutually used signs and a diachronic analysis of the use of signs across the process are considered for investigating students' semiotic bundles in learning processes. One diachronic relationship is given by the so-called *genetic conversion*, which is identified when one semiotic set is converted into another one (Arzarello, 2006, p. 281). This is given, for example, when the reference made by a gesture is fixed so that the inscription arises from a former gesture and by this, enlarges the semiotic bundle.[9] Concluding from a study carried out in a 11th grade calculus class, Arzarello et al. (Arzarello, Robutti, Paola, & Sabena, 2009) emphasized that "the synchronic and diachronic analyses show the complex intertwining of gestures, speech, and inscriptions in learning mathematics. These ingredients jointly support the thinking processes of students in a unitary way" (ibid., pp. 106-107). Arzarello and his colleagues observed that students create personal signs to express mathematical meaning. These can become endowed with meaning within social interaction by being used repeatedly in similar contexts. Those signs are called *basic signs* (Arzarello & Paola, 2007) and their use can be initiated and supported by the teacher, playing a *semiotic game* with the students. This allows the teacher "[to tune] with the students' semiotic resources and [to use] them to guide the evolution of mathematical meaning" (Arzarello, Bikner-Ahsbahs, & Sabena, 2009, p. 1547). Arzarello, Robutti, Paola, and Sabena further note that "in order that such opportunities can become concretely realized, the teacher must be aware of the role that multimodality and semiotic games can play in teaching" (Arzarello, Robutti, Paola, & Sabena, 2009, p. 107).

2.3 The Epistemic Role of Gestures Against The Background of Different Theoretical Frameworks

The abovementioned applications of the semiotic bundle analyses have been carried out against the background of a semiotic-cultural approach (Radford, 2003). That is, they take into account the students' semiotic-cultural background, referring to Radford's process of objectification, described etymologically as "a process aimed at bringing something in front of someone's attention or view" (Radford, 2002, p. 14). For example, gestures referring to a graphical representation may function as a means to make something visible or indicate something that has not been apparent before. Representations are one specific form of what Radford calls *semiotic means of objectification*, generally denoted as any "objects, tools, linguistic devices, and signs that individuals intentionally use in

---

[8] Also other semiotic sets, for example artefacts used by the students may be part of a semiotic bundle.
[9] The model of the semiotic bundle will be described more in detail in chapter 4.3.

social meaning-making processes to achieve a stable form of awareness, to make apparent their intentions, and to carry out their actions to attain the goal of their activities" (Radford, 2003, p. 41). Furthermore, he sees great potential in gathering insights into the process of the social construction of knowledge by investigating learners' use of semiotic means of objectification. Radford has been one of the first to appreciate the role of different signs as they play together to constitute mathematical meaning, and to establish a theory about it. Against the background of this theory, Sabena (2007) investigated the role "gestures play in the semiotic bundle intervening in processes of knowledge objectification in Calculus" (Sabena, 2007, p. 257). Among others, she found that a gesture may contribute to a blended character of the semiotic bundle, i.e. to the possibility of merging two related meanings in a gesture by using it in two different, but related contexts. In Sabena's example, a gesture referred to as the Δ-gesture[10] is used in a numerical and in a geometrical context: Using this gesture in a geometrical environment but using language related to Numerics, the two meanings are blended within the gesture (ibid., pp. 259-260). Sabena also suggests that the less contextualized the character of the mathematical activity becomes, the more conventionalized are the gestures used in the discourse as carriers of a certain meaning. In other words, they get a symbolic character developing somehow in classroom interaction (ibid., pp. 261-262). Objectification, and with that "learning, [...] can be theorized as those processes through which students gradually become acquainted with historically constituted cultural meaning and forms of reasoning and action" (Radford, 2010, p. 3). This approach to learning considers the teacher as participating and mediating a mathematical culture within the mathematics classroom. While this becomes an important component in the actual application of theories for teaching practice, it leaves aside how, and what kind of knowledge is constructed *by the students*. However, getting a better understanding of students' epistemic processes is considered important in order to become aware of conditions that foster or hinder them. Since gestures constitute a part of the social interaction in which the epistemic processes take place, my study addresses the conditions shaped by gestures.

Just recently, a case study of networking two approaches demonstrated that the epistemic bundle model can be adopted to provide deeper insights into epistemic processes: Two research groups, each focusing on different aspects of students' learning processes, collaborated to add a semiotic perspective to analyzing an epistemic process (Dreyfus, Sabena, Kidron, & Arzarello, 2014). With this, they sought to answer *whether* gestures contribute to the *individual* construction of mathematical knowledge and of course, also the ways in which they contribute to it. For this purpose, both groups decided about a

---

[10] The shape of the gesture can be described as follows: The thumb and index finger are held as if they were holding something, like an imaginary line segment, between them. The fingers indicate a starting point and end point of this segment. Depending on the mathematical context, i.e. depending on whether the context is geometrical or numerical, narrowing the fingers means a line segment getting shorter or a number that is decreasing.

piece of data together: Dreyfus and Kidron (in the following referred to as the 'epistemic group') searched for episodes that bear a "potential for the emergence of constructing knowledge" (ibid., p. 131), based on an a-priori-analysis of the task. The 'semiotic group' (Arzarello and Sabena), however, preferred to analyze episodes in which gestures were used in order to communicate. The amount of data interesting for both groups resulted to be very small (ibid., p. 129). Therefore, their initial plan to compare two parallel analyses of the same episode was changed in favor of integrating gesture analysis to the epistemic analysis of an episode in which the process of knowledge construction has been found to be hardly accessible from the verbal utterances alone. This way they aimed to enrich the theoretical and methodological framework of the epistemic group. The semiotic perspective on the epistemic process allowed the epistemic group to attain better comprehension of the students' construction of knowledge by considering the gestures as an individual and a social means of expression to "better understand the interplay between the social and the cognitive dimensions" (ibid., p. 149). In fact, they found that gestures contribute to the process of constructing knowledge in different ways, for example to "illustrate or clarify to [the students] themselves the mathematical objects and their properties rather than to communicate to one another" (ibid., p. 146). Dreyfus et al. call gesture that fulfil such a function *epistemic gestures* (ibid. p. 146). Given another example, they identified a gesture that recurred during the epistemic process. The semiotic group interprets these to recall meaning as embodied in a reasoning situation, while the epistemic group considers it as an additional sign for consolidating constructed knowledge, confirming what they have already reconstructed from "the progressively more elaborated language" (ibid., p. 136).

While at first sight this seems to deal with exactly the same issue as does the study presented here, crucial differences are evident when taking a second look: The epistemic group focuses on *individual* processes of constructing knowledge. That is, the reconstructions of the epistemic processes concern the individual constructs of the students as they develop in social interaction through communication as cognition-driven. It does not take into account the *social* constructs developing in social interaction. Furthermore, a case study cannot provide in-depth insights into such a phenomenon as epistemic gestures, but can indicate their existence and significance.

However, this case of networking shows that the integration of gestures in the analysis of epistemic processes can supplement an approach that had been mostly restricted to the interpretation of verbal means of expression to this point.

2.4 The Epistemic Role of Gestures in This Project

The current study understands knowledge as constructed as shared among the participants. To analyze this, the epistemic action model developed by Bikner-Ahsbahs will be used (Bikner-Ahsbahs, 2005; 2006, see also chapter 4.1). Bikner-Ahsbahs obeserved

2.4 The Epistemic Role of Gestures in This Project

how interest-dense situations (Bikner-Ahsbahs, 2004; Bikner-Ahsbahs, 2005, pp. 119-149) emerge in the mathematical classroom and describes characteristics of such situations in which mathematical knowledge is constructed (Bikner-Ahsbahs, 2005). As a consequence, her findings lead to conclusions about conditions that foster and hinder the emergence of interest-dense situations and with that, of the construction of mathematical knowledge. Her epistemic action model embraces the three epistemic actions of gathering mathematical entities, connecting them and seeing mathematical structures, such as patterns or generalities (GCSt). Knowledge in this regard is understood as shared structural relation generated through connecting and structure-seeing in social interactions (Bikner-Ahsbahs, 2005, pp. 200-204, see also chapter 4.1). The model has been developed based on reconstructions of the social interaction tracing the development of knowledge in a *semiotic sequence analysis* (Bikner-Ahsbahs, 2005; 2006; see also section 5.2.3.1) . However, the reconstruction was framed by speech act analysis and non-verbal signs have been taken into account only when needed for the interpretation of the speech act. Observations of social interactions in the mathematics classroom and also the research presented in this chapter show that non-verbal signs cannot be seen as merely supporting verbal language, but also as a particular part of social interactions in which epistemic processes take place. Hence, excluding them from analysis might lead to overlooking important aspects. This study will focus on gestures as specific kinds of signs to get insights about how their use can foster the construction of mathematical knowledge, but also how it may hinder it.

The review of literature leads to focussing on two aspects of gestures in mathematical epistemic processes:

1. **The communicational aspect:** As a social means of expression, gestures support the verbal utterance and facilitate the expression of non-elaborated ideas so that they are offered to the shared pool of ideas. Gestures may help to compensate for lacking terms (Reynolds & Reeve, 2002), to simplify the verbal expression by using deictics (Bjuland et al., 2008), or to test ideas or approaches without going too much into detail (Sabena, 2007, pp. 260-261). All of these communicative functions are suggested to lighten the cognitive load[11] by allowing verbal utterances to be imprecise up to a certain point while still giving reference to the core of the idea. This unburdening of speech when words are hard to find can also

---

[11] In cognitive science, the cognitive load described the cognitive effort that has to be overcome in order to make progress in learning. Sweller, as one of the first ones to highlight this topic, investigated students' problem solving strategies and how they are influenced by the instructional design (Sweller, 1988). He established Cognitive Load Theory to investigate the factors that influence this effort, integrating the concept of the schema thinking pattern to release this effort: "[...] in contrast to a huge long-term memory, working memory is very limited. Working memory can store and process no more than a few discrete items at a given time" (Sweller, 1994, p. 299). The research just cited suggests that gesture can reduce the number of items and processes stored in order to communicate information by means of words.

be regarded as benefitial from the theory of interest-dense situations: According to Bikner-Ahsbahs (2004), one of the characteristics of interest-dense episodes lies in the progressive dynamics of the epistemic process[12] in which knowledge is socially negotiated. This can be disturbed "by the unwillingness to accept indistinct meanings" (Bikner-Ahsbahs, 2004, para. 6). Her findings reveal that a hindering condition for the emergence of interest-dense situations can be caused by the teacher when "forcing the students to use precise words before the interaction process may continue. This preciseness of words disturbs the flow of ideas and prevents the students from presenting their own ideas" (Bikner-Ahsbahs, 2004, para. 6). The possibility to express ideas in a premature stadium is thus essential for social learning processes and may be benefited by using a communicative function of gestures. In this regard, gestures may ground the shared understanding on visual means of expression rather than on verbal ones.

2. **The representational aspect:** Gestures can themselves provide visual representations of a mathematical object that are not or cannot be represented in inscription. Studies of more advanced learners and of teachers give evidence that the conceptualization of more sophisticated mathematical ideas may be revealed by gestures in the gesture space in front of a person (Edwards, 2009; Marghetis et al., 2014; Sabena, 2007). This can also be endowed with mathematical meaning to be used as shared in social interaction, as described by Yoon, Thomas and Dreyfus (2011). It allows to collaboratively elaborate more general ideas by detaching the reference from the concrete inscription and provides the potential for using the gesture space as an experimental space for trying out and representing ideas (ibid., p. 390). The difference to the communicational function as mentioned above is seen here in the potential to represent something to work on and not to compensate speech, as mentioned above. This has been seen also in the study of Chen and Herbst (2013), where gestures have been used to complement inscription in order to carry out or test hypotheses: "[Iconic] gestures represented virtual mathematical objects [...] or mathematical relationships [...]. Students also gestured to animate the diagrams" (ibid., p. 302). Gestural representation can thus be considered to potentially bring forward the epistemic process by representing the unrepresented.

Nevertheless, also possible pitfalls are to be mentioned: Yoon et al. (2011) point out the aspect of inaccuracy of gestures that "may lead to significant miscommunications and misconceptions" (ibid, p. 390). They suggest that there need to be moments of synchronizing details, for example by producing inscriptions. Furthermore, and probably of more importance, the meaning of gesture is not given per se. It needs to be elaborated by the students together with the knowledge of the object it refers to

---

[12] Literally, she refers to this as *'progressive Erkenntnisdynamik'* (Bikner-Ahsbahs, Kidron, & Dreyfus, 2011) in German and "dynamic of the epistemic process" in English (Bikner-Ahsbahs, 2004).

2.4 The Epistemic Role of Gestures in This Project 17

in order to express *shared* meaning in social interaction and to be potentially have a beneficial effect for all the participants.

This study aims at investigating how both, the communicational as well as the representational aspect of gestures may play a part in fostering and in hindering social processes of constructing mathematical knowledge. An analysis of the epistemic process carried out by epistemic actions according to the GCSt-model shall thus be combined with an analysis of gestures within the semiotic bundle, shedding light on gestures' contribution to social epistemic processes.

3 Shaping the Semiotic Approach
An Introduction of Semiotic Concepts and Notions to be Used

This chapter will provide the semiotic basics that this study is grounded on. First, a concept of signs suitable for analyzing social processes of learning mathematics will be presented. This concept provides the framework in which the focus of this study, the use of gestures in the learning of mathematics, becomes embedded.

3.1 Signs: The Peircean Semiotics

When dealing with semiotics, there are three main traditions that need to be considered for defining what is meant when talking about a sign: the one founded by de Saussure, a Peircean tradition, and the Vygotskyan school (Radford, 2006). The Peircean concept of signs provides some striking features that enable describing and understanding social processes of learning mathematics in social interaction as carried out by means of signs. While de Saussure understood signs as defined by the dyadic relationship between the *signifier* (the 'corpus' of the sign) and the *signified* (the mental concept or 'meaning' referred to by the sign), he predominantly studies linguistic signs. The Peircean comprehension is broader and includes everything that becomes a sign by being interpreted. The two concepts differ in how the 'meaning' of a sign is understood: For Saussure, a sign has a conventional meaning within the larger frame of a language (de Saussure, 1995, p. 145). The concept offered by Peirce comprises that the meaning of a sign is not given per se, but depends on different factors, such as the interpreter, context, and interpretation background. This is crucial for the observation of mathematical learning processes, in which the objects referred to by signs are not directly accessible: Considering the factors just mentioned, the same sign can be understood differently, and the mathematical meaning of a sign needs to be negotiated to be shared in social interaction. Vygotsky focuses on the sign's effect on the individuals mind rather than as the sign as a means to interact with others. He described how individual meaning develops in a process of internalization from its social use so that its function changes from an interpersonal (between persons) function to an intrapersonal one (within the mind of a person) (Vygotsky, 1978, p. 57). This understanding makes it difficult to describe how social interaction takes place by means of signs as becomes observable within the mathematics classroom.

The Peircean concept of a sign not only includes how meaning becomes negotiable rather than given; it also makes the epistemic process a social development of 'mathematical signs'[13] that goes hand in hand with the social development of mathematical knowledge. These considerations will be described thoroughly in the following section.

3.1.1 Definition of Signs and the Process of Semiosis

Following Peirce, a

> sign, or representamen, is something which stands to somebody for something in some respect or capacity. It addresses somebody, that is, creates in the mind of that person an equivalent sign or perhaps a more developed sign. That sign which it creates I call the *interpretant* of the first sign. The sign stands for something, its *object*. It stands for that object, not in all respect, but in reference to a sort of idea, which I have sometimes called the *ground* of the representamen.
>
> (Peirce, CP, 2.228)[14]

Hence what is termed 'sign' always needs to be seen as being related to an *object* that is represented, inducing an *interpretant* that is itself a sign that represents the object with respect to a certain perspective. This triadic relationship can be represented as follows (Fig. 3.1, following Hoffmann, 2001, p. 3):

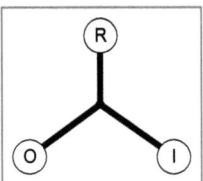

Fig. 3.1: **A Peircean understanding of signs as a triadic relation between representamen (R), object (O), and interpretant (I)**

The definition by Peirce also includes 'signs in one's mind' that is 'thought signs'. In social interaction, we consider only external, perceivable signs. The *object* can be anything the sign may signify, such as perceivable objects, abstract concepts, notions, ideas, and underlying meaning that organizes social interaction, as well. The *interpretant* can now be understood as the reaction on the representamen, constituting a new sign and by this, making it a sign as being interpreted.

---

[13] This means signs being used with reference to mathematical ideas.
[14] According to conventions established in literature on semiotics, it is referred to works of Peirce in terms of '(opus, section/page)'. CP stands for 'Collected Papers' and 2.228 for 'Volume 2, paragraph 228'.

With the *ground*, the presented concept of 'sign' comprises the background against which the interpretation refers to an object in a global sense, meaning everything that influences the shaping of the object. Hoffmann and Roth (2005) describe this as follows:

> Here we take 'ground' like in 'background', that is, as a certain horizon that specifies a certain triadic relation due to the fact that (a) any object can be represented by many different signs and (b) any sign can have many different interpretants – both according to different 'respects' of the object to be represented.
>
> (Hoffmann & Roth, 2005, p. 112)

In social interaction, the shared understanding of any signs is thus always dependent on the shared ground that may change and develop in time in a process that Peirce calls *semiosis*:

> A sign is anything which is related to a Second thing, its Object, in respect to a Quality, in such a way as to bring a Third thing, its interpretant, into relation to the same Object, and that in such a way as to bring a Fourth into relation to that Object in the same form, ad infinitum. If the series is broken off, the Sign, in so far, falls short of the perfect significant character. It is not necessary that the interpretant should actually exist. A being in future will suffice.
>
> (Peirce, CP 2.92)

This quotation shows how the interpretant is again a new representamen representing the object and being related to a new interpretant, "ad infinitum". The object represented by the interpretant is always an "immediate object" (Hoffmann, 2005, p. 51), the object in its current formation. The actual, *"dynamic object"* (ibid.) stands in the 'end' of the infinite process of semiosis. Hookway explains the distinction between the understanding of these two perspectives on the object as the difference between "two answers to the question: what object does this sign refer to? One is the answer that could be given when the sign was used; and the other one we could give when our scientific knowledge is complete" (Hookway, 1985, p. 139). This mentioned completeness is, in fact, a pure, unreachable ideal. Randell writes in this regard:

> [T]he immediate object is the object as it appears at any point in the inquiry or semiotic process. The [dynamic] object, however, is the object as it really is. These must be distinguished, first, because the immediate object may involve some erroneous interpretation and thus be to that extent falsely representative of the object as it really is, and, second, because it may fail to include something that is true of the real object. In other words, the immediate object is simply what we at any time suppose the real object to be.
>
> (Randell, 1977, p. 169)

In mathematics, there are no real objects but substituting the "really" in the second sentence by "ideally" and the "real" in the last sentence by "mathematical", this description can easily be adapted. The quotation explicates that the immediate object is not *identical* to the mathematical object and indicates problems possibly arising from this distinction, hampering a mutual understanding of mathematical knowledge.

The description of the semiotic process, of the *semiosis*, also expresses the inseparable character of the triadic relation between the three relata of the sign: No two of them can be considered as dyadic but what constitutes a sign always includes all three of them.

3.1.2 The Trichotomy of the Object: Iconic, Indexical, and Symbolic

Peirce subdivides the three dimensions of the triad (representamen, object, and interpretant) into three categories each, depending on their integration within the triadic relation. This integration is related to the idea that a sign is "anything which is so determined by something else, called its Object, and so determines an effect upon a person, which effect I call its Interpretant, that the latter is thereby mediately determined by the former" (SS, 80-81). To investigate how mathematical objects are negotiated in social interaction by means of signs, the integration of the object becomes especially important, concerning the instantaneous meaning of a sign in development. According to Peirce, an object can determine a sign as an *icon*, an *index*, or as a *symbol*:

> A sign is either an icon, an index or a symbol. An icon is a sign that would possess the character which renders it significant, even though its object has no existence; such as a lead-pencil streak as representing a geometrical line. An index is a sign which would, at once, lose the character which makes it a sign if its object were removed, but would not lose that character if there were no interpretant. Such, for instance, is a piece of mould with a bullet-hole in it as sign of a shot; for without the shot there would have been no hole; but there is a hole there, whether anybody has the sense to attribute it to a shot or not. A symbol is a sign which would lose the character which renders it a sign if there were no interpretant. Such is any utterance of speech which signifies what it does only by virtue of its being understood to have that signification.
>
> (Peirce, CP 2.304)

Furthermore, he terms 'icon', 'index' and 'symbol' as 'three kinds of signs':

> It has been found that there are three kinds of signs which are all indispensable in all reasoning; the first is the diagrammatic sign or *Icon*, which exhibits a similarity or analogy to the subject of discourse; the second is the *Index*, which like a pronoun demonstrative or relative, forces the attention to the particular object intended without describing it; the third or *Symbol* is the

general name or description which signifies its object by means of an association of ideas or habitual connection between the name and the character signified.

(Peirce, CP, 1.369)

Hoffmann summarizes the three ways of integrating the object within the triadic relationship giving a special remark to its role concerning the development of *mathematical thinking*. He considers the trichotomy of the object as *"the* essential instrument of [Peirce'] epistemology and phenomenology" (Hoffmann, 2005, p. 10)

The Iconic Integration:
An icon evokes a certain similarity in terms of 'relational structures' within an object. A representation of a mathematical object can be iconic, insofar as it refers to specific aspects that characterize the object in a specific representational register. Thus, a graph as well as an algebraic term can function as iconic signs associated with a mathematical object. Diagrams are special kinds of icons that carry out relations between other represented entities.[15]

The Indexical Integration
An index refers to something and, because of this, stands in a dyadic relationship to this something. It has no intrinsic meaning and its sole function is to refer to what it forces attention to. Examples are pointing gestures, deictic phrases or highlights within an inscription.

The Symbolic Integration
Symbols signify objects by habit or convention. This is the case when something is perceived as something regularly so that certain prognoses of its being can be made based on experienced rules, regularities and laws. Examples of symbols are road signs or, in the case of mathematics, certain representations like the relational numerator-denominator structure found in fractions, or the conventionalized $f(x)$-notation of functions.[16]

This distinction may be quite hard to grasp from a theoretical point of view. Giving an example already introduced by Peirce and elaborated by Hoffmann (2001) may render the whole idea more comprehensible. For this, it is important to mention that a sign never *is* an icon, an index or a symbol because what it is seen as always depends on the ground

---

[15] "A *diagram* is a representamen which is predominantly an icon of relations and is aided to be so by conventions. Indices are also more or less used. It should be carried out upon a perfectly consistent system of representation, one founded upon a simple and easily intelligible basic idea." (Peirce, CP 4.418, 1903).
[16] These two examples show as well that symbols can be identified as symbols only related to a certain background. Road signs are not necessarily international and one has to be introduced to the knowledge about them to let them become symbols. One does not know the conventionalized meaning of a road sign naturally and the same holds for representations of mathematical objects.

(see page 21). What is a symbol for someone may be an icon for someone else. Furthermore, signs in general are not that simple but form a relational unit together with other signs. Hoffmann considers this in the following example referring to Peirce:

> He takes the sentence 'Ezekiel loveth Hulda',[17] which can be understood as a diagram insofar that there is an icon linked with indices and a conventional symbol according to the definitions. It may be the case, that we already know the persons mentioned in the sentence from somewhere, than by the names 'Ezekiel' and 'Hulda' would additionally be given symbols which we interpret in a special sense, based on habits, as signs for the persons we already know. But primary for Peirce is that 'Ezekiel and Hulda have to be indices or contain such, because without indices it is impossible to denote what is spoken about' (SEM I 211f., 1895). Even if we do not know them, our attention is directed towards two persons and that is exactly the function of indices.
>
> At the same time, the function of a symbol as that one of an icon fulfills the word 'loveth' in this example. It is iconic insofar that it evokes the idea of two persons in love and by that represents a certain relation between the indices 'Ezekiel' and 'Hulda', and it is symbolic because we assign this relationship on the convention of our language with the sequence of letters l-o-v-e and none else. In this sense the convention 'supports' the iconic function.
>
> (Hoffmann 2001, p. 13, my own translation)

With special remark to mathematics, he adds:

> With symbolically supported icons we are dealing especially in mathematics; for Peirce an algebraic formula like $a + b = c$ is an icon insofar that it evokes a particular relation between a, b and c, which is, at the same time, determined by the conventions of commutativity, associativity and distributivity.
>
> (ibid., p. 13, my own translation)

This characterization of a sign is far from being clear-cut. To the contrary: a sign often reveals iconic, indexical and symbolic features, depending on the larger context in which it occurs. To give an example: A pen placed in a diagram is not per se an icon for a straight line. By sharing the characteristic of being straight, it can be an icon for a straight line and becomes so by being embedded into the specific setting. The easiest way to evoke this iconic relation is perhaps to specify the situation verbally and to refer to the pen as 'a straight line'. Nevertheless, in the surroundings of the diagram, it also may function as an index if the specific location within the diagram is also deemed to be important. This, again, has to be judged in the specific situation. A repeated use of the

---

[17] Hoffmann's sentence "'Hesekiel liebt Hulda' is the translated quote from "Ezekiel loveth Hulda" in the original by Peirce.

pen as a straight line may constitute a convention and, thus, *pen as a straight line* can become a symbol in the course of interaction.

Considering the learning of mathematics, this distinction between index, icon, and symbol has another significance: These signs used in the mathematical classroom or in textbooks are often called, and considered to be, mathematical *symbols*. However, it is by far not taken for granted that this symbolic relationship is as clear for the learner as it is for the teacher.

3.1.3 The Representational and the Epistemic Function of Signs

Hoffmann states: "[O]nly the representation of epistemic qualities inherent in the sign enables further development of the sign and with this a further development of epistemic qualities" (Hoffmann, 2001., p. 1, my own translation). According to him, this develop-ment becomes possible by two functions that are inherent in every sign, a representa-tional and an epistemic function.[18] The representational function is related to the sign standing for something it represents, the epistemic function relates to the possibility to see "through the sign" (Hoffmann, 2005, p. 37) and to constitute knowledge from what is represented. The main difference between the functions a sign can fulfill lies in its familiarity: The representational function comes to the fore when the sign itself becomes a knowledge object and its meaning needs to be made familiar. The epistemic function on the other hand concerns using the sign to refer to a *meaning behind it* that in turn may lead to further insights into the object. Hoffmann and Roth distinguish the kind of expli-cation in this regard depending on whether the mathematical meaning of the sign inter-preted is "knowledge in focus" (representational function) or "collateral knowledge" (epistemic function) (Hoffmann & Roth, 2007, p. 107). Concerning the epistemic func-tion of signs, they write: "We have to have collateral knowledge of a meaning of a sign before we can use it to make distinctions, specify objects and relations, and structure our observations and experiences" (ibid., p. 108). Nevertheless, both functions interact in social processes of learning:

> In communication, obviously both these functions of a sign are realized. The speaker uses signs as means to signify something, and for the hearer the same sign represents something. At the same time however, our semiotic model makes clear that communication works *only if* both partners share the collateral knowledge necessary to use and to understand signs. Misunderstanding

---

[18] Hoffmann and Roth call this function 'epistemological' while their description of this function suggests that they refer to an 'epistemic function': While the term 'epistemological' is used to refer to the theory of the process of coming to knowledge, 'epistemic' means 'concerning knowledge constitution'. Since I investigate gestures in *epistemic* processes, using an epistemic action model to describe these processes, I will stick to the notion of *epistemic* function here as well.

and conflicts result when the collateral knowledge on both sides is different, or not similar enough.

(ibid., p. 108)

The following section will now specify the focus of this study, the use of gesture.

3.2 Gestures: A Framework for this Study

3.2.1 Understanding of Gestures in this Study

As gestures, I consider "idiosyncratic spontaneous movement[s] of the hands and arms accompanying speech" (McNeill, 1992, p. 37), "being done for the purposes of expression rather than in the service of some practical aim" (Kendon, 2004, p. 15). This excludes touching and personal habituated behavior such as playing with rings or a necklace or hair-patting from the definition of gestures. A main aspect of the definition concerns the positioning of the gestures considered within *Kendon's Continuum* (McNeill, 1992, p. 37):

> Gesticulations → Language-like Gestures → Pantomimes → Emblems → Sign Language

McNeill named this typology of gestures in honor of Kendon, who proposed it first (1988). The continuum represents how "(1) the obligatory of presence of speech declines, (2) the presence of language properties increases, and (3) idiosyncratic gestures are replaced by socially regulated signs" (McNeill, 1992, p. 37), each considered from left to right. Kendon furthermore writes:

> At one end of the continuum gesture is used in conjunction with speech, it is global and holistic in its mode of expression, idiosyncratic in form and users are but marginally aware of their use of it. At the other end, gesture is used independently of speech, it is compositional and lexical in its structure and organization and users are fully aware that they are using it.
>
> (Kendon, 2004, pp. 104-105)

Taking into account the aspects of 'obligatoriness of speech', 'awareness of usage', and 'idiosyncrasy', Kendon's continuum can be represented more comprehensively as follows:

3.2 Gestures: A Framework for this Study   27

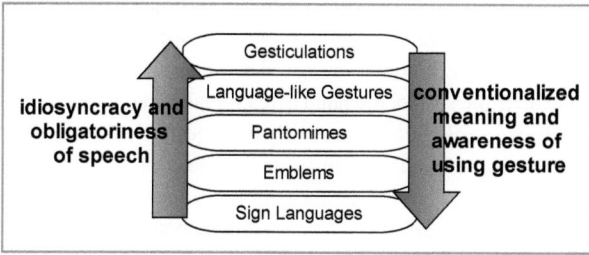

Fig. 3.2:   Kendon's Continuum

The gestures examined in this study are located in the upper part of the continuum, taking into account gesticulations and language-like gestures. While gesticulations mean gestures co-occurring with speech, language-like gestures complement speech within verbal utterances. In this regard, gestures are understood as movements of the hands and arms having a meaning contextually embedded in verbal discourse. These are what Bavelas, Chovil, Lawrie, and Wade (1992, p. 470) call *illustrators*, referring to a recent classification by Ekman and Friesen (1969). Bavelas et al. furthermore propose to distinguish between *topic gestures* and the less frequently occurring *interactive gestures*: "Topic gestures depict semantic information directly related to the topic of discourse, and interactive gestures (a smaller group) refer instead to some aspect of the process of conversing with another person." (Bavelas et al., 1992, pp. 472-473). In this study, only topic gestures related to the topic of mathematical discourse are considered.

Following Kendon, a gesture normally passes three phases (McNeill, 1992, p. 25, referring to Kendon, 1980; Kendon, 2004, pp. 114-115): First, a phase of *preparation*, in which the hand is raised and positioned in the space where the actual gesture takes place. This often is the space in front of the body, framed by the shoulders and the hips, respectively the table while sitting, and is called the *gesture space*[19] (see Fig. 3.3).

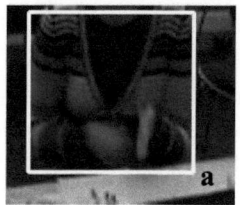 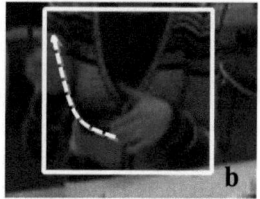

Fig. 3.3:   Gesture space when sitting:
            In front of the body, framed by shoulders and the table.
            a) preparation phase of a gesture, b) stroke of a gesture

---

[19] For a broader description of the gesture space see McNeill (1992, pp. 86-91).

Secondly, what is conventionally agreed on as being the main part and movement of the gesture, the *stroke* is performed (Fig. 3b). The third phase is that of *recovery*, in which the hand is moved to where it comes to rest. The only obligatory phase is that of the stroke. Before and after the stroke, the hand often remains in one position. These phases are called *pre-* and *post-stroke hold* respectively. One *gesture unit* may consist of several strokes between which the hand is held in preparation for the next stroke. '*Gesture phrase*' is defined as one part of the gesture unit meaningfully assigned to one part of the verbal utterance (Kendon, 2004, p. 112). A gesture phrase is constituted by the *stroke* and a *preparation phase* and/or a *poststroke hold*. Fig. 3.4 represents a gesture unit that contains two strokes.

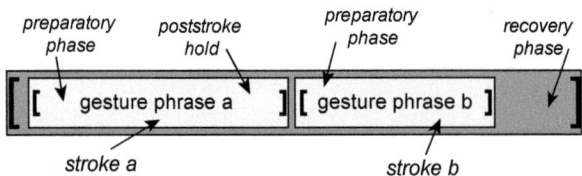

Fig. 3.4:   **Exemplified gesture unit, containing two gesture phrases**

The gesture unit concerns the "entire movement excursion" (Kendon, 2004, p. 124) and since there are two strokes, there are also exactly two gesture phrases within this gesture unit. In this example, the first gesture phrase embraces a preparatory phase, 'stroke a', and a poststroke hold. The second gesture phrase starts with a new preparatory phase which is followed by 'stroke b'. Generally, it depends on the concrete movement whether the poststroke hold of one gesture phrase contains the preparatory phase of the following gesture phrase. In the gesture unit presented in Fig. 3.4, the second stroke is prepared separately.

The synchrony of the phases to the co-occurring speech helps to value the role of what is considered as its meaning (Kendon 2004, p. 112; McNeill 1992, pp. 26-29). The relation between speech and gesture will be described in more detail in the following section.

3.2.2   Gesture and Speech: An Integrative System

Speech, in spoken or in written form, is considered to be the primary channel for communication since its production as well as its comprehension is subject to conventionalized rules concerning syntax, semantics and pragmatics. Producing speech may be the conscious mode of expression we are most familiar with. However, while speaking, people gesture. People do not only use gestures to represent something for others, but also when there is nobody else to see the gestures, for example when they are talking on the phone (de Ruiter 1995; 2000, p. 290). Even blind people use gestures when they are talking to another blind person (Iverson & Goldin-Meadow, 1998; 2001, p. 421). From this it has been concluded that gestures not only have a communicational aspect in the sense that they are accomplished to give information, but that their use also has an influence on cognition, both on an individual and on a social level. Underlying in all the research on gestures is the assumption that gesture and speech are co-expressive: McNeill sees "gesture and the spoken utterances as different sides of a single underlying mental process" (McNeill 1992, p. 1) elaborating on Kendon's work "Gesticulation and speech: Two aspects of the process of utterance" (Kendon, 1980). McNeill claims that "speech and gesture must cooperate to express the person's meaning" (McNeill 1992, p. 11), "each can include something that the other leaves out" (ibid., p. 79). This he bases on unique semiotic properties that are specific for each, speech and gesture[20] (McNeill & Duncan, 2000, pp. 143-144): "Each modality, because of its unique semiotic properties, can go beyond the meaning possibilities of the other, and this is the foundation of our use of gesture as an enhanced window into mental processes."

Nevertheless, the same syntactic and semantic rules that help us understand each other by the use of speech become linguistic stumbling blocks when it comes to communicating ideas that are not fully elaborated. This may concern the lack of adequate notions as well as the requirement of constructing grammatically correct sentences in order to be understood. In chapter 2, the results from Yoon et al. (2011) and Sabena (2007) already pointed out the positive effects gesturing may have on cognitive load. They adopt the *Information Packaging Hypothesis (IPH)* (Kita, 2000, p. 163), assuming gestures to play an important role in reducing cognitive load by organizing spatio-motoric knowledge. Through this, gestures assist in verbalizing this knowledge:[21] "According to the Information Packaging Hypothesis, gesture is involved in the *conceptual planning* of the message to be verbalized, in that it helps speakers to 'package' spatial information into units appropriate for verbalization" (Alibali, Kita, & Young, 2000, p. 593, italics in the original). By the use of gestures, information is parsed into entities that are more convenient to put into words. "[C]onsequently, the collaboration between the two modes [gesture and speech] provides speakers with wider possibilities to organize thought in

---

[20] The lack of rules for gesture-production allows gestures to be produced idiosyncratically while their co-expressiveness to speech allows them to bear meaning as interpreted together with the verbal utterance. Furthermore, the constitution of a meaningful unit of speech or gesture is substantially different. Gestures are *global* and *holistic* in contrast to the *analytical* and *segmented* character of verbal language: The whole gesture determines the meaning of its parts while in verbal language, a sentence derives its meaning from the meaning of the words. A further property of gestures is that they can "combine many meanings" (McNeill 1992, p. 19); they are *synthetic*.
[21] The three conjectures stated in the IPH as proposed by Kita (2000) are:
"1. The production of a representational gesture helps speakers organize rich spatio-motoric information into packages suitable for speaking.
2. Spatio-motoric thinking, which underlies representational gestures, helps speaking by providing an alternative informational organization that is not readily accessible to analytic thinking, the default way of organizing information in speaking.
3. Spatio-motoric thinking and analytic thinking have ready access to different sets of informational organizations. However, in the course of speech production, the representations in the two modes of thinking are coordinated and tend to converge." (Kita, 2000, p. 163)

ways suitable for linguistic expression" (Kita, 2000, p. 180). Furthermore, this packaging also explains situations in which information is first expressed in gesture before a verbalization takes place. This may be the case when words or even whole parts of a sentence are replaced by gestures, providing visual access to an idea by referring to a non-verbal representation of an object. When different information is conveyed in speech and in gesture – when there is a *mismatch* in Goldin-Meadow' notion – gesture is considered to be the path-breaker (McNeill, 2005, p. 137) and gives a more genuine while more spontaneous view on the concept present in a student's mind. Goldin-Meadow distinguishes two kinds of gesture-speech relationships: In a gesture-speech-*mismatch*, gesture conveys a version of a situation differing from that expressed in speech (Goldin-Meadow, 2003, pp. 25-29). According to Goldin-Meadow, the grade of conformity between speech and gesture while expressing thoughts on a mathematical concept reflects to what extent different, but not necessarily competing perspectives on that concept are present in a student's mind. She highlights the cognitive potentials of mismatch-situations, claiming that a propensity to learn can be seen in the students' embodied representation of a new mathematical idea (Goldin-Meadow, 2010, pp. 1-2). This in turn may be useful for the teacher to help the students activate their implicit knowledge, made apparent in gestures (Broaders, Cook, Mitchell, & Goldin-Meadow, 2007).

Another consideration that allows inferring from how a thought is expressed in speech and gesture is described by McNeill's concepts of *Growth Point* (GP) and *catchment* (McNeill, 1992, pp. 219-220; McNeill & Duncan, 2000). He defines the *Growth Point* as

> the theoretical starting point, in a microgenetic sense, of a speech-gesture combination – 'growth' in the sense that it is the seed out of which speech and gesture grows. A growth point is a combination of imagery and linguistic categorical content. It is a single idea unit. It is a unit, moreover, that holds not only content but also implies the immediate context of speaking from which it has been differentiated.
> (Montredon et al. 2008, p. 173, citing from a correspondence with McNeill from January 2003)

Furthermore, it links imagistic and linguistic representation, combining an idiosyncratic meaning system (gesture) and a socially constituted meaning system (verbal language) in a minimal idea unit (Röpke, 2011, p. 2). The 'differentiation of context' mentioned in the citation above was described as "novel departure of thought from the presupposed background" in McNeill (McNeill, 1992, p. 220). This background or (semantic) 'context' can be disclosed from an observation of the *catchment(s)* involved in its constitution (McNeill, 2005):

> A catchment is recognized when two or more gesture features recur in at least two (not necessarily consecutive) gestures. The logic is that the reoccurrence of an image in the speaker's thinking will generate recurrent gesture features. Recurrent images suggest a common discourse theme.
>
> (ibid., p. 116)

These features are related to the appearance of the gesture and can be revealed in form, movement or dynamics of the gesture. Further, he writes:

> A catchment is a kind of thread of visuo-spatial imagery that runs through a discourse to reveal the larger discourse units that emerge out of otherwise separate parts.
>
> By discovering the catchments created by a given speaker, we can see what this speaker is combining into larger discourse units – what meanings are being regarded as similar or related and grouped together, and what meanings are being put into different catchments or are being isolated, and thus are seen by the speaker as having distinct or less related meanings.
>
> (ibid., pp. 116-117)

Following this, the identification of catchment-gestures makes it possible to determine *which* contexts are related for the speaker (contexts in which the gesture appears). Furthermore, taking into account which feature reoccurs, it becomes interpretable *what* the relation consists of. The larger discourse units mentioned in the last paragraph frame the contextual background in which the growth point emerges. It can then be identified as a starting point of a minimal idea unit, tracing catchments also in retrospect.

3.2.3 Gesture Classification

The meaning of a gesture always needs to be inferred in relation to the semantic content of speech that provides a frame for interpretation. McNeill constitutes a classification of gestures distinguishing *iconics, metaphorics, beats, (cohesives),*[22] and *deictics,* all standing in a certain relationship to the semantic content of speech (McNeill, 1992, pp. 12-19):

Iconics: Iconic gestures "bear a close formal relationship to the semantic content of speech" (McNeill, 1992, p. 12), where this semantic content refers to a *concrete* object, event or action.

Metaphorics: According to McNeill, metaphoric gestures are like iconic gestures but differ in the kind of references to which they are affiliated: The semantic content of the

---

[22] The cohesive gestures belong to the original categorization of gestures. With the development of the concept of 'catchments', McNeill abandoned cohesive gestures and narrows down to an "Iconic-Metaphoric-Deictic-Beat Quartet" (McNeill 2005, p. 34).

accompanying speech refers to an *abstract* idea, event, object or action. Hence, the metaphorical character of those gestures arises solely from their relation to metaphorical speech.[23]

Beats: Beat gestures are quick and short movements of the hands, reminiscent of hitting a musical beat as it is related to prosody, the melody of speech. They typically have two movement components (such as 'up-down') and mostly lack a preparation phase. McNeill states that "the semiotic value of a beat lies in the fact that it indexes the word or phrase it accompanies as being significant, not for its own semantic content, but for its discourse-pragmatic content" (McNeill, 1992, p. 15).

Deictics: These are the familiar pointing gestures, used to indicate concrete objects and events as well as abstract ones. The latter deictic gestures also have a metaphorical character. For example pointing to some point on the left or behind when the utterance refers to some event in the past. One property that is explicitly mentioned for deictics is that they also can be executed by other parts of the body, like the head or the nose, and by manipulating artifacts, like pointing with a pen or even by the positioning of a cursor in a computer environment.

(Cohesives: McNeill speaks of a cohesive when a gesture accompanies speech and is, after a temporal discontinuation, repeated, building a thematic link between these two utterances. These are the gestures that establish catchments such that in his later work, McNeill replaces the cohesive gestures with catchments instead (McNeill, 2005, p. 41).)

One single gesture can show aspects of different characteristics, it is very rarely clearly attributed to one class of gestures. For example, an iconic gesture can also have a deictic character when it locates the represented object in relation to others. This can only be decided by considering the verbal utterance and the situation and this affordance again makes apparent the deep connection to speech. In his later work, McNeill refrained from the idea of categorization and began to talk about these characteristics as 'dimensions' (McNeill, 2005, pp. 41-42).

While beats and cohesives particularly have a discursive function, iconics, metaphorics, and deictics are related to the representation or indication of objects, ideas, or concepts. The discursive function is similar to what Kendon grasps as 'pragmatic function', "by which we mean any of the ways in which gestures may relate to features of an utterance's meaning that are not a part of its referential meaning or propositional content" (Kendon, 2004, p. 158). Iconics, deictics, and metaphorics are those gestures he describes as bearing a 'referential function' (ibid., pp. 159-160). In this regard, these three dimensions of a gesture can be seen as related to the three dimensions of the sign in a Peircean sense,

---

[23] The concept of metaphors is discussed thoroughly in section 4.2.1.

described in chapter 3.1.2.[24] The considered reference (or 'semantic meaning') of the gesture correlates to the dimension of the object of the sign in the following sense: An iconic gesture/sign refers to something according to similarity. An indexical gesture/sign (deictic gesture) refers to something by a dyadic relationship, that is by pointing at something else it refers to. The correspondence between 'metaphoric gesture' and 'symbol' is based on the assumption that the understanding of metaphors needs to be carried out based on their conventionalized meaning,[25] just as the meaning of symbols.

A Classification Suitable for this Study
The terming of 'metaphorics' as a gesture dimension (see pages 31-32) becomes problematic in mathematics since mathematical objects are abstract per definition. In a narrow sense, there are no iconic gestures referring to mathematical objects but only to other representations of them. To highlight this, Edwards (2009) proposes a distinction between *iconic-physical* and *iconic-symbolic* gestures: "Rather than referring to a concrete object in and of itself, the [iconic-symbolic] gesture refers to a symbolic, written inscription, which in turn represents a specific mathematical entity or procedure" (ibid., p. 138).[26] Iconic-physical gestures on the other hand refer to actual, concrete material so that its use or they themselves are being represented by the gesture. *In mathematical talk, I thus consider those gestures metaphoric that represent a mathematical idea without being iconic-physical or iconic-symbolic.*

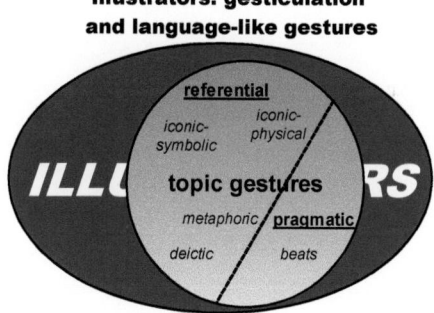

Fig. 3.5: **Overview on relations between gestures examined in this study**

---

[24] The existence of such a correspondence between these three dimensions of gestures and the iconical, indexical, and symbolic integration of the object in a sign has already mentioned by Sabena (2007, p. 42).
[25] See chapter 4.2.1 for further explanations on metaphors.
[26] The use of the term „symbolic" is considered as problematic when considering gestures as signs in a Peircean sense: For Peirce, signs are symbols when they are interpreted according to rules or conventions (see chapter 3.1.2). The mere iconic reference to an inscriptive representation of a mathematical object does not imply a reference to the mathematical object. It is rather seen as an iconic reference to a mathematical notation, regardless of whether the mathematical object is connected to the representation or not.

4 Perspectives on Learning
How Mathematical Learning is Seen in This Study

Learning, in its general sense, means to acquire knowledge so that when something has been learned, it is known more than before. Rather than as a *product* of learning, knowledge is considered as temporal, always being in a state of development that potentially never reaches an end. Nevertheless, this *process* of learning cannot proceed in a vacuum but needs an external stimulus. In this chapter different approaches towards learning mathematics will be presented, providing different possibilities of how such an external stimulus may look like. Together, they shape a more comprehensive picture on mathematical learning processes as they can be observed in the mathematics classroom.

4.1 Learning as a Social Process: An Epistemic Action Model

In the mathematics classroom, learning takes place through interaction with peers and with the teacher. In this regard, learning is understood as a social event in which knowledge is constructed socially among individuals. As Bauersfeld (1992) writes:

> From a constructivist perspective, again it is crucial that the children develop their constructive competence through creative experimenting, commenting about their images, ideas, and expectations, and particularly through adapting to adequate issues in interaction with their teacher and their classmates.
>
> (ibid., p. 472)

That is, knowledge needs to be experienced as shared in social mathematical practice, negotiated as consistent for the participants of the interaction.

This process of knowledge construction in social interaction can be analyzed either by focusing on the individuals or by considering the knowledge "taken as shared" as referred to by Voigt in one of his studies (Voigt, 1995): "Through their discussion, the students and teacher constituted an explanation that perhaps neither would produce individually. They arrived at knowledge taken as shared" (ibid., p, 183). The first perspective has been chosen to be analyzed by Dreyfus, Hershkowitz, and Schwarz (2001). They shaped an *individual* approach on socially constructed knowledge by introducing the epistemic action model RBC+C (Dreyfus, Hershkowitz, & Schwarz, 2001; Hershkowitz, Schwarz, & Dreyfus, 2001; Schwarz, Dreyfus, & Hershkowitz, 2009).[27] On the contrary, the *social* approach is considered by Bikner-Ahsbahs (2004; 2005) in her investigations on the emergence of situated collective interest, referred to as *interest-dense situations*. These situations are characterized by the three following features in collective activities during classroom discussions:

- one after the other students get involved in the activity (**involvement**).
- one after the other students construct continuously farther-reaching meanings (**dynamic of the epistemic process**)
- the value of the situation is concerned with maths (**mathematical valence**)

(Bikner-Ahsbahs, 2004, third page, bold in the original)

The GCSt-Model for Social Epistemic Processes

To identify conditions that foster or hinder the epistemic process within interest-dense situations, she developed an epistemic action model that can be applied to reconstruct and analyze the students' process of constructing mathematical knowledge as carried out in social interaction. This model embraces the three epistemic actions of *gathering* mathematical entities, *connecting* them and *seeing* mathematical *structures* such as patterns or generalities (GCSt). Within this approach, knowledge is considered as structural relations generated through the actions of connecting and structure-seeing in social interaction (Bikner-Ahsbahs, 2005, p. 196; 2006, p. 166).

- Gathering: The students collect mathematical entities that seem to be helpful in order to fulfill a need. Priwitzer found that this gathering of information may be *directed* towards a certain approach to follow, *associative* (or *open*) in order to get an overview of the possibilities to develop an approach, or *resumptive*, when knowledge is repeated that has already been negotiated (Priwitzer, 2010, pp. 54-55).
- Connecting: Connecting can be observed when relations between gathered entities are identified such that some are linked.
- Structure-seeing: This takes place when students recognize a certain generality or pattern. Here, "a new entity is built or a known entity is re-built in a new context" (Krause & Bikner-Ahsbahs, 2012, p. 19). Situations in which a structure is seen can be identified in verbal markers such as "always", "all" or "everywhere", that is by linguistic expressions of generality.

---

[27] Understanding knowledge as 'Abstraction in Context' (AiC, Schwarz, Dreyfus, & Hershkowitz, 2009), knowledge construction according to the RBC+C-model starts with a *need for a new construct* (NNC), encompasses the three nested epistemic actions of recognizing, building-with and constructing, and then becomes consolidated (see also (Dreyfus, 2012) for a comprehensive overview).

Furthermore, these structures can be concretized by means of examples or linked to other aspects in order to reason the structure. When structures are applied or continued in such a way, a gathering or connecting action *extends the structure*.

Within the social interaction, epistemic actions are carried out by means of sign. Their use allows communicating observations and insights, and organizing collaboration in social working processes (Seeger, 2006, p. 267). So far, the epistemic process has been reconstructed mainly by interpreting speech as the most conscious and most conventionalized means of expression. The students' verbal utterances are thus considered to provide the context that makes it possible to interpret, understand and share less conventionalized signs like gestures or inscribed sketches.

4.2 Embodied Thinking and Learning: How our Body Shapes our Mind

As has just been described, this study is guided by an understanding of learning as constructing knowledge in social interaction with peers. In this regard, a hypothesis counts as *knowledge* when it proves to be viable in social interaction and finds its expression in it (Knoblauch, 2010, pp. 158-159), while *meanings* can develop individually. These meanings are seen to be coined by the experience with the environment, in other words, *by the bodily existence and the being in the physical world* (Núñez, Edwards, & Matos, 1999, p. 53). Embracing also unconscious modes of perceiving and understanding abstract concepts, Lakoff and Núñez (2000) emphasize the role of bodily experience as source for all conceptual understanding:

> Human concepts and human language are not random or arbitrary; they are highly structured and limited, because of the limits and structure of the brain, the body, and the world.
>
> (Lakoff & Núñez, 2000, p. 1)

This integration of the body in the way of thinking or, in other words, of the *embodied mind*, has entered cognitive science, especially considering a "mind-based mathematics, limited and structured by human brains and minds" (Lakoff & Núñez, 2000, p. 4):

> Mathematics, as we know it is human mathematics, a product of the human mind. Where does mathematics come from? We create it, but it is not arbitrary – not a mere historically contingent social construction. What makes mathematics nonarbitrary is that it uses the basic conceptual mechanisms of the embodied human mind as it has evolved in the real world. Mathematics is a product of the neural capacities of our brains, the nature of our bodies, our evolution, our environment, and our long social and cultural history.
>
> (Lakoff & Núñez, 2000, p. 9)

The role of the embodied mind in the acquaintance of mathematics is founded on three basic premises stated by Núñez (2000, p. 6): The first one concerns the abovementioned functioning of our physical bodies in the physical world and the experiences brought out by that, as already mentioned above (*"the embodiment of the mind"*). The second hypothesis states that we are not aware of all our thinking. Indeed, "most cognitive processes is [sic] unconscious" (ibid., p. 6) (*"the cognitive unconscious"*). The third aspect concerns the ability to think metaphorically.[28] Its major importance becomes apparent by the following description of what Lakoff and Núñez call the *metaphorical thought*:

4.2.1 The Role of Metaphors in the Conceptualization of Mathematics

Lakoff and Johnson describe the notion of a metaphor in a general way: "The essence of metaphor is understanding and experiencing one kind of thing in terms of another" (Lakoff & Johnson, 1980, p. 5). Taken as a starting point to deal with conceptualization in mathematics, Lakoff and Núñez motivate and define the term of *conceptual metaphor* as follows:

> For the most part, human beings conceptualize abstract concepts in concrete terms, using ideas and modes of reasoning grounded in the sensory-motor system. The mechanism by which the abstract is comprehended in terms of the concrete is called *conceptual metaphor*.
>
> (Lakoff & Núñez, 2000, p. 5, italics in the original)

The relevance of these aspects, especially for the learning of mathematics as science dealing with objects not directly tangible, is standing to reason considering that access to mathematical objects and ideas is always approached from concrete representation. They can only depict specific cases or situations; the generality of a mathematical concept is itself not representable. Núñez (2008) encapsulates the essential meaning of metaphorical thought for the comprehension of mathematics: Mathematical concepts

> don't exist in any real perceivable world. They are metaphorical in nature. It is important to understand that these conceptual metaphors and metonymies

---

[28] The notion of metaphor originally stems from linguistics and is a rhetorical means by the use of which one conceptual domain is mapped into another one. By expressing something in terms of something else, a whole system of relations is mapped, such that an entire conceptual environment can be understood via another conception that may be more familiar or more illustrative. The domain referred to is called *source domain* and conveys how the relationships in the *target domain* shall be understood (Lakoff & Johnson, 1980). As a notational convention, the conceptual domains of the target and source are written in capitals. A case in point is the famous metaphorical expression "Love is a battlefield" referred to by Lakoff and Johnson (ibid.): "Love" is the target from target domain LOVE and "battlefield" the source, from source domain WAR. A whole system of associated terms can now be understood by the suggested mapping: For instance, if love is understood as a battlefield than lovers can be seen as soldiers. Of course, love is not literally a battlefield. By using this metaphor, a certain view on the target is revealed by the reference to the source domain.

4.2 Embodied Thinking and Learning: How our Body Shapes our Mind

are not simply embellishments added on top of formalisms, or 'aids' to understand these formalisms. They are in fact constitutive of the very bodily-grounded forms of sense-making that make mathematical ideas possible.

(Núñez, 2008, p. 103)

According to Lakoff and Núñez (2000), there are two different types of conceptual metaphors to consider in mathematics: *grounding metaphors*, which map everyday experience onto abstract concepts, and *linking metaphors*, concatenating different mathematical domains:

- *Grounding metaphors* allow fundamental mathematical concepts to be based in real world experience, such as addition or subtraction are grounded in adding and taking away things from a collection. While the source domain, in which the everyday illustrative concept comes from, is the real world, the target domain is inside mathematics.
- *Linking metaphors* provide the possibility to treat one mathematical concept in different mathematical domains. This includes the switching between different representational registers, for example between the graphical representation of a function and its function equation in the algebraic register. Lakoff and Núñez refer to ideas approached by linking metaphors as more sophisticated mathematical ideas (Lakoff & Núñez, 2000, pp. 52-53).

Grounding metaphors that rely on experiences of spatial relation use *image schemata*[29] (Johnson, 1987, p. xiv). Examples for these are the *container schema* (Lakoff & Johnson, 1980, pp. 29-32; Lakoff & Núñez, 2000, pp. 30-34) and the *source-path-goal schema* (Lakoff & Núñez, 2000, pp. 37-39). The container schema concerns spatial relations like 'in' and 'out' and becomes applicable in a mathematical context in concepts like *basic arithmetic* ('cardinal numbers as quantity of objects in a collection: A collection is a container and numbers can be conceptualized as collections) (Lakoff & Núñez, 2000, pp. 54-65) and also *substitution* ('putting something *in* for something else '). The source-path-goal schema "is the principal image schema concerned with motion" (Lakoff & Núñez, 2000, p. 37) and it is considered to be "ubiquitous in mathematical thought" (ibid., p. 38) due to its verbal manifestation in *fictive motion*: In fictive motion, concepts or objects are incarnated so that they get ascribed abilities of movement or behavior linked to motion (Talmy, 2000, pp. 103-104). For example, speaking of "The Equator [sic] *passes through* many countries" or "The fence *stops* right after the tree" (Núñez, 2008, p. 104, italics in the original) as if a dynamic process is accomplished by the

---

[29] Johnson defines image schema as "a recurring, dynamic pattern of out perceptual interactions and motor programs that gives coherence and structure to our experience" (Johnson, 1987, p. xiv). They are characterized by the three following aspects: "(1) recurrence across many different experiences; (2) a relatively small number of parts or components; (3) an internal structure that supports inferences" (Dodge & Lakoff, 2005, p. 59)

(abstract concept of the) Equator or the fence. Following Núñez (ibid.), fictive motion needs a *trajector*, which is the object or concept that 'moves' and a *landscape* in which the movement takes place. The important property of the trajector is that it, even though it does not move itself, should theoretically have the capability to move or to give something else the opportunity to move (i.e. to move on it, to move along it, etc.). Investigating how mathematics and mathematical thought is constituted from metaphors, Núñez combines the metaphorical nature of the mathematical world with the fictive motion of the entities in this 'metaphorical reality' (Núñez, 2008). For mathematics, as being metaphorical in nature as mentioned above, "the trajector has always a metaphorical component. That is, the trajector as such can't be literally capable or incapable of enacting movement, because the very nature of the trajector is imagined via metaphor" (ibid., p. 104). Núñez claims that in the case of mathematics, the metaphorical understanding of the landscape and the trajector has to be investigated first to examine how "fictive motion operates on a network of precise *conceptual metaphors*" (ibid., p. 105, italics in the original).[30]

The correspondence between source domain and target domain in metaphorical thinking does not necessarily need to be captured in the verbal modality. Moreover, it turned out to be useful integrating the idea of conceptual mapping into a *multimodal* approach on thinking:[31]

> Metaphors are central instruments in cognition, and do not only manifest themselves in language but also in pictures and sounds. What constitutes a metaphor, however, is partly affected by the medium in which it occurs. In the case of metaphors involving one or two domains that are pictorially represented, such (a) domain(s) is/are inevitably rendered in highly concrete ways, involving specific forms, textures, and colours, all of which may play a role in the mapping. In addition, the manner of representation and the material used (see Kress & Van Leeuwen 1996: chapter 7; Forceville (1999) is a critical discussion of this important book) to render it may influence the construal of the metaphor as well as its interpretation.
>
> (Forceville, 2009a)

---

[30] Núñez exemplifies his train of thought by imagining a point moving from one location within the Cartesian plane to another one: A point is defined by its coordinates and changing its location would mean to change its defining properties. Thus it refuses all possibility to enact movement. The conceptual metaphor NUMBERS ARE LOCATIONS IN SPACE now allows to understand points as spatial locations and lines that are given by functions as traces of their movement (Núñez, 2008, pp. 104-105).
[31] While perceptual aspects of multimodality will be discussed in chapter 4.2.2, the multimodality referred to at this point concerns sensory channels – sight, hearing, touch, taste, and smell. In this regard, Kress defines multimodality as the "idea that communication and representation always draw on a multiplicity of semiotic modes of which language may be one" (Kress, 2001, pp. 67-68).

Metaphors in which the source and the target domain are represented in different modalities are called *multimodal metaphors* (Forceville, 2009b, p. 24). In linguistic metaphors, the word "is" indicates the conceptual equality, and with that the mapping between two conceptual domains (Forceville, 2009b). Since at least one of the domains is non-linguistic in the case of a multimodal metaphor, the conceptual link between them needs to be explicated differently, as has been described by Forceville (2009b, pp. 31-32). According to him, the following possibilities of establishing a conceptual mapping can occur in isolation or in combination:

- "perceptual resemblance" (Forceville 2009b, p. 31) (same color, shape or style of representation that evokes a certain association of similarity, for example shaping a frame in the air by gesture while verbally referring to a thematic frame),
- "filling a schematic slot unexpectedly" (ibid.) (e.g., a dog making the sound of a sheep, suggesting DOG IS SHEEP and with that, for example that the dog is harmless),
- and, as a third possibility, "simultaneous cuing" (ibid.) (suggesting the analogy by the simultaneous use of two modalities in a bold manner[32]) (Forceville, 2009b, pp. 31-32.).

In the gestural-linguistic case, that is gesture representation as source or as target domain and speech as the other one respectively, Mittelberg and Waugh (2009) do not consider metaphors only but point out the importance of metonymy to build metaphors on. Similar to metaphors, metonymic relationships function by referring to one thing by means of another one. Different to metaphorical meaning, this metonymic relationship is not established by analogous similarity but by reference evoking some association by contiguity. One example of a metonymic relationship is the PART FOR WHOLE-relationship that can be identified when saying "This is my mother" while showing a photograph of the face of a person. It is only the part of the mother showed which then stands for the whole mother. Actually, this is an example for a connection for metonymic and metaphoric interpretation in series: Metonymically, the face stands to represent the whole person, not only the face. Metaphorically, the picture is understood to "be" the mother, not to represent her; the target (the represented person) is understood in terms of the source (photography as representation). According to Mittelberg and Waugh (2009), the process of interpreting a multimodal speech-gesture-metaphor often requires a two-step interpretation like the one just presented: first a metonymical part of the gesture has to be identified before the metaphorical meaning of the utterance can be interpreted.[33] In everyday life, these links are made automatically, due to conventions of communication experiences and developed when interacting with others. In

---

[32] As an example for this, Forceville proposes to think of a picture of a kiss in combination with the sound of a car crash that may suggest that KISS IS DESASTER (Forceville, 2009b, pp. 31-32).
[33] For further explanation, see Mittelberg and Waugh (2009).

mathematics, such 'chains of interpretation'[34] need to develop from experience as well to become linked to meaning. For example, segments of graphs of functions are seen to represent the entire function in a PART FOR WHOLE-relationship. In addition, the object "graph" is conceptualized in a metaphoric way when the graphical representation is considered the object itself.[35]

Concerning fictive motion, Lakoff and Núñez discuss the example of "the function approach[ing] zero" (Lakoff & Núñez, 1997, p. 33) where the "function" works as a substitute for 'the y-value of the function' (metonymy). In a second step, fictive motion can be identified as a 'movement' of the y-value of a function for increasing x-value. To think of a multimodal metaphor, this covariational aspect of the function 'movement' of y-value' in dependence of x-value increasing, detectable in fictive motion, may be represented through the dynamism of a gesture.[36]

In sum, the use of metaphors and metonyms is ubiquitous and often unconscious in our everyday life and reveals itself in multivarious modes of expression:

> Our ordinary conceptual system, in terms of which we both think and act, is fundamentally metaphorical in nature. […] If we are right in suggesting that our conceptual system is largely metaphorical, then the way we think, what we experience, and what we do every day is very much a matter of metaphor.
>
> (Lakoff & Johnson, 1980, p. 3)

4.2.2 Multimodality and Perceptuo-Motor Experiences in Learning Mathematics

To recall an important consideration mentioned above, the way in which a metaphor is expressed influences its interpretation: "In addition, the manner of representation and the material used […] to render it may influence the construal of the metaphor as well as its interpretation" (Forceville, 2009a). Taking into account the just described link between metaphoric and mathematical thinking guides, this consideration can base the perspective on multimodal modes of expression and perceiving. As Nemirovsky points out, acting, perceiving, and understanding are crucial issues on the construction of knowledge (Nemirovsky, 2003, pp. 106-109). With respect to this, he states "[t]hree

---

[34] From one aspect of an object to the idealized representation of the whole object in a first step, and from this to the mathematical idea referred to by the representation of the object in a second step.
[35] Another metonymic relationship often used in mathematical context (although not necessarily consciously) is the PRODUCT FOR PROCESS-relationship (Lakoff & Johnson, 1999, p. 203), used in the question 'How is the thesis going?': Despite the fact that nothing is really "going" (This is also a case of 'fictive emotion'.), the final result, or 'product of the writing process' (the thesis), stands for the process of writing itself.
[36] This has been confirmed by Sabena in a study carried out in calculus class (Sabena, 2007), providing further results on the dynamism of gestures related to students' reference to mathematical functions.

4.2 Embodied Thinking and Learning: How our Body Shapes our Mind

conjectures concerning the relationship between body activity and understanding mathematics":

(1) [M]athematical abstractions grow to large extent out of bodily activities having the potential to refer to things and events as well as to be self-referential.

(ibid., p. 106)

(2) While modulated by shifts of attention, awareness, and emotional states, understanding and thinking are perceptuo-motor activities; furthermore, these activities are bodily distributed across different areas of perception and motor action based in part, on how we have learned and used the subject itself.

(ibid., p. 108)

(3) [I]n connection to the previous statement [conjecture 2], we add here that that [sic] of which we think emerges from and in these activities themselves.

(ibid., p. 109)

Following these conjectures allows the suggestion that the thinking process seems to be deeply influenced by our perceptuo-motor experiences. Furthermore, the peculiarity of our thinking and understanding is assumed to be bodily-based in how we perceive and act. On a neuronal level, this multimodality in information processing as biological basis for the "embodied mind" is discussed as being linked to a specific class of neurons, mirror-neurons, located in the human's Broca's area of the brain (Di Pellegrino, Fadiga, Fogassi, Gallese, & Rizzolatti, 1992; Gallese, Fadiga, Fogassi, & Rizzolatti, 1996). Rizzolatti and Craighero suggest that mirror neurons are linked to imitation since they fire not only when a grasping-action is executed but also when it is perceived (Rizzolatti & Craighero, 2004, pp. 172-174; Rizzolatti, 2005). Moreover, they claim that "[...] in humans, in addition to action understanding, the mirror-neuron system plays a fundamental role in action imitation" (Rizzolatti & Craighero, 2004, p. 169). Gallese and Lakoff explicate the consideration of different modalities concerning action and perception in the context of conceptual knowledge:

> Mirror neurons and other classes of premotor and parietal neurons are inherently 'multimodal' in that they respond to more than one modality. Thus, the firing of a single neuron may correlate with both *seeing* and *performing* grasping. Such multimodality, we will argue, meets the condition that an action-concept must fit both the performance and perception of the action.
>
> (Gallese & Lakoff, 2005, pp. 457-458)

Emanating from the conclusions on the grasping-action Gallese and Lakoff furthermore suggest that this result holds for all concrete concepts (ibid., pp. 468-469). While all of these hypotheses are inferred from research on monkeys, the possible transfer to humans stands to reason. However, the neuronal model is suggested to support theories on the embodied character of mathematical conceptualization as presented in this chapter, shortly termed as 'embodied mathematics' (Edwards, Ferrara, & Robutti, 2013, section on 'neural multimodality'). It shall not be seen as an explanatory model, but more so as a link between individual, brain-based and social, interactional aspects of multimodal learning processes; as a link between the perceptual and the sensual. For example, Arzarello takes this into account when emphasizing on the multimodal character of social learning processes. His approach is described in the following subchapter.

4.3 Learning as a Multimodal Process

Arzarello argues "that learning processes happen in a multimodal way, namely in a dynamically developing bundle, which enlarges through genetic conversions and where more and more semiotic sets are active at the same moment" (Arzarello, 2006, p. 283). Together with colleagues, he developed a model that aims to describe the simultaneous and progressive interplay of different semiotic resources as parallel and serial dynamics among the handling and genesis of signs in processes of objectification (Arzarello, Ferrara, Paola, Robutti, & Sabena, 2005, p. 129).[37] The striking feature of his model concerns the integration of gesture interpretation, since gestures show core significance in embodied thinking. This model allows for a semiotic perspective on knowledge construction, observing the parallel and serial dynamics in epistemic processes.

4.3.1 The Basic Idea of the Semiotic Bundle

Ernest (2006) introduces the notion of *semiotic systems* that allows different semiotic resources to be grouped according to certain rules. These rules concern the *type of 'materiality'* of the signs, *rules for sign production* and the possibility for interpretation based on an *underlying meaning structure*:

> The term semiotic system is here used to comprise three necessary components. First, there is a set of signs, each of which might possibly be uttered, spoken, written, drawn, or encoded electronically. Second, there is a set of rules of sign production, for producing or uttering both atomic (single) and

---

[37] They use the terms 'signs' and 'semiotic resources' interchangeably for "anything that 'stands to somebody for something in some respect or capacity'" (Arzarello, Robutti, Paola, & Sabena, 2009, p. 99), referring to the Peirce (see also chapter 3.1). Even though in general, this use will be retained in this book, the term 'semiotic resource' will be used with a connotation regarding the intentionality of its use as a resource in the working process, directed towards doing something, such as communicating. This is supported by a definition provided by van Leuwen: "Semiotic resources have a meaning potential, based on their past uses, and a set of affordances based on their possible uses, and these will be actualized in concrete social contexts where their use is subject to some form of semiotic regime" (van Leuwen, 2004, p. 285).

4.3 Learning as a Multimodal Process

> molecular (compound) signs. [...] Third, there is a set of relationships between the signs and their meaning embodied in an underlying meaning structure.
>
> (Ernest, 2006, pp. 69-70)

This definition constrains the variety of semiotic resources that can actually be observed in social interactions in the mathematics classroom. For example, it excludes sketches or extra-linguistic signs such as eye-movements or non-linguistic verbal expressions. Particularly, this definition does not embrace sets containing gestures since there are no 'molecular' gestures produced according to specific rules or algorithms. Arzarello broadens the concept to include gestures by forming the notion of *semiotic sets*, satisfying the following three affordances, slightly modified from those claimed by Ernest:

> a) A set of signs, which might possibly be produced with different actions such as uttering, speaking, writing, drawing, gesticulating, and handling an artifact;
>
> b) A set of modes for producing signs and transforming them; such modes may possibly be rules or algorithms but may also be more flexible action or production modes used by the subject.
>
> c) A set of relationships among these signs and their meanings embodied in an underlying meaning structure.
>
> (Arzarello, 2006, p. 279)

Sabena gives a succinct overview of how these three conditions are realized in a semiotic set of gestures: Gestures

> i) can be considered as signs, in the Peircean sense, since they can 'stand to somebody for something in some respect or capacity';
>
> ii) are endowed of underlying meaning structures, which are highly context-bounded;
>
> iii) are produced by bodily enactments, in a balance between idiosyncratic features and conventional character that does not follow precise rules nor algorithms.
>
> (Sabena, 2007, pp. 71-72)

With the notion of semiotic sets it is possible to structure the semiotic resources in processes of knowledge construction and to observe how they are related. For this, Arzarello defines the concept of the *semiotic bundle* as being constituted of

> (i) A collection of semiotic sets.

and

> (ii) A set of relationships between the sets of the bundle.
>
> (Arzarello, 2006, p. 281)

According to Arzarello, "a semiotic bundle is a dynamic structure which can change in time" (ibid.). That is, the components of the sets as well as the relationships between them are not static but change in time as the students act by means of signs.

4.3.2 The Analysis Of and Within the Semiotic Bundle

The relationships within the semiotic bundle are approached from two angles: A *synchronic analysis* gives insights about what Arzarello et al. call the *parallel process of objectification*[38] (Arzarello, Ferrara, Paola, Robutti, & Sabena, 2005, p. 129). In this case, the relationships between the sets of the bundle are shaped by the interplay of different semiotic resources and their simultaneous usage. They are considered in a *synchronic* analysis of the semiotic bundle in mathematical discourse (Arzarello, 2006, p. 283). The *development* of this interplay is brought out by observing the *serial process of objectification* (Arzarello et al. 2005, pp. 129-130). Arzarello refers to the analysis of this development of relations as *diachronic analysis* (Arzarello, 2006, p. 283).

The dynamic structure of the semiotic bundle, and with this the diachronic analysis, becomes an important approach to investigate how the use of signs develops in epistemic processes. To interpret the relationships, Arzarello (2006) refers to the transformational functions of *treatment* and *conversion of signs:*

> The transformational function consists in the possibility of transforming signs within a fixed system or from a system to another, according to precise rules (algorithms). For example, one can transform the sign $x(x+1)$ into (x^2+x) within the algebraic system (register) or into the graph of a parabola from the Algebraic to the Cartesian system. Duval (2002, 2006) calls treatment the first type of transformations and conversion the second ones.
>
> (Arzarello, 2006, p. 272)

Treatment thus means the modification of a sign within a semiotic register while *conversion* denotes a change of register while keeping reference to the same object. Duval claims that the transformational flexibility of students is a core aspect in their learning process:

> Thus, it appears that the thinking processes in mathematics are based on two quite different kinds of transformations of representations. Even if a single representation register is enough from a mathematical point of view, from a cognitive point of view mathematical activity involves the simultaneous mobilization of at least two registers of representation, or the possibility of changing at any moment from one register to another. In other words, conceptual comprehension in mathematics involves a two-register synergy, and

---

[38] Arzarello et al. (2005) refer to Radford's notion of semiotic means of objectification as described in chapter 2.3.

sometimes a three-register synergy. That is the reason why what is mathematically simple and occurs at the initial stage of mathematical knowledge construction can be cognitively complex and requires a development of a specific awareness about this coordination of registers.

(Duval 2006, pp. 126-127)

This idea is rendered especially important taking into account that any two representations of a mathematical object each provide different access to it, prompting an accentuation of different aspects of it. According to him, this is a crucial consideration when analyzing mathematical learning processes (Duval, 2006):

(1) [Mathematical activity] runs through a transformation of semiotic representations, which involves the use of some semiotic systems.

(2) For carrying out this transformation, quite different registers of semiotic representations can be used.

(3) Mathematical objects must never be confused with the semiotic representations used, although there is no access to them other than using semiotic representations.

(ibid., p. 126)

Furthermore, he cautions against underestimation of the diversities between the representations and claims that this reduces them to the object they represent:

Dismissing the importance of the plurality of registers of representations comes down to acting as if all representations of the same mathematical object had the same content or as if the content of one could be seen from another as if by transparency!

(Duval, 2002, p. 14)

But referring to the coordination of representations in processes of knowledge construction, Duval claims to **not** assume this transparency:

In other words, conceptual comprehension in mathematics involves a two-register synergy, and sometimes a three-register synergy. That is the reason why what is mathematically simple and occurs at the initial stage of mathematical knowledge construction can be cognitively complex and requires a development of a specific awareness about this coordination of registers.

(Duval 2006, pp. 126-127)

The negligence of this necessary coordination of different representations is referred to as the *Piaget effect* by Arzarello (2006, p. 285). In addition, he introduces the *Wittgenstein effect* as the reduction of expressing mathematical thoughts in formal terms and symbolic signs and considering the use of signs in semiotic sets as irrelevant for mathematical epistemic processes (ibid.). He claims that these two effects seen in the mathematics classroom crucially restrict possibilities in processes of learning mathematics (ibid.). One of these possibilities concerns the development of representations across different semiotic sets carried out as relationship within the semiotic bundle from a diachronic analysis. This is given by the so-called *genetic conversion*, which is identified when "one semiotic set is generated by another one, enlarging the bundle itself" (Arzarello, 2006, p. 281) and becomes identifiable from recurring aspects represented across the semiotic sets (Arzarello, 2006, p. 281).[39]

4.3.3 Implications for the Mathematics Classroom

As one application of his results, Arzarello introduces the concept of *basic signs* (Arzarello & Paola, 2007). These are signs that receive a symbolic character from being used repeatedly and in combination with other signs within the social interaction; they become conventionalized and may be used at later points in time in similar contexts without further explanation needed. While they provide resources for constructing knowledge individually and socially, they may serve as a powerful heuristics for teaching (Arzarello & Paola, 2007, p. 23; Arzarello, Robutti, Paola, & Sabena, 2009, pp. 107-108). Arzarello makes use of this in a *semiotic game* between teacher and students: The teacher may help a student to express himself by refining a link between the students' gesture and his verbal utterance. If the teacher recognizes a gesture-speech-mismatch[40] in a students' utterance of a thought in classroom interaction, he may detect a fruitful approach to an idea although this idea does not need to be conscious to the student. If the verbal formulation is not yet elaborated, the teacher can paraphrase the idea, simultaneously using the gesture introduced by the student. Assuming that gesture and speech are co-expressive, the idea represented implicitly in gesture is linked to the verbal formulation and the gesture becomes enriched with mathematical meaning. On the one hand, the student may now be able to express his idea more explicitly in speech and gesture so that the idea could be elaborated further in social interaction. On the other hand, the introduction of the gesture as means of expression may be beneficial for all participants of the interaction, prompting them to use gesture as means to express themselves as well.[41]

---

[39] This has already been referred to in section 2.2.
[40] See chapter 3.2.2, p. 26 for Goldin-Meadow's concept of gesture-speech-mismatches.
[41] Besides being a shared means of referring to the mathematical idea, Goldin-Meadow refers to the cognitive benefits also from a neuroscientific perspective. Confirmed in an empirical study she proposes that "[s]eeing gesture makes learners gesture, which leads to learning" (Goldin-Meadow, 2010, p. 4). That is, seeing the teacher or peers gesture encourages students to gesture themselves and it is what has been found to promote learning.

4.4 Summarizing the Theoretical Frame for This Study

The study is based on an understanding of *learning as constructing knowledge in social interaction*. While each participant may construct meaning individually, socially constructed knowledge is considered being constituted by those conjectures that are made explicit in social interaction and prove themselves viable in it (Knoblauch, 2010, pp. 158-159). This approach is based on Bauersfeld's comprehension of "learning [as] characterized by the subjective reconstruction of societal means and models through negotiation of meaning in social interaction" (Bauersfeld, 1988, p. 39). Furthermore, this study sees the "subjective reconstruction", as Bauersfeld calls it, as crucially *influenced by the experiences as human being in the physical world*:

> Meaning is in many ways socially constructed, but, it is *not arbitrary*. It is subject to constraints which arise from biological embodied processes that take place in the ongoing interaction between mutually constituted sensemakers and the medium in which they exist.
> (Núñez, Edwards, & Matos, 1999, p. 53, italics in the original)

This consideration provides an explanation why knowledge is not constructed arbitrarily, and why it is possible that different individuals can construct shared knowledge in social interaction in the first place.

Adopting these theoretical approaches, the study is framed within the background theories of *social constructivism* and *embodied cognition*. While social-constructivism concerns the social-communicative aspects of learning, embodied cognition adds individual aspects that are considered being shared as based on bodily experience. In addition, the results presented from recent research on the analysis of gestures in the mathematics classroom, and from neuroscientific studies suggest a complementation by a *multimodal approach to learning*, as proposed by Arzarello (2006). Multimodal means that the signs considered do not only concern one mode of expression, such as speech, for example, but the *interplay between different kinds of signs* that are used in the mathematics classroom for communication, expression and representation. This study aims to investigate gestures' contribution to epistemic processes. Because of that, a focus is set on gestures' role within this interplay with different signs, analyzed within the semiotic bundle emerging and developing during these processes. In particular, the relationship between the semiotic sets of *speech, gesture*, and *inscriptions* are considered. As inscriptions are understood those "signs that are materially embodied in some medium, such as paper or computer monitors" (Roth & McGinn, 1998, p. 37). This includes written language, sketches, printed diagrams, but also representations that become visible in the environment of a Dynamical Geometry System (DGS). To keep distance from gestures as embodied means of expression, the term 'embodied' as used in the description by Roth and

McGinn is thus suggested to be replaced, such that the defining feature of an inscription is being '*fixed* in some medium'.[42]

In this regard, gestures are considered as specific kinds of *signs in a Peircean sense* (see chapter 3.1), deriving their meaning from the context given by their relationship to speech and inscription. Following the model of the semiotic bundle, these relationships between verbal, inscriptive and gestural signs can be diachronic or synchronic, the latter concerning simultaneously used signs. These simultaneously used signs will be considered to form a *semiotic composition,* that is a composition of signs of different modalities, together forming one representamen. For example, the verbal utterance "this is the function $y = 2x$" together with simultaneously pointing at a straight line within an inscribed coordinate system constitutes a semiotic composition: All three signs, the verbal sign, the gesture sign and the inscription, play together to shape the reference of the utterance, the immediate object. Consequently, this semiotic composition provides the representamen of what I understand as a *multimodal sign* (see Fig. 4.1). Its triadic relation in the Peircean sense is shaped by the interrelated appearance of gesture, speech and inscription. While the three modalities are deeply intertwined in their use and their interpretation, they will be distinguished only for the sake of analysis.

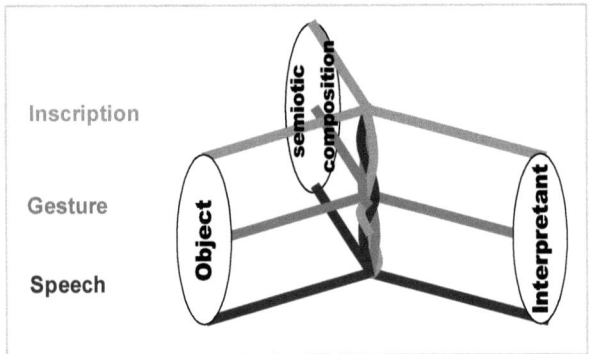

Fig. 4.1: Multimodal sign shaped by the synchronous relationship between the three semiotic sets speech, gesture, and inscription. The 'semiotic composition' forms the representamen of the multimodal sign.

---

[42] Roth and McGinn also give a list of characteristics of possibilities provided by the use of inscription, as well as of its conditions (Roth & McGinn, 1998, pp. 37-38). While this shall not be given in full extent here, the necessity of interpreting the meaning of the use of inscriptions in relation to other signs is an issue that also concerns their interplay with gestures: "The meaning of any inscription is […] not an inherent property but, similar to the meaning of a word, arises from the context of its use (Wittgenstein, 1994/1958). That is, an inscription does not have meaning in and of itself; rather, its meaning arises in the context of other inscriptions and sign forms (e.g. language)" (Roth & McGinn, 1998, p. 38).

Part II
The Study

As Pike says, "[v]erbal and nonverbal activity is a unified whole, and theory and methodology should be organized or created to treat it as such" (Pike, 1967, p. 26). Part I aimed at clarifying the terms, notions, and concepts taken into account in this study, thus the language of theory used in this book. The following second part of the book is dedicated to the methodological concerns as well as to the methods used for conducting the study.

5 Methodological Assumptions
Underlying Research Logic

After pinpointing the research goal, this chapter starts with explicating the methodological considerations that underlie this study. This concerns its qualitative character first: It seeks to give insights about a question of *how*, and with that about *qualities* of empirically gained phenomena. Second, these phenomena are observed within social interactions, suggesting an interpretative approach. Third, both aspects just mentioned need to be considered from an interactionist theory as well as from the perspective of gesture studies.

5.1 The Goal of This Study: Focus, Aims, and Questions against the Theoretical Background

Ensuing from the initial research interest in gestures' contribution to learning processes, the temporally stated aim as developed from the review of recent research in chapter 2 is recalled as follows:

> This study aims at investigating how both, the communicational as well as the representational aspect of gestures may take part in fostering and in hindering social processes of constructing mathematical knowledge.
>
> (Chapter 2.4)

A specific focus shall thus be given to the performance of epistemic actions, as they constitute the epistemic process according to the GCSt-model (see chapter 4.1). Since social interaction takes place by means of signs, the epistemic actions are identified by tracing the signs used. These allow for the reconstruction of the immediate mathematical object referred to and its development in the process of knowledge construction. So far, this has mainly been carried out by interpreting the speech acts as the most conscious, explicit and conventionalized means of expression. Since gesture and speech are co-expressive, the reference given in speech frames the context to understand the gesture, but gestures' influence on the interpretation of the whole utterance is often undervalued due to its implicit nature. Considering the (analytical construct of the) multimodal sign unfolding in social interaction when a gesture is performed, the following question is aimed to be answered:

How can gestures contribute in shaping the immediate mathematical object in social epistemic processes?

This question concerns the feature of gestures (considered as signs) to represent the mathematical object "in some respect or capacity" (CP 2.228). Following Hoffmann and Roth (2007), gestures then fulfill a *representational function* as described in chapter 3.1.3. Their *epistemic function* thus concerns their use directed towards generating knowledge that is, to "distinguish objects, to structure our experiences, to organize interaction, and so on" (Hoffmann & Roth, 2007, p. 118). Relating this consideration to the GCSt-model, the research question is complemented by another corollary question:

How can gestures contribute to acting epistemically? What beneficial ways of using gestures can be identified in this? What possible pitfalls can emerge?

As Bavelas (1994) claims in a more general context: "We need to move away from taxonomic approaches, with their apparent goal of classifying gestures and focus instead on function: *What does this gesture do and how?*" (ibid., p. 202, italics in the original). In this study, this question will be posed within the specific context of collaborative mathematical working processes. In this regard, the functions of gestures are then given "by their value with respect to a larger whole" (Nöth, 2000, p. 199, my own translation). That is, there are considered different "larger wholes", depending on the kind of function looked at:

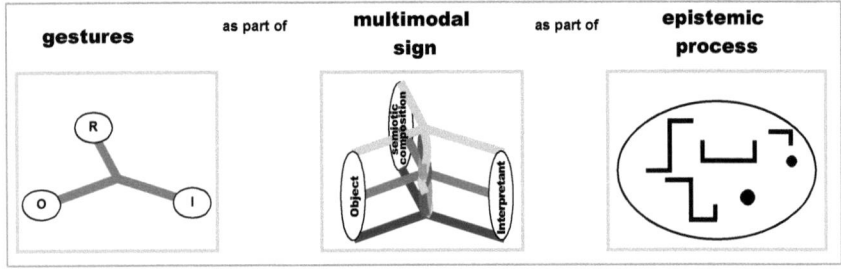

Fig. 5.1: 'Larger whole' in which the functions of gestures are investigated[43]

The *representational* function describes a gestures' value as a sign within the multimodal sign as the "larger whole" with respect to the immediate mathematical object that is shaped by this multimodal sign. The *epistemic* function describes its value for the multimodal sign, with respect to the epistemic process as the "larger whole". This also fits the idea of functions followed in social sciences, as "specific meaning of a part for the

---

[43] The picture on the right hand side, representing the epistemic process displays token that are used to represent epistemic actions in a condensed way. These token will be explained in detail in section 6.2.3.1.

whole, for example the kind in which one element contributes to the preservation or the 'functioning' of a system" (Nöth, 2000, p. 199, my own translation). Remember that the investigation of this "contribution" is considered the overarching aim of this study (see chapter 1.2), in which the subquestions for the "representational contribution" and the 'epistemic contribution' are distinguished for structuralizing reasons.

The functions identified shall not describe single phenomena that did occur once, but moreover shall *describe* gestures' contribution to the epistemic process. This description shall provide a building block for a theory that describes and explains students' epistemic processes within interest-dense situations.

This building block shall be formed by

1. Characteristic ways in which gestures can represent mathematical entities, hence modes of representational functions of gestures.
2. Characteristic ways in which gestures can help acting epistemically, that is modes of epistemic functions of gestures.

5.2 The Interpretative Approach

"Interpretative research assumes that people develop their social world through their mutual, interpretative actions" (Jungwirth, 2003, pp. 189-190, my own translation). The analysis of social interactions demands for a reconstruction of the interpretation process among the participants. They build their reality as a construction of first degree which is reconstructed by the researcher as a construction of second degree (Schütz, 1971, p. 68). Therefore, the hypotheses that are gained in this interpretative empirical research are not hypotheses that concern any objective "reality" (if existing) per se but specific aspects of it that always have to be seen as related to the specific setting and situation. A constant comparison shall bring forth the core and characterizing features to describe the phenomena reconstructed concerning the investigation of the research question.

5.2.1 Handling the Data Within the Interpretative Approach

The fixation of data discharges the researcher from the pressure of immediate action. The textual form of data[44] makes it possible to slow down the interpretation by 'zooming into the discourse'. This makes it possible to consider various possibilities of how the partner in interaction may interpret the utterance before the actual reaction is known, or

---

[44] This does not exclusively refer to texts in form of written words. Bikner-Ahsbahs (2005) refers to Ricoeurs understanding of texts and its extension by Beck and Maier (1994). According to this, text is everything which "documents meaning in the form of signs" (Bikner-Ahsbahs, 2005, p. 62, my own translation), including audiovisual data and pictures as well.

to compare sequences of interaction (Przyborski & Wohlrab-Sahr, 2010, p. 163). This is an important feature used to carry out sequential analyses that understand action and reaction in social interaction as "reason-consequence-relation" (Przyborski & Wohlrab-Sahr, 2010, p. 249). Furthermore, it facillitates the documentation of the research process and the presentation of the results. This is obvious for inscriptive primary data since extractions of it can be directly used for printed publication. Audiovisual data demands the production of secondary data in the form of transcripts. In these transcripts, verbal utterances are written down literally (also considering fluency, prosody, emphases and so forth) and non-verbal actions are described as neutral as possible. Although the production of the transcripts is accomplished according to specific rules,[45] strict neutrality can never be guaranteed and it is never possible to capture all aspects of non-verbal behavior.

5.2.2 Transcript versus Video Data: Dealing with Gestures

As has just been described, the data collection in interpretative research is typically carried out by video or audio data capturing the investigated situations (i.e. interviews and written examinations), and by documents, such as inscriptions produced by the observed persons, questionnaires or curricula (Jungwirth, 2003, pp. 190-191). The precise formation of the type of data depends on the phenomenon that is studied and, hence, at first on the research interest.

The textual fixation of the data offers advantages for the analytical process, as well as for the documentation and the publication of the results. Since the publication usually appears in printed form, the transcription of the non-inscriptional data becomes an issue. To make it comprehensible, one should include everything that could be relevant for the interpretation of the object of interest. Concerning this matter, the transcription of the verbal expressions is considered as particularly relevant, since speech is regarded as "a socially objectivated system of meanings, detached from human subjectivity" (Jungwirth 2003, p, 191, my own translation). Furthermore, Jungwirth refers to Ricoeur by mentioning the fixation of the "outlasting, linguistic based meaning of a speech act" (ibid., p. 191, my own translation[46]) as a core benefit of the transcription of speech acts. This allows being aware of a "surplus of meaning" (ibid., p. 191) that can be considered

---

[45] These transcription rules have to satisfy quality factors such as 'practicability', 'expandability and flexibility concerning the data', 'clarity', ensuring their neutrality on the one hand and their intelligibility on the other hand (Przyborski & Wohlrab-Sahr, 2010, pp. 89-90).
[46] Literally, she writes „Sie [(die Sprache)] ist ein von der Subjektivität des Menschen abgelöstes, sozial objektiviertes Bedeutungssystem." This „systematisation" fell victim to the translation but becomes important regarding the aspects to which it is related: The "Bedeutungssystem" is systemized by "typifying schemas built on experience [('Erfahrungsschemata')] that form a part of human beings 'given social apriority' (Schütz & Luckmann 1988, 282)" (Jungwirth, 2003, p. 191). This systematization provides a point of reference for the interpretation.

in a subject-detached interpretation of a speech act, and makes it possible to reconstruct and analyze situations with respect to a specific aspect.

For this study, this reconstruction becomes important since the GCSt-model so far has been mainly carried out by speech analysis. Speech is widely considered to be a resource having conventional meaning in discourse and allows researchers to reconstruct the social interaction. The semiotic perspective added in this study requests to make the integration of non-verbal means of expressions explicit, giving a special remark to the description of gestures: Gestures are *not* conventionalized, but are idiosyncratic and spontaneous movements. Especially complex gestures, in which a gesture unit may comprise more than one stroke, possibly also including iconic non-deictic features, are hard to describe from a neutral perspective. Gestures are performed in three dimensions of space and their movement can only be described from a certain perspective. Since gesture is ephemeral in nature, it leaves no fixed product. The movements cannot be reconstructed based on such products but have to be described by adjectives that themselves may suggest a pre-interpretation of the movement.[47] Moreover, every taxonomy or coding scheme (see, for example, McNeill's dimensions of gestures as described in chapter 3.2.3) is more than a pure description of the movement; it classifies the gesture in some prescribed way to capture the aspects considered as significant. For this reason, every purely verbal description of gestures, fulfilling the condition of being neutral for the purpose of analysis, appears to be confusingly detailed. Therefore, it is not functional to carry out the analysis based on verbal descriptions of the gesture. A pictorial representation also does not settle the claim to reflect all aspects of a gesture since it can only give a glimpse of the dynamical movement and its timing in relation to speech. The dynamic character as well as the lack of socially based systematization of meaning of representational gestures (in contrast to speech as semiotic system) demands for their fine-grained analysis in relation to the other semiotic resources within the multimodal sign. This can be carried out by constant parallel use of the video data, analyzing the gestures frame by frame. Within a micro-analysis a way to describe and decode gestures' contribution to social interaction, as well as to the epistemic process, can be developed.

To present the results in printed form, other methods need to be carried out in order to make the interpretations comprehensible to the reader. This demands for a notation of gestures that respects the co-timing to speech as well as its specific formation. One way to carry this out is to integrate pictures taken from the data. To *'quote a gesture'*, additional means can visualize its dynamics and the co-timing to speech can be marked in the transcript.

---

[47] For example "parabola-shaped movement": This choice of words already suggests the curve of a parabola as represented by the gesture and may influence the researcher's interpretation.

5.2.3 Analyzing Interaction by Tracing the Development of Signs

In social interaction, meaning is thought to develop among the participants by ongoing mutual interpretation of each other's actions. These actions in forms of signs can be traced to investigate the interaction, based on an ethnomethodological point of view. That is, it allows understanding the "process of subjective allocation of meaning not as an inner 'private' act of consciousness, but from the outset as a social, 'public' event" (Bergmann, 2004, p. 75). Furthermore, as Soeffner states:

> Human behaviour, as an observable form of human action – be it linguistic or non-linguistic – is interpretable by and for human beings because, in addition to many other properties, it always displays that of being (proto-)symbolic. From the gesture to the 'significant' symbol from the token and symptom to the constructed and unambiguously defined mathematical sign, from the body and facial expression to clothing, from the natural impression to the human product, we attribute to ourselves and our environment the qualities of signs, and with these we constitute the human interpretative horizon (Wundt 1928; Bühler 1934; Mead 1934). Here the different types of signs and their varying semantics and associations also correspond to different interpretative procedures (cf. Schütz and Luckmann 1989: 131-147).
>
> (Soeffner, 2004, pp. 96-97)

With this, Soeffner expresses a fundamental assumption underlying the interpretational approach: The participants in social interaction are in a steady process of interpreting each other, based on the signs they 'send out', consciously or unconsciously. Within this process, they construct their social reality. In turn, this assumption provides the methodological base for researchers to reconstruct the social reality students construct within their social interaction as construction of second degree (Schütz, 1971, p. 68). Nevertheless, the reconstruction needs to be carried out methodologically controlled to be distinguished from everyday-interpretations.

5.2.3.1 A *Re*constructivistic Approach, Based on a Peircean Understanding of Signs

Analyzing sequentially means to analyze one unit of meaning[48] after the other, assuming the "reason-consequence-relation" already mentioned before (see chapter 5.2.1). Reconstructing the construction of meaning is accomplished by interpreting a meaning unit to

---

[48] "Bedeutungseinheit" in German can be interpreted as an expression to which a meaning is assigned. The terms "meaning unit" and "meaning" need to be crucially distinguished since a "meaning unit" refers to parts of an action and reaction that are detectable in textual form, while "meaning" means an underlying idea reconstructed from these parts.

5.2 The Interpretative Approach

get as many variations as possible of what could be meant and in turn exclude improbable meanings by including the consequence. This in turn unfolds a new space of possible interpretations (Przyborski & Wohlrab-Sahr, 2010, pp. 249-250). This becomes more comprehensive if one considers the specific case of face-to-face interactions: The action of one participant can be seen as a meaning unit that in turn provokes a reaction of the other participant. The interpretation would begin by considering what could be meant by the action and concluding the most probable meanings from the reaction. This again is considered another action in the sequential process of the social interaction.

To reconstruct the social processes of constructing mathematical knowledge, Bikner-Ahsbahs developed a method that takes into consideration both the sequential and the sign-based character of social interaction (Bikner-Ahsbahs, 2006). Her *semiotic sequence analysis* of the transcript encompasses three steps in which the data becomes more and more aggregated: The first step concerns the reconstruction of the social processes of knowledge construction through tracing the development of mathematical objects within the social interaction. This way, the three epistemic actions of *gathering, connecting* and *structure-seeing* (see chapter 4.1) have been identified. In a second step, the results of this analysis are coded by means of signs (see Tables 6.2 and 6.3 in section 6.2.3.1) and then represented in a condensed way in *compressed process diagrams*. Thirdly, phases of similar structure with respect to the epistemic process can be identified from the compressed process diagrams and represented by means of *pictographs* (Bikner-Ahsbahs, 2006). In the first step of Bikner-Ahsbahs' semiotic sequence analysis, from here on referred to as *semiotic reconstructive analysis (SRA)*, she considers students' verbal utterances as signs and reconstructs the corresponding immediate object. Since social interactions are accomplished through signs (Seeger, 2006, p. 267. See also Soeffner, 2004, pp. 96-97 as quoted above.), action and reaction can be understood as a representamen (sign) and interpretant in a Peircean understanding of the sign as a triadic relation.[49] The *immediate object* takes the place of what is called *meaning* in the above considered sequential analysis. The main idea of this semiosis becoming visible in the students' utterances is illustrated by the following Fig. 5.2:

---

[49] See chapter 3.1.1.

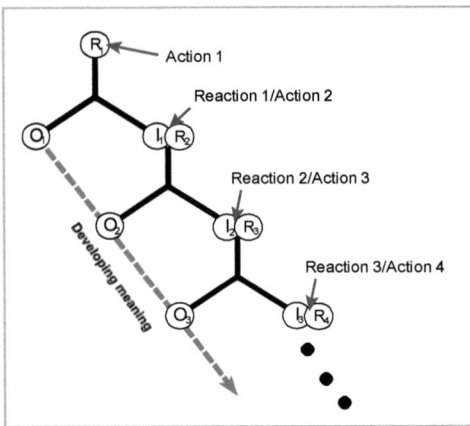

Fig. 5.2: The idea of the first step of the semiotic sequence analysis to reconstruct meaning as developing in social interaction (following Bikner-Ahsbahs 2005, pp. 197-198, p. 208)

This analogy was used by Bikner-Ahsbahs in her investigation of social classroom interactions (2005, 2006). Tracing the action-reaction relation and by that the development of the immediate mathematical objects gave her insights into the process of social construction of mathematical meaning (Bikner-Ahsbahs 2006, pp. 161-162). For this, the verbal signs are analyzed according to Austin's speech act analysis on three levels (Austin, 1962):[50]

- On the *locutionary level*, the utterance is regarded in its purest form, as saying something for conveying meaning through the content of speech. I understand it as the level of semantic content.
- On the *illocutionary level* one considers uttering as an action through which something is performed, such as a question, a request, a proposal etc.
- The *perlocutionary* level is the level of 'what is done through saying something' and is understood as the actual effect of e.g. asking, requesting.

The distinction between the *illocutionary* and the *perlocutionary* level is important. This may become clear when one thinks about requesting something on the perlocutionary level without actually formulating a request on the illocutionary level. For example, the statement "We don't have any more bread for breakfast." may be an objective observation about the amount of bread, a reproach that somebody ate all of the bread, or a request to buy new bread. This concerns the illocutionary level. In turn, the perlocutionary act is

---

[50] Although this has been briefly described in the introduction, it shall be explained here in more detail.

revealed in the interaction as the reaction to the illocutionary act. To stay with the example, somebody might answer "Yes, I know", "I ate it yesterday", or "I will buy new bread". All three answers suggest different perlocutionary effects of the speech act.

For the analysis, the students' verbal utterances are the main sources to get an understanding of a situation and constitute the units of analysis. Speech is considered to be the most consciously used way to communicate since one can rely on its underlying meaning structure which is shared in the common ground among the participants of the social interaction and, important for the interpretative approach, also shared in the common ground with the researcher. This allows for the interpretation of each utterance as a speech act on the three levels, in order to generate hypotheses on possible interpretations of it. Bikner-Ahsbahs refers to this unfolding of possible interpretations as *interpretation space (IS)* (Bikner-Ahsbahs, 2005, p. 189). Within the ongoing social interaction, some of these become more probable than others so that the interpretation space of one utterances frames the interpretation space in the following interaction and with that the reconstruction of the immediate mathematical object. The now presented example will give an idea of how a situation can be reconstructed based on the speech acts (and the non-verbal actions as referred to in the verbal utterances). First, an interaction analysis (Krummheuer & Naujok, 1999, pp. 68-71) of the episode will be given to clarify the situation. Then, the semiotic reconstructive analysis will be presented to show how the mathematical object develops as reconstructable from the speech acts, and how these shape epistemic actions.[51] To make the example more comprehensible, the analyzed episode is subdivided in several scenes.

Example 1.1: A Scene Taken from the Elaboration of a Task, Dealing with the Parabola as Geometrical Locus[52]
The two students Tim and Mike work together on a task that deals with the parabola as geometrical locus. They have been asked to prepare a piece of a paper according to instructions that includes folding the paper in a specific way and producing inscriptions to mark specific points. The instructions provide the construction rules for one step, that is, the construction of one point: 'Choose any point on the lower edge of the given sheet, on which a point M has been marked. Fold the chosen point so that it comes to rest on point M. Draw a perpendicular to the lower border through the chosen point, and mark the intersection point of this perpendicular and the folding line with a cross.' This step shall be repeated until the students recognize a curve. The folding lines can be identified

---

[51] The example is chosen from an episode in the very beginning of the elaboration of a task. This minimizes the background knowledge that is needed while the situation will be outlined only in its briefest features in order to provide access to the students' elaboration of the task.
[52] The task will be presented in detail in chapter 6.1.3.1. The presented scene is the very first scene of the so-called PA5-data set (see chapter 6.2.2 or footnote 63 for the naming) and starts with transcript-line 49.

as tangent lines that envelop a parabola. The marked intersections are the points on the parabola. Preceding the following scene, the students constructed a third point.[53]

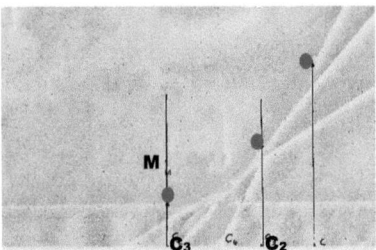

Fig. 5.3: Tim and Mike's folding sheet after constructing three points (M is the point of reference for performing the folding)

---

49. T: *(Makes an arc-shaped movement above the sheet, linking the constructed points)* I can actually nearly see the curve now already.

50. M: yes

51. T: do we want to do another one here' *(points at a point on the bottom edge between C_2 and the left point marked on the bottom edge)* ,then we have roughly the same distances. (.)

52. M: in fact it would actually have to be exactly the same here. *(points to and fro between the left and the right hand side of point M)* well then *(briefly points to the sheet)* ,we don't need to do any on the other side.

53. /T: *(successively indicates the red dots on the paper with the pen)* well no actually when one connects these three points by a good curve *(briefly points to the sheet)* one can theoretically predict it already. (.) ,approximately.

54. M: *(briefly points to the sheet)* never mind *(points at a point between C_2 and C_3)* let's do another one between

55. T: *(sets the pen on the paper, then changes the pen, continues to draw, mumbles)* not red ,that would be silly *(draws a point between C_2 and C_3, labels it C_4 and folds)* (17s) that one there *(draws a perpendicular with the bottom edge through C_4, takes the red pen and draws the intersection of folding line and perpendicular)* (12s) shall I now try to draw this *(follows the red dots with the pen)* curve approximately'[54]

---

[53] See Table 6.1 in chapter 6.2.1 for the transcription rules.
[54] Within utterance number 55, two long speech pauses mark interruptions within the turn. The first scene is considered to end with the first pause, the second scene is considered to start after the second pause. The light grey color is used to fade out the part of the turn that is not considered part of the respective turn.

Interaction Analysis of the Scene

Tim mentions that the curve is visible for him (49). Since the constructed folding lines constitute an envelope curve, it does not become clear whether he sees the curve by looking at these lines, or rather as a connection of the constructed points. Nevertheless, Mike confirms Tim (50: "yes"). This confirmation might be considered an agreement to the idea of not constructing any more points expressed by Tim on an *illocutionary* level: The curve is seen and thus they can stop constructing any more points. Tim however explicitly suggests to construct another point and uses nonverbal means of expression to make the reference of the imprecise verbal deictic phrase "here" (51) more precise. He indicates a point on the lower edge in the middle between the two left points, reasoning the choice of this position verbally: ",then we have roughly the same distances." (51). Mike neither agrees nor disagrees, but mentions that there is no need "to do any on the other side." (52). He bases his suggestion on the assumption that "it would <u>actually</u> have to be exactly the <u>same</u> here." (52), giving more information about the "here" by alternately pointing left and right of point M. Nevertheless, he stays unspecific regarding what he means with "it" and with "the same". Seen as a response to Tim's previous proposal regarding the construction of a point between two others, "the same" may refer to the idea that it does not matter where to construct a point, it is "the same" (situation) anywhere. At the same time, he may start a new line of thought, referring to the situation to be constructed being "the same" on the left hand side of point M as on the right hand side. Both interpretations are compatible with his suggested redundancy of constructing points "on the other side." (52). Tim's reaction does not clarify the situation. It seems that he refers rather to his former utterance than to Mike's suggestion although he also considers parts of it: "well no actually when one connects these three points by a good curve one can theoretically predict it already. (.) ,approximately." (53). He may agree on a redundancy of constructing another point at all, probably understanding Mike's statement in line 52 as disagreement on his idea of constructing another point (51). Adding the word "approximately" after one second pause, he qualifies the shape of the curve as not necessarily precise, possibly *illocutionary* seeking for support in elaborating it. Mike gives up on his thought ("never mind", 54) and he himself proposes to "do another one between" (54), most probably referring to another point between two others. Tim sees this *perlocutionary* prompt as invitation, starting to accomplish the construction of another point without hesitation (55).

---

55. T: (*sets the pen on the paper, then changes the pen, continues to draw, mumbles*) not red ,that would be silly (*draws a point between C2 and C3, labels it C4 and folds the paper*) (17s) that one there (*draws a perpendicular to the bottom edge through C4, takes the red pen and draws the intersection of folding line and perpendicular*) (12s) shall I <u>now</u> try to draw this (*traces the red dots with the pen*) curve approximately'

56. M: (*looks at I*) are we allowed to connect those'

57. I: you can draw it in there.

58. T: yes we can begin there (*moves the pen from* $[MC_4]$ *to the right*) or we simply draw a curve in a different color. (*looks at I*) or do we need the other colors later on.

59. I: you can- (2s)

60. T: (*takes a green pen*) do you want'

61. M: (*takes the pen from T*) (incomprehensible)

62. T: leave it to you.

63. M: (*moves the sheet to his side*) (5s) the points. (.)

64. T: these of one ‚approximated curve.

65. M: (*turns the sheet by 90°, draws a connection between the red dots with the green pen*) (15s) wow. (*laughs*) (..)

Following the construction of the fourth point, Tim proposes to "draw this curve approximately'" (55). With this, he *illocutionary* asks for Mike's permission to fix the curve. Adding the word "approximately" recalls the planned drawing to be a sketch and not necessarily precise. However, it can also be an *illocutionary* request directed to Mike to participate in shaping the curve in a more precise manner. Mike questions the allowance of manipulating the folding sheet when not required by the instruction (56). This can be seen as *illocutionary* agreement to Tim's proposal of drawing the curve. Mike's question suggests that he considers the folding sheet as a product of the instruction rather than as a representation that can be explored. After getting the interviewer's permission (57), the students fix the curve as a connection of the constructed points after clarifying which of them will draw the line (59-65).

Semiotic Reconstructive Analysis

The SRA of the scene is presented in Fig. 5.4 and Fig. 5.5. It provides a more detailed and in-depth analysis of the scene. Furthermore, *it facilitates the tracing of the development of the immediate mathematical object and the reconstruction of the (epistemic) actions*:

On the right hand side, a variety of interpretations (interpretation space) IS_k of the interpretant I_k is noted. That is the interpretant determined by the representamen R_k, related to the object O_k. The index k denotes the respective utterance/line in the transcript. This is reasonable, since the unfolding of the IS is accomplished by making use of speech act analysis, considering a verbal utterance as a meaning unit. There is no claim for having found all possible interpretations and however, for spatial reasons only the most accessible ones are written down here. The O_k's on the left hand side make it possible to trace the immediate objects as they develop in the course of interaction. Those O_k's described in a bolt-framed box are immediate objects that are considered sufficiently elaborated from the perspective of the students; they mark the actions within the epistemic process

5.2 The Interpretative Approach

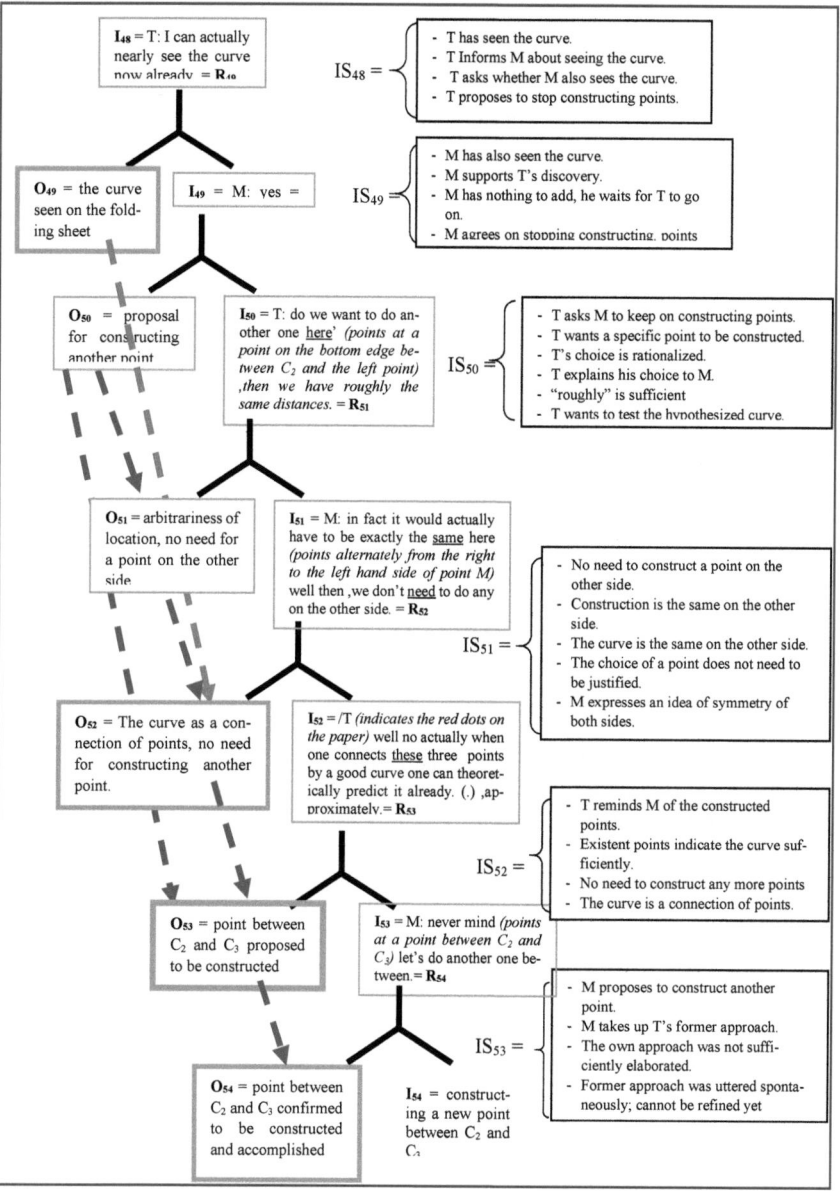

Fig. 5.4: Semiotic reconstructive analysis of the first part of the first scene in the elaboration of the parabola-task by the students Mike (M) and Tim (T): A first suggestion on the shape of the curve that becomes visible in the folding product is stated and shall be tested by constructing an additional point.

5 Methodological Assumptions

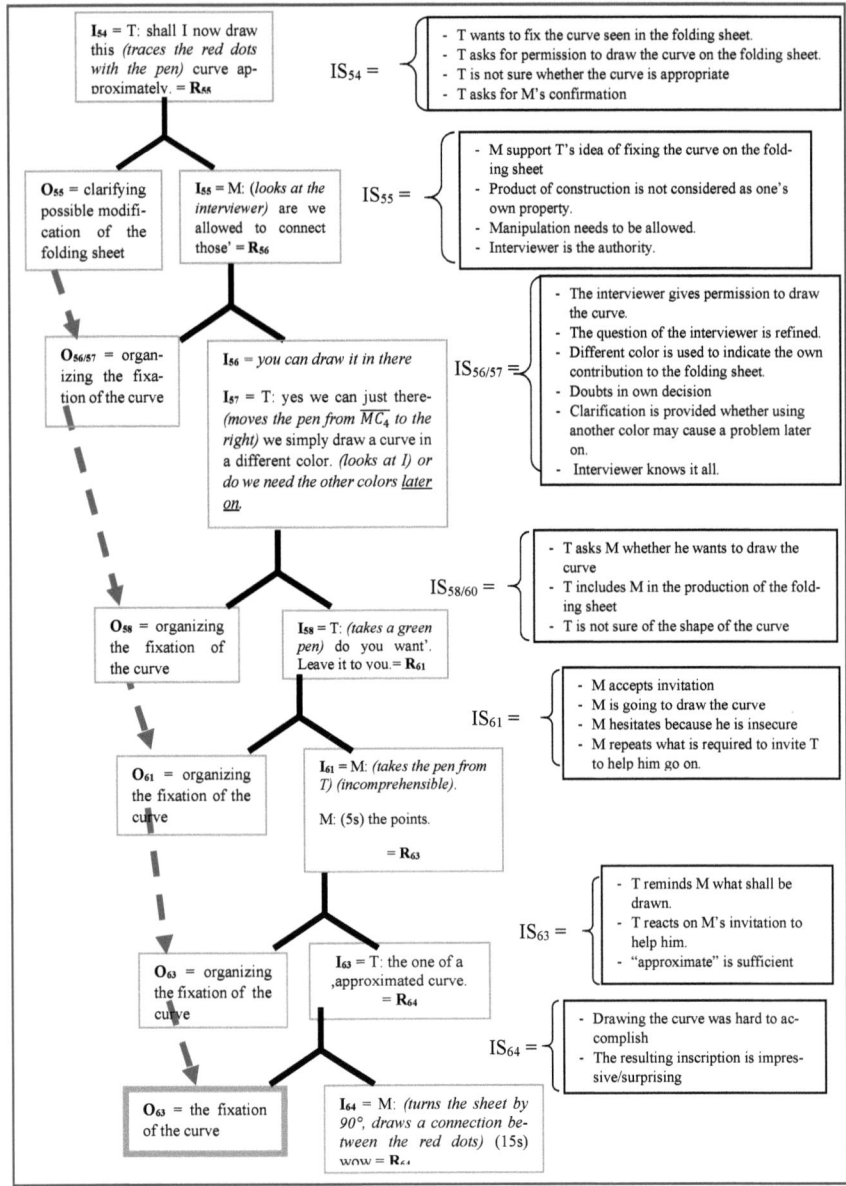

Fig. 5.5: Semiotic reconstructive analysis of the first part of the second scene in the elaboration of the parabola-task by the students Mike and Tim: The additional point confirms the first hypothesis and the suggested curve becomes fixed within the folding sheet.

(even though not all of them are epistemic actions). In this specific case, the curve is considered a **seen structure** (49) since its continuous shape needs to be imagined within the folding sheet; or as an abstraction of the envelope curve, or as 'smooth' connection of the constructed points. Mike's utterance in line 52 is not clear from speech alone. He suggests something to be "exactly the same" and his pointing suggests that he refers to the left and the right hand side of the folding diagram. With this he might start **extending the seen structure** (connecting action) but Tim does not react to his suggestion. He also **extends the seen structure** (connecting action) as well in line 53, making explicit that the curve is given as the connection of the constructed points. The last action accomplished (55) affects the working process: The students agree on constructing another point between the points $C2$ and $C4$ on the lower edge. The immediate object 'shape of the curve' is seen as a structure within the folding sheet, and then it becomes elaborated as a connection of points. Mike's suggestion of another extension, a suggested symmetry, is not confirmed as shared by both students. Moreover, they decide to test the assumed structure by constructing another point.

The second part of the scene deals exclusively with the modification of the folding sheet, drawing the curve as a connection of points. This is first clarified in participation with the interviewer, and then organized between the students.

This illustrative example shall help to understand how the development of objects can be traced as immediate objects of the signs produced verbally. This way of analyzing the social interaction by means of speech acts reveals how the immediate object develops in social interaction. However, it does *not* disclose the role of the inscriptions and gestures used by the students nor how this is intertwined with speech; in short, it does not disclose how the semiotic bundle shapes the epistemic process.

5.2.3.2 Considering a Multimodal Approach by Using the Model of the Semiotic Bundle

The model of the semiotic bundle has already been described in chapter 4.3 as a model, allowing for a description of the multimodal nature of learning processes. Within this model, gesture cannot be extracted as a single modality to look at, but it can be focused on while examining its synchronic and diachronic relations to speech, inscription and possibly other semiotic resources, such as Dynamic Geometry Software or artefacts. Taking these relationships into account, the meaning of a gesture can thus be disclosed by taking into account the following components:

- its simultaneous interplay with other signs (co-expressivity of speech and gesture, indexical reference to inscriptions etc.)

- its development from previous signs (for example through *generic conversion*[55] or evoking iconic reference by similarity of some features such as shape, orientation etc.)
- a potential metaphorical meaning

Furthermore, the interpretation of gestures needs to be carried out within the concrete situation, in which the context is influenced by various factors, such as the mathematical topic, the affordance of the task, and so on. In turn, the gesture itself as part of the social interaction needs to be considered as influencing the situation in the subsequent interaction. Fig. 5.6 summarizes these factors of gesture analysis within the semiotic bundle.

The upper right (green) circle concerns the interplay of the gesture with inscription and speech in a semiotic composition,[56] respecting what the gesture does and how inscription and speech may synchronically influence its interpretation within the social interaction. The upper left (blue) circle takes into account which former signs may influence the meaning assigned to the gesture. The lower (yellow) circle is related to possible metaphorical meaning that is not necessarily explicated, but that may influence how the gesture becomes interpreted by the students to express meaning in social interaction.[57]

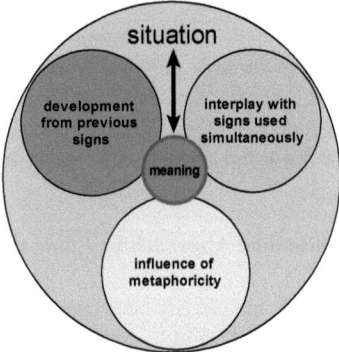

Fig. 5.6: A three-component-model for analyzing the meaning of gestures within the semiotic bundle

---

[55] See chapter 4.3.2.
[56] See chapter 4.4.
[57] Being grounded in everyday concepts, the metaphorical mapping of the gesture often does not develop in social interaction and its situated reference rarely becomes explicit between the participants.

To illustrate this idea, the meanings of the gestures performed in example 1.1 are analyzed in the following, using the background of the social interaction as disclosed in chapter 5.2.2.1:

a) The first gesture in this scene is performed by Tim in anticipation to his verbal utterance in line 49, where he announces that he "can actually nearly see the curve <u>now</u> already."[58]

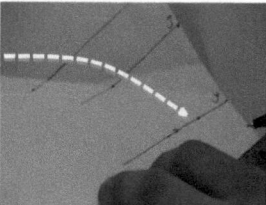

Fig. 5.7: Tim's gesture in anticipation to his utterance in line 49

The meaning of the gesture is constituted by the interplay with the inscribed diagram, taking into account the situation constituted by the task of recognizing a curve based on constructing points according to the instruction:

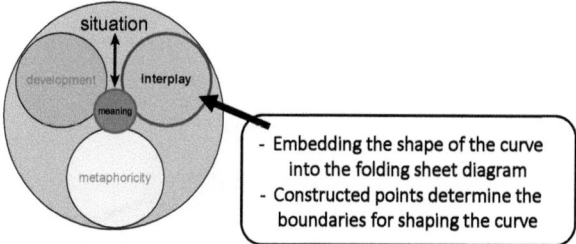

Fig. 5.8: Components influencing the interpretation space for disclosing the students' interpretation of the gesture 49

The iconic gesture shapes the right side of the curve as embedded in the folding sheet. The extent of the gesture is restricted by the positioning of the points that have already been constructed: The shaping of the curve begins above the point farthest to the right and stops shaping at the point constructed beneath point M. By this, the style of the curve is sketched ephemerally to make its shape visible such that *the performance of the gesture slightly changes the situation*: The

---

[58] In the following will be referred to the gestures by the line number of the co-occurring utterance and, in cases that there are several gesture phrases co-occurring to an utterance, by adding letters to the numbers.

entities within the folding diagram are *visibly connected* with each other, providing an idea of how the curve looks like for Tim, but also as shared between the students.

In turn, Tim mentions that he "can actually see the curve now already." (49) and Mike agrees (50).

b) The next gesture is again performed by Tim, complementing his question in line 51, "do we want to do another one here'", to be semantically precise:

Fig. 5.9: **Gesture accompanying the utterance in line 51**

Tim indicates a point on the bottom edge of the folding sheet,[59] located in the middle between the points $C2$ and $C3$. The gesture starts co-timed to the word "here" and clarifies to which location is referred to as being "here". The gesture is held during the entire remaining utterance in line 51 and is thus considered as being co-expressive to it.

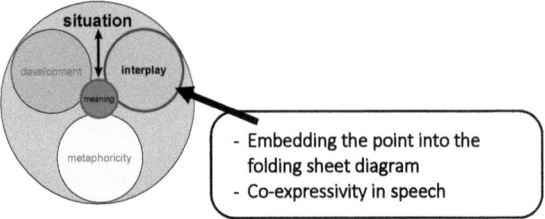

Fig. 5.10: **Components influencing the interpretation space for disclosing the students' interpretation of the gesture 51**

The gesture needs to be interpreted within the situation of constructing points according to the instruction, where points are chosen on the bottom edge as a basis for one construction step. By indicating a location on the bottom edge, Tim

---

[59] Note that the camera was directed on the students' writing process from their upper left. From that perspective, the students are positioned on the right side and the bottom edge of the paper laying in front of them appears to be on the upper right part of the picture.

5.2 The Interpretative Approach

thus refers to another point for *starting* a construction step. This contrasts to another point as a *result* of the constructing step.[60] Furthermore, the hypothetical point as embedded within the folding diagram is placed in the middle of the two construction points $C2$ and $C3$, giving visual access to the "same distance" mentioned right after.

c) In line 52, Mike adds a gesture to his verbal utterance, "in fact it would actually be exactly the <u>same</u> here.":

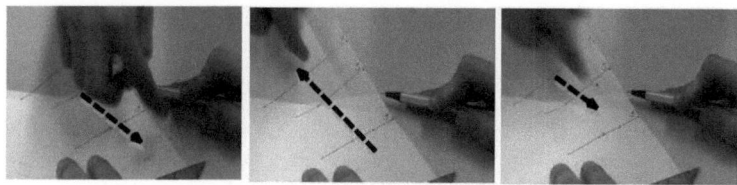

Fig. 5.11: Gesture co-timed to "exactly the same" as part of Mike's utterance in line 52

Mike moves his hand to the left, to the right and again a bit to the left above the folding sheet, crossing the left perpendicular that goes through the point M. This is co-timed to a part of his utterance, "exactly the same", completed in anticipation to the expression that requires specification, "here". Again, the interpretation of the gesture becomes possible by considering the interplay with inscription and speech within the situation.

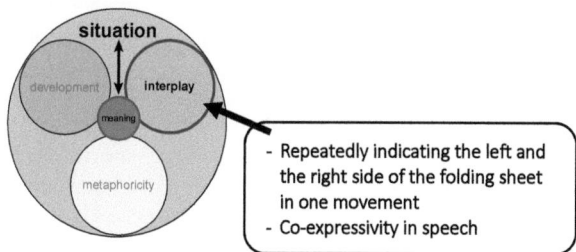

Fig. 5.12: Components influencing the interpretation space for disclosing the students' interpretation of the gesture 52

---

[60] In the following, I will refer to the points on the lower edge, used to start a construction step as *construction points*, and to those that are the product of a construction step as *constructed points*. While the constructed points constitute the curve, the notations of "construction" and "constructed" points shall make the reference to the folding process explicit and shall highlight that the students not necessarily refer to the mathematical object "point on the curve".

The interpretation of the gesture is not clear. It is considered to be co-expressive to the utterance "in fact it would be exactly the same here", but does not provide information about what is referred to by "it". The alternating pointing gesture, indicating the left hand side, the right hand side, and the left hand side again, may clarify the "here" by relating the left hand side of the diagram to the right hand side.

He concludes that "well then" they "don't need to do any on the other side." (52), seemingly justified by the relation between the left and the right hand side, expressed by the use of gesture.

d) While Mike is still speaking, Tim starts an ephemeral connection of the constructed points by a gesture similar to the one in line 49, while Mike still finishes his utterance in line 52. This is directly merged into successively indicating the three constructed points one after the other in the opposite direction while Tim starts to express his observation "well no actually when one connects these three points by a good curve one can theoretically predict it already. ‚approximately." (53).

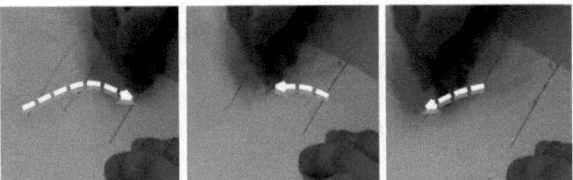

Fig. 5.13: Tim's gesture starting while Mike still speaks (52/53)

As in line 49, the curve is embedded in the diagram without Tim accompanying it by words. Then, co-expressive to him saying "well no actually when one connects these three points by a good curve", the movement changes to an indication of the three constructed points, starting with the point on the left hand side.

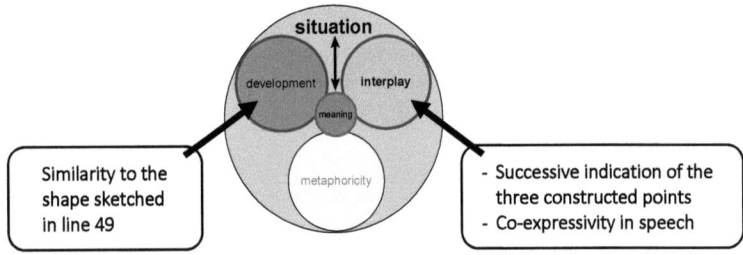

Fig. 5.14: Components influencing the interpretation space for disclosing the students' interpretation of the gesture 52/53

The first embedding of the shape of the curve (from right to left) is reminiscent of the initial shaping in line 49, and the successive indication of the three points comes along with a strong iconic component, automatically shaping the curve from left to right. Co-expressively to speech, it refines the curve to be the connection of points and furthermore, suggests the extent of the curve to be bounded by the three constructed points.

Tim's gesture thus provides a reference to what can already be 'predicted theoretically' (53).

e) Mike's response in line 54 is accompanied by a pointing gesture.

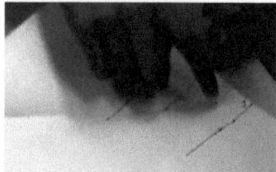

Fig. 5.15: Gesture accompanying Mike's proposal in line 54

While saying "let's do another one between", he indicates a location between the two perpendicular lines related to the construction points $C2$ and $C3$ on the lower edge. Furthermore, a diachronic relationship to Tim's gesture in line 51 can be identified, not in iconic similarity, but with respect to the location it indicates:

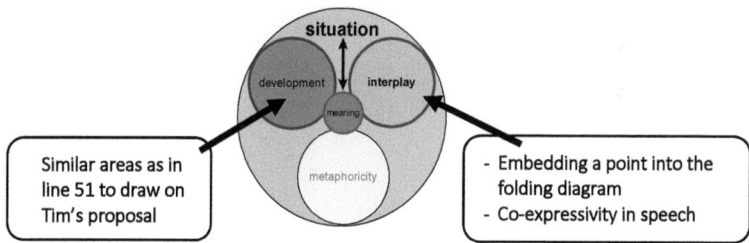

Fig. 5.16: Components influencing the interpretation space for disclosing the students' interpretation of the gesture 54

From the similarity to Tim's reference in line 51, pointing to the same area, it is suggested that Mike takes up Tim's proposal made in line 51 when he refers to "do another one between".

In this first part of the scene, it can already be observed how the analysis of gestures may influence the students' interpretation of the utterance and with that, of the social interaction: The students' gestures add further references and may provide a better understanding of the epistemic processes. Tim's utterance in line 49, "I can actually nearly see the curve <u>now</u> already" can be seen in another light since he actually made the shape of the curve visible by ephemerally embedding it within the folding sheet by using a gesture. The seeing of a structure thus does not come out of nowhere but is prepared by a **visual connecting action**, expressed in a gesture. Assuming that Mike's agreement (50) refers to Tim's gesture-speech utterance rather than to the verbal utterance alone, a shared understanding of the shape of the curve can be suggested. Furthermore, with respect to the diachronic relationships, gestures may recall references to former ideas in a kind of *catchment*[61] (e.g. line 53), also interpersonally, referring to a partner's former expression (54).

Taking into account the development of the semiotic bundle, another component is added in the second part of the scene:

f) After constructing a fourth point (55), Tim proposes to fix the curve within the folding sheet as it has been shaped in gesture so far:

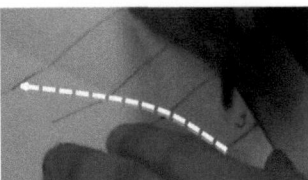

Fig. 5.17: Tim's gesture, performed from left to right: "shall I now try to draw this curve approximately' " (55)

The gesture recalls the gestures performed in lines 49 and 53, again shaping the curve within the folding sheet as has already been agreed on.

---

[61] See chapter 3.2.2 for the concept of catchment. Note that the catchment needs to be seen in a wider sense for the correspondence of the gesture in line 54 to the one in line 51. It can only be identified as catchment when respecting that both gestures are related to the same proposal of constructing another point.

5.2 The Interpretative Approach

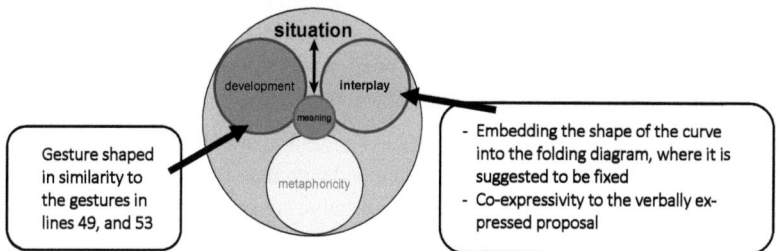

Fig. 5.18: Components influencing the interpretation space for disclosing the students' interpretation of the gesture 55

The gesture recalls the curve as a connection of points in a catchment (diachronic relationship), making it visible within the folding diagram once more.

Mike does not reject the idea but turns towards the interviewer to make sure that it is permitted to add an inscription that is not requested to be added by the instructions (57).

g) Refining the idea of fixing the curve as drawing it "in a different color." (58), Tim repeats the embedding of the curve ones more:

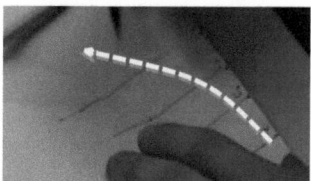

Fig. 5.19: Gesture sketched from right to left in line 58

The gesture sketches the curve above the folding sheet in a hasty movement. Rather than shaping the curve precisely, it only *recalls* the shape as referred to before (49, 53, 55).

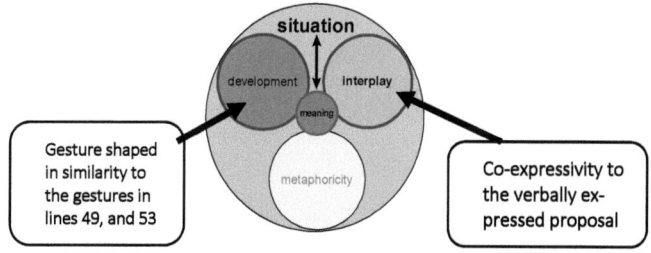

Fig. 5.20: Components influencing the interpretation space for disclosing the students' interpretation of the gesture 58

The similarity to the former gestural formations of the shape of the curve allows for its identification as a catchment even though it is not accurately performed. It is thus more so the diachronic relationship to former signs that shapes the possible reference than the concrete shaping itself.

After clarifying who performs the fixation of the curve, it is drawn in the folding diagram and enlarges the semiotic set of inscription and with that, the semiotic bundle by generic conversion (ephemeral gesture leads to inscriptive sign).

5.2.3.3 Combining Both Approaches

As we have just seen, the different signs in use influence the students' understanding of their meaning not only simultaneously, but also in the process of meaning making. This is not an entirely new insight, recalling that the phenomenon of genetic conversion, in which an inscription is added to the semiotic bundle as derived in similarity to a former gesture has already been mentioned by Arzarello (2006, p. 281).[62] However, while the 'semiotic reconstructive analysis' (SRA) so far mainly focusses on the analysis of speech acts as signs, the analysis of the semiotic bundle lacks the fine-grained sequential character of the SRA. I thus plan to combine both analytical approaches in moments when gesture occurs and claim that this enlightens the relations between the signs as they interplay in the semiotic composition and develop within the process. Furthermore, I also suggest that this combined analysis allows for a more comprehensive description of the epistemic process. A similar approach has already been considered by Dreyfus et al. (2014), as has been described in chapter 2.3. The present study however considers gestures not only when experiencing limits of speech act analysis but *generally* as one modality influencing the epistemic process. I assume gestures to have influence on the way verbal utterances are interpreted in social interactions. Therefore, every utterance is considered in its multimodal form as semiotic composition of speech, inscription and gesture.

In the previous two sections, two different phenomena have been reconstructed for the same scene as complementing each other: First, the epistemic processes as observable from speech acts; second, the meanings reconstructed to be represented by gestures within the epistemic process.

Combining both, the SRA can be refined to a *multimodal* SRA.

---

[62] See also chapter 4.3.2.

5.2 The Interpretative Approach 77

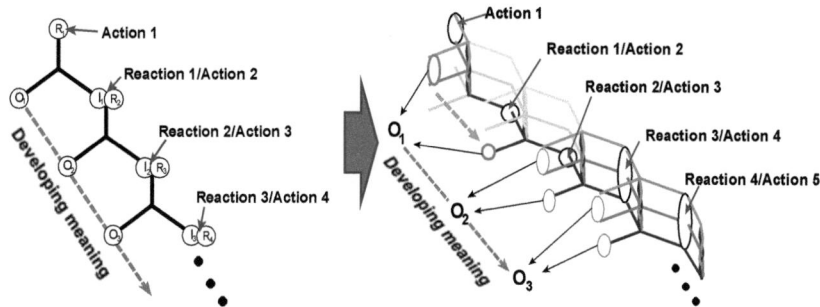

Fig. 5.21: Schematic representation of the SRA (left hand side) in comparison to the schematic representation of its refined, multimodal version (right hand side)

Fig. 5.21 represents both, the original (left side) and the multimodal (right side) versions of representing interaction on which the semiotic reconstructive analysis can be based. Due to its complexity, the new, multimodal version will be explained more in detail: The darker 'limbs' of the multimodal signs in the representation on the right hand side represent signs that are simultaneously activated. These signs form one semiotic composition (See chapter 4.4, page 50, for the explanation of the multimodal sign.). While the main idea considered in the original version of the SRA stays the same for the multimodal one, a core difference concerns the shaping of the immediate mathematical object. This becomes visible on the left sides of each representation: In the original model, the immediate object is shaped by the verbal utterance. The multimodal model now makes it possible to consider several immediate objects, related to parts of the semiotic composition of which the verbal utterance is one part. The immediate object of the entire multimodal sign now appears to become analytically decomposed. This approach makes it possible to answer the question of how gestures take part in shaping the immediate object.

In the following Fig. 5.22, a multimodal SRA of the first utterances of the yet familiar scene PA5e1.1[63] is presented to illustrate a fine-grained semiotic analysis that makes possible to zoom on single meaning units as decomposed for the analysis.[64] The immediate object represented on the left side (grey box) is the one shaped by the whole utterance, while those in the ellipses may refer to parts of the utterance. As can be seen in

---

[63] The episodes and scenes are each labeled with a code: The first two letters and the following number identify the data set, the 'e' stands for 'episode', and the following number concerns the episode and the scene. This scene is thus taken from the data set 'PA5', episode 1, scene 1 (See section 6.2.3.1 for a more comprehensive description.).
[64] A multimodal SRA of the first part of the scene, lines 49-55 is given in the attachment on page 339 as fitting an entire page.

O_{49}, an immediate object can also be assumed to develop within one utterance. Here, the gesture that gives visual access to the shape may lead to actually seeing the curve and to Tim's verbal expression.

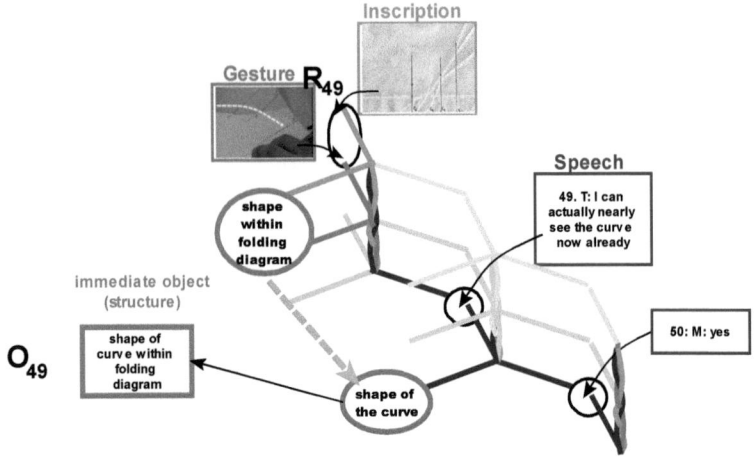

Fig. 5.22: **Multimodal SRA of the first utterance of scene PA5e1.1**

This example (and the extended diagram attached in the appendix) also shows that it is far from being practicable to conduct such a multimodal SRA; it serves as a *mental model* that can be explicated to get a more in-depth understanding of the social interaction that constitutes the epistemic process in a multimodal way. For example, the 'splitting' of the utterance in line 49 provides a better comprehension of what may cause the structure-seeing action that otherwise would be considered being accomplished spontaneously, leaving open the question about whether the students see the same shape of the curve within the product of the folding process.

In total, integrating a multimodal approach in the investigation of the epistemic process reveals how gestures contribute in preparing epistemic actions (49). In this case, it explains the seemingly 'spontaneous' structure-seeing. Furthermore, it is also shown how gestures contribute in establishing non-verbal representation of mathematical objects and by that shape a visual reference in social interaction.

Therefore, this integration of a multimodal process is adopted for this study: Utterances are not seen as being shaped only by verbal means of expression. Moreover, I assume utterances to be shaped by the entire semiotic composition (see chapter 4.4) whose components are distinguished only for reasons of analysis.

5.3 Theoretical Sampling as a Focusing Strategy

Not the entire amount of data needs to be of interest for investigating the research question. Considering it may not only be uneconomic regarding the aspect of time but may also lead to great biases if a hypothesis shall be tested by means of a random sample (Kelle & Kluge, 2010, p. 41). Qualitative research considers a small amount of cases but aims to make statements about their underlying structure, which may show aspects of great diversity while others are comparable. The choice of data considered for the analysis is as important as the collection of data itself and needs to be substantiated, driven by the concrete research aim. In this regard, Lamnek refers to the "quality of the sample" (Lamnek, 2005, p. 189) that ensures that the considered cases are chosen in a suitable manner to investigate the research question.

Lamnek proposes *theoretical sampling* as introduced by Glaser & Strauss in their Grounded Theory-approach (Lamnek, 2005, p. 191) as a possible sampling-strategy. For theoretical sampling, relevant episodes are selected based on criteria that are grounded in theoretical considerations. This way, the data can be reduced to focus on those samples, in which the appearance of the investigated phenomenon becomes most probable or most relevant from a theoretical perspective. The samples chosen are then considered to represent the developed theory (Kelle & Kluge, 2010, pp. 47-49). Grounded Theory aims to develop new theories from data and the theoretical considerations and the sampling are linked and develop as intertwined within the research process. For building on, that is refining or extending an already existing theory, the theoretical considerations can be grounded in this theory. Based on a concrete research aim, additional theoretical input shall substantiate the choice of the criteria that underlie the theoretical sampling.

5.4 Forming the Building Blocks for a Theory on Gestures' Contribution to Social Epistemic Processes

As has been made clear in chapter 5.1, the results aimed to be derived from the analysis are *representational* and *epistemic* functions of gestures. Both kinds of functions need to be approached differently since each of them is related to a different "larger whole" (see chapter 5.1).

5.4.1 Identifying the Gestures to be Analyzed

An important issue to be considered is determining the cases relevant for analysis (Kelle & Kluge, 2010, p. 42). This relevance highly depends on the research question that shall be answered by the analysis of the cases.

The representational functions of gestures are related to the shaping of the immediate mathematical objects. Therefore, their identification desires an analysis of those multimodal signs that contain gesture as one modality. With this it can be answered how the immediate object of the gestural sign contributes to shaping the immediate object of the multimodal utterance. The search for relevant cases is thus *gesture-driven*. Epistemic functions of gestures on the other side seek towards answering the question of "How can gestures contribute to acting epistemically?" Their identification is orientated on the epistemic actions reconstructed from speech act analysis, searching for a possible answer to the question "How does gesture contribute to the establishment of *this concrete epistemic action*?" The indication of relevant cases is *speech-driven* here.

5.4.2 Elaborating Functions of Gestures as Empirically Based Categories

This study aims to provide fundamental research on gestures in epistemic processes. This issue comes along with a lack of predefined categories to which the gestures can be assigned in a subsumptive manner. Consequently, the categories need to be developed from the data. This can be accomplished by *abduction*, a strategy to construct a new category. As Kelle and Kluge write with reference to Peirce:

> The sudden occurrence of an unexpected phenomenon can prompt the researcher to construct another class, or to find a new rule. PEIRCE calls *abduction* such a conclusion from an unexpected event that serves as an explaining rule. Hypotheses resulting from an abductive conclusion are thus hypotheses on a new general rule that explains a surprising phenomenon.
>
> (Kelle & Kluge, 2010, p. 25, my own translation, italics in the original).

The research questions concern matters of *"how* gestures contribute" to two different events; representing and acting epistemically. The hypothesized rules will thus lead to *descriptive* categories. Nevertheless, these categories develop in the course of opening the data. Each new case is used to test the hypotheses ("Does the category describe the function in this specific case?") in order to reject or to refine it. In case it is rejected, a new catgeory may be found to describe the respective function of the gesture. In case it is refined, the description of the category is elaborated. This requires a constant reinterpretation of recent cases in order to check whether they still fit the refined category. By that, the construction of new theoretical knowledge on the phenomenon is made possible. Reichertz writes in this context:

> An abductive discovered order, therefore, is not a (pure) reflection of reality, nor does it reduce reality to its most important components. Instead, the orders obtained are *mental constructs* with which one can live comfortably or less comfortably. For many purposes particular constructs are of use, and for other purposes different constructs are helpful. For this reason the search for order is never definitively complete and is always undertaken provisionally. So long as the new order is helpful in the completion of a task it is allowed to remain in force; if its value is limited, distinctions must be made; if it shows itself to be useless, it is abandoned. In this sense

> abductively discovered orders are neither (preferred) constructions nor
> (valid) reconstructions, but *usable* (re-)constructions.
>
> (Reichertz, 2004, p. 163, italics in the original)

From this also follows that there is no claim that the categories to be found are exclusive or generally valid for mathematical epistemic processes but that they have been reconstructed from the data. The scope of the results is widened by the variety of the tasks worked out by the students, being different in mathematical areas and affordances on semiotic and mathematical activity.

5.4.3 The Scope of the Theoretical Building Blocks

The pieces of theory this study seeks to find are of **middle range** (Krummheuer & Brandt, 2001, pp. 198-200; Merton, 1968, pp. 39-72). That is, they do not attempt to make statements that are considered as general truth but moreover to yield contextually bound descriptive explanations as developed in close relation to the phenomenon or object under investigation.

> It is intermediate to general theories of social systems which are too remote from particular classes of social behavior, organization and change to account for what is observed and to those detailed orderly descriptions of particulars that are not generalized at all. Middle-range theory involves abstractions, of course, but they are close enough to observed data to be incorporated in propositions that permit empirical testing.
>
> (Merton, 1968, p. 39)

As has already been mentioned, the theoretical building blocks developed in this study shall contribute in enhancing an already existing theory on social mathematical epistemic processes.

5.5 Quality Criteria and Methodological Concerns from Gesture Studies

5.5.1 Quality Criteria Considered in Qualitative Research

Denzin (2009) gives a comprehensive discussion of affordances examined in comparison to those related to quantitative study. Nevertheless, as the methods used to conduct a qualitative study need to fit the research questions and aims, so do the standards the study shall comply with. Their disclosure helps the reader to evaluate the study (Hammersley, 2007, pp. 288-291) by revealing the measurement of quality underlying the research process. Furthermore, they serve the researcher as grounds from which his own research can develop during the process of inquiring about a given issue, framing the affordances (s)he claims to be relevant within the research process. Besides documenting the affordances considered for the study, this allows forming conclusions on limits and constraints of the results that can be gained. Hammersley summarizes these considerations as follows:

> While I am rejecting the idea of a finite set of explicit and exhaustive criteria that can substitute for judgement, [sic] or render its role minimal, the above discussion indicates my belief that criteria, in the form of guidelines, can play an important role in the work of researchers. They may facilitate reflections on previous judgements that enables us to learn from our own experience, and from one another. Such learning may also involve exploring the implications of applying locally used criteria in new contexts, and considering the extent to which there are commonalities across fields. This is important because such reflections can lead us to conclude that locally used criteria need to be changed or developed, even as regards their original context. [...]
>
> In a much more mundane way, but no less importantly, criteria, in the form of guidelines, can remind researchers of what they ought to take into account in assessing their own and others' research. This is not a small matter. When engaged in any complex activity, it is easy to overlook what one would, in other circumstances, routinely take into consideration.
>
> (Hammersley, 2007, p. 289)

Furthermore, Hammersley argues that

> it is not possible for researchers to make their judgements transparent, in the sense of fully intelligible to *anyone*, irrespective of background knowledge and experience. Indeed, there are limits to the extent to which these judgements can be made intelligible even to fellow researchers, because of the situated nature of judgement. Certainly, it is the case that such intelligibility is an *achievement*, it is not automatic: speakers need to be able to formulate the situation, the reasons for making the judgements that they did, and so on, in ways that facilitate understanding; and, equally importantly, the audience must be able to draw the right inferences from what is said, on the basis of the background resources they have. [...]
>
> A central theme here is 'transparency', for example it is demanded that the basis on which professionals work should be made explicit, so that the lay people who use their services can judge the quality of what is provided.
>
> (Hammersley, 2007, p. 291, italics in the original)

Steinke (2004) refers to this as *"inter-subject comprehensibility"*, *"indication of the research process"*, *"empirical foundation"*, and the explication of *"limitation"* of the results (ibid., pp. 186-190), and adds the criteria of *"coherency"*, *"relevance"* and *"reflected subjectivity"* (ibid., p. 190) . These criteria substitute the criteria that are traditionally adapted from quantitative study, modified to fit the affordances, which interpretative, qualitative research shall fulfill; *objectivity, reliability,* and *validity* (Jungwirth, 2003, p. 196).

Inter-Subjective Comprehensibility
It is due to the nature of interpretative research that its results cannot be considered as being objective: The interpretative reconstruction of social interaction always requires taking into account multiple variations of possible interpretations of the participants' contributions under investigation. Nevertheless, it cannot be guaranteed that all possible variants are ascertained (Bikner-Ahsbahs, 2005, p. 91). According to Steinke, intersubjective comprehensibility encompasses the explication of the choice of research methods, their documentation as well as their application. Furthermore, it embraces the background the study is embedded in, the gathering of the data, and the interpretation methods applied for the analysis of the data. That is, all kinds of information explicitly influencing the study leading to its results needs to be documented (Steinke, 2004, p. 187). These affordances are all dedicated to fulfil the claim of making the research process as transparent as possible.[65]

In addition, interpretation can be conducted in groups or "discussed with colleagues who are not working on the same project" (Steinke, 2004, p. 187) to provide an overlap in inter-subjectivity. A third method to consider concerns utilizes "codified procedures" (Steinke, 2004, p. 188): If no formerly established procedure has been applied to carry out the analysis, a detailed description of the steps followed within the analysis, providing its intelligibility, needs to be given.

Indication of the Research Process
This criterion concerns the issue of having chosen a suitable approach towards answering the research question. Besides addressing the qualitative approach, Steinke refers to four main aspects to consider (Steinke, 2004, pp. 188-189):

1. The *choice of methods* related to gathering and analyzing the data;

2. The *transcription rules* used to fix the data in inscriptive form. This is considered to be crucially linked to the choice made, concerning the kind of data used for conducting the analysis. That is, transcription rules depend on the concrete research question and the phenomenon under investigation, the exclusiveness of the use of the transcription as a resource for the analysis, as well as for the grade of

---

[65] Due to the reasons described in chapter 5.2.2, the analyses of gestures have been carried out directly from the video data and not from a verbatim description of the gestures. To respect the anonymity of the students, video data is only made accessible on demand, in anonymized form and in presence of the author.

subjectivity that shall be provided by the transcript. Nevertheless, the rules should be practicable; it should be possible to easily write and read the transcript.

3. The *sampling of the data* for reduction and choice of the analyzed data;
4. The indication of the *evaluation criteria*: Do the results answer the research question?

Empirical Foundation

The results shall be grounded in the data, derived from the data considered for analysis. As Steinke writes:

> The formation *and* testing of hypotheses or theories in qualitative research should have an empirical foundation (or grounding), that is, it should be based on the data. *Theory-formation* should happen in such a way that there is a possibility of making new discoveries, and questioning or modifying the investigator's prior theoretical assumptions. Theories should be developed close to the data (for example, the informants' subjective views and modes of action) and on the basis of a systematic data analysis. For *theory-testing* implications or prognoses are derived deductively from the theories, and these are verified or falsified on empirical data. While verification looks for confirmation of the theory with the data, falsification – as a tougher criterion – tests the theory by attempting to reject it (on the latter, cf. Seale 1999a: 73ff.).
>
> (Steinke, 2004, p. 189, italics in the original)

She proposes various ways to give evidence of the empirical foundation of the results. For this study, the application/development of codified methods can be considered appropriate, as well as what Steinke calls the "textual evidence" (Steinke, 2004, p. 189). This is closely related to the following criterion:

Explication of the Limitation of the Results

Empirical results are rarely generally applicable but depend on the concrete circumstances and conditions in which they have been carried out. It is the researcher's job to extract the main points of generalization by abstracting from the specific cases up to a certain degree. It is also his work to make this degree explicit as a degree of generalizability of the derived hypotheses.

> If all of the (very specific) conditions of the investigation must be fulfilled for the results to be transferable, then the results can hardly be claimed to be transferable. There must also be a clarification of what conditions must be fulfilled, as a minimum, for the phenomenon described in the theory to occur. At the same time, aspects that are incidental, and – from the point of view of the theory – irrelevant, are filtered out.
>
> (Steinke, 2004, p. 190)

This can be carried out by searching for deviant and extreme cases of the phenomenon under investigation, as well as for cases with minimal difference to a prototypical case.[66] By comparing and contrasting these cases, the typical aspects of the phenomenon can be pointed out as constituting its specificity.

Coherency and Relevance
Finally, two criteria need to be particularly related to the theory on mathematical epistemic processes of interest-dense situations, as it already exists:

By coherency, Steinke refers to the coherency of a developed theory in itself. The present study seeks to provide building blocks that fit a preexisting theory concerning the understanding of students' social processes of constructing mathematical knowledge. The coherency thus needs to be seen in this regard, as a criterion of *fitting the preexisting theory*.

The criterion of relevance concerns the question whether the building blocks aimed to be built actually provide any additional value at all. Why is it important to address this research question? How does the (in this case preexisting) theory benefit from answering the question? Are the results applicable for other research?

5.5.2 Criteria Considered in Gesture Studies

Although the criteria described in the preceding chapter do not explicitly refer to the interpretation of verbal utterances, some requirements for the analysis of gestures need to be made explicit. In the following sections, some methodological concerns from gesture analysis are explicated to justify some parts of the proceeding that may differ from what is common in the analysis of spoken or written utterances. Since methods need to fit the research objects and the question under investigation, the specificity of gestures as a research object has to be considered for methods and methodology. Gullberg provides a list of "desiderata"[67] for gesture analysis (Gullberg, 2010, p. 88)[68] that will be discussed in the specific context of this study.

---

[66] The term 'prototypical case' is used here as synonym for 'illustrative example'.
[67] While generally, "research desiderata" means topics desired to investigate, Gullberg uses the word "desiderata" here moreover to refer to 'conditions a study on gestures shall fulfill.
[68] More in detail, Gullberg refers to "Desiderata for gesture studies in SLA [(Second Language Acquisition)] and bilingual studies" (2010, p. 88). However, I do not see any reason why her list shall be restricted to SLA-related studies since she writes the following in the introduction of the respective section: "As for any other domain of inquiry, replicability and methodological rigour [sic] is vital. It is essential to be explicit about theoretical assumptions, tasks and procedures, issues of baselines, units of analysis, coding and reliability. This is equally important in qualitative and quantitative approaches" (ibid.).

On the Issue of Developing Functions of Gestures

Gullberg sets the starting point that has to be made clear for any study that concerns gestures as "the question of what gesture dimensions are relevant for a given research question: form, meaning, function, timing, contextual use, presence/absence, etc." (Gullberg, 2010, pp. 86-87). For the present study, this has been determined in chapter 5.1.

Although gesture studies in general have a strong quantitative orientation, Gullberg mentions that "[c]ompanion qualitative analyses are [...] needed to examine how individual, social and contextual factors interact with the phenomenon under study" (ibid., p. 87). In this study, this idea can be related to the contextualization of the gestures as contributing to the epistemic process in social interaction. It is thus the approach more common in gesture studies that suggests the proceeding of first coding gestures and then grouping them for developing the description of the code by comparing the elements of one group with respect to the just mentioned factors.

"Desiderata" on Gesture Analysis

a) *Theoretical Assumptions*: "It is crucial to specify what gestures are supposed to index (a concept, lexical access, etc.), and how the link between gesture and speech is viewed" (Gullberg, 2010, p. 88). In this study, this issue concerns the gesture being a component of the semiotic composition, having its link to speech as well as inscription within the semiotic bundle.

b) *Tasks and Procedures*: Task and procedure concern the design of the setting so that it initiates gesture production. In this study, this has not been considered for the gathering of the data, since the setting has been designed to yield fruitful epistemic processes and make them observable in social interaction. While in this regard mainly the observation of verbal utterances has been in mind, the setting may have also elicited gestures as means of communication "since gestures are more frequent in dialogue face-to-face interaction" (Gullberg, 2010, p. 88, referring to Bavelas et al. 2002 and Bavelas et al., 2008).

c) *Coding*: "Studies should be explicit about (1) how gestures are identified and (2) the unit of analysis (e.g., the 'gesture phrase' between major resting positions or the more fine-grained gesture stroke; Kendon 1980)" (Gullberg, 2010, p. 89). This depends on the phenomenon that is investigated. Consequently, different kinds of units are considered for representational functions and for epistemic functions of gestures (see chapter 5.4.1).

d) *Distinguish Form and Function*: Gullberg distinguishes "coding for form" and "coding for function, meaning and co-expressivity of speech and gesture" (Gullberg, 2010, p. 90). While she claims that the former one shall be coded with sound turned off "in order to avoid an influence from meaning in concomitant speech" (Gullberg, 2010, p. 90), the latter one seeks just for this kind of meaning, answering questions like, "What does the gesture do/mean?" and "What is expressed where?" (Gullberg, 2010, p. 90). The diverseness of gesture and speech (see footnote 20 in chapter 3.2.2) causes that "[s]peech and gesture information is [...] rarely 'identical' even if closely related" (Gullberg, 2010, p. 90) so that it needs to be made explicit which one of both modes of expression is considered first for analysis: "Analyses can take speech or gesture as their starting point depending on the research question. [...] It is equally possible to first identify a relevant stretch of speech and then investigate gesture co-occurring with it (Gullberg, 2003). Both approaches are equally valid depending on the research questions and provided the approach is clearly described" (Gullberg, 2010, p. 90). However, this 'desideratum' is closely linked to the units of analysis mentioned in the preceding point, related to the coding of gestures: To determine possible representational functions, the gestures have been the leading marker to in turn take a closer look at the relation to speech. Differently, the investigation of the epistemic function of gestures considered only those gestures which are temporally related to an epistemic action that has been reconstructed from speech act analysis.

e) *Interrater Reliability* (also Intercoder Reliability): Different coders will most likely derive different results when coding. The result worked with will be a measure of all coders' results. In this study, one aim is to develop such codes as functions and describe their value for the epistemic process. Following this, intercoder reliability is not an issue at this point but may become one in upcoming studies in which the same functions are used as codes. Their characteristics must thus be made clear to enhance the best possible intercoder reliability.

f) *Qualitative vs. Quantitative Analyses*: "The most fruitful approach is to combine some degree of qualitative analysis with a quantitative approach to allow for a clear understanding of the phenomenon under study as well as generalizability" (Gullberg, 2010, p. 91). This study seeks to identify and describe the phenomenon of epistemic function of gesture that is rooted in qualitative work. Although the derivation of functions of gestures is undoubtedly based on the qualitative analysis of the data, a basic quantitative enumeration provides a more comprehensive view of similarities and differences between different data sets.

6 Methods
The Course of the Study

This chapter aims to provide an overview of the study, starting from considerations concerning the gathering of the data, that is, the setting. The project is affiliated to a more comprehensive research program that deals with the investigation of processes of constructing mathematical knowledge, conducted at the University of Bremen. It makes use of data that has been gathered before, but extended and elaborated in the context of this study with respect to the present research perspective. The study focuses on a specific phenomenon that shall be investigated within epistemic processes, namely students' use of gestures. The specific way of collecting data made it possible to align the preparation of the data so that the integration of a semiotic perspective on epistemic processes is respected. For that, this chapter will clarify (i) how the data fits the semiotic approach examined in this study, (ii) the extensions that had to be made concerning the preparation of the data and above all, (iii) the concrete process of data analysis.

6.1 Setting

The term 'setting' has a twofold meaning for this study: The data itself has been gathered within a setting, comprising the students, the tasks, and the situation of dyadic collaboration in a laboratory setting. This setting however has already been 'set' in the beginning of the study. In the following section, the setting of the study will be described in more detail in both senses.

6.1.1 General Aspects of the Study and on Gathering the Data
The data used in this study stems from the German contribution to a project that aimed to connect a social and an individual perspective on knowledge construction by 'networking' (Bikner-Ahsbahs & Prediger, 2014) two theories, namely the theoretical approach of 'interest-dense situations' (IDS, Bikner-Ahsbahs, 2005, see chapter 4.1) and the investigation of 'abstraction in context' (AiC, Hershkowitz et al., 2001).[68] The tasks

---

[68] "Effective knowledge construction in interest-dense situations" (funded by the German-Israeli-Foundation, grant 946-357.4/2006).

used for gathering the data have been constructed and tested in a pilot study in order to prompt the construction of mathematical knowledge. The variation of the semiotic material provided by the three tasks, as well as the consideration of three perspectives when videotaping the students' working processes allows the epistemic processes to be investigated from a semiotic perspective, specifically on the use of gestures: A front perspective captured the gestures produced in the gesture space, and a second perspective focused on the inscriptions produced and referred to by the students. When Dynamic Geometry Software was used, a third camera was directed to the screen.

6.1.2 The Students

Three pairs of high-performing students from 10th grade were invited to take part in teaching experiments in which they each worked together on three mathematical problems. The students were recommended by their teachers who considered them as particularly driven and talented in dealing with mathematical problems. Furthermore, they were chosen as pairs that were experienced in working together. These conditions took part to provide a setting in which fruitful social processes of knowledge construction were expected to be observable. The mathematical topics of '*Parabola as geometrical locus*', '*Continued fraction*', and '*Induction*' were new to the students, so the underlying concepts had to be constructed within the session. The students brought along adequate previous knowledge to develop a way to handle the mathematical task.

Two of the pairs are composed of male students, a third one consists of two girls.

During the working process the students sit side by side at a table, rendering it possible for both of them to have same access to the given material. This placement is particularly important considering the use of gestures since it provides a similar spatial orientation and with this facilitates establishing a common gesture space (Yoon et al., 2011, p. 390).

The Role of the Interviewer and the Observers
During the gathering of the data, up to four observers, including an interviewer, were present in the field. The interviewer's role was to present the task to the students, to clarify upcoming questions regarding its comprehension, and to give hints in case they got stuck. Based on an a-priori-analysis of the task, auxiliary cards were prepared for those situations. These could give suggestions to stimulate the working process of the students. The interviewer was trained to react spontaneously to the students' approaches and to give support. The interviewer only participated in the social interaction when there was no other possibility. At most two additional persons were needed to take care of the videography, each controlling the perspective of one camera.[69] They did not interact with the students. One observer sat in the back to make notes about the students' elaboration

---

[69] The third camera was statically directed to the students from a left front perspective. A static right perspective was used for those elaborations in which the screen was not recorded.

of the task. At times, this observer also intervened and acted as a second interviewer, although he mostly stayed in the back.

During the gathering of the data, I did not take the role of the interviewer. This research setting made it best possible for me as the researcher to keep a neutral attitude towards the data. This becomes important for the analysis from a methodological point of view: Not being involved in the students' learning processes allows the analysis to be carried out more objectively.

6.1.3 The Tasks

As mentioned before, each pair of students worked out three different tasks, each of these tasks having different qualities concerning the semiotic design[70] and the mathematical topic. In the following, a summarized overview of the tasks and their solutions will be given as an important part of clarifying the setting. The results in chapters 7 and 8 are illustrated by examples taken from the data which are difficult to understand without knowing the mathematical and semiotic background the students deal with. The tasks have been chosen from different mathematical fields in order to investigate whether and how the use of gestures differs between them. For reasons of clarity and comprehensibility, the concrete formulation of the task as given to the students, as well as a detailed mathematical solution adding an analysis from a pedagogical content perspective is given attached in the appendix rather than at this point (see pages 277-307).

6.1.3.1 Geometric-Algebraic: The Parabola as Geometrical Locus (PA)[71]

A geometric-algebraic task deals with the parabola as geometrical locus. Several diagrams become available to the students to work with. The first one needs to be produced by the students following a folding instruction. For this, the students are asked to choose any point on the lower border of a sheet, to mark it as point C and to bend the sheet so that the chosen point comes to lie exactly on a point that is already marked on the sheet (point M, see Fig. 6.1).

---

[70] Semiotic design means everything which is explicitly provided to work with as semiotic resource. This includes, for example, the input and output registers (input concerns the formulation of the task, output the formulation of the solution) and the semiotic means provided to work with (e. g., tables, coordinate systems, environment given by a Dynamic Geometry System,...)
[71] This task has been developed in the project "Effective knowledge construction in interest-dense situations". It is a revised version of a task designed by Nava Gilboa (Gilboa, Dreyfus, & Kidron, 2011) with specific contribution of Julia Cramer. It was used for several publications (Bikner-Ahsbahs, Sabena, Arzarello, & Krause, 2014; Bikner-Ahsbahs, Sabena, & Arzarello, 2014; Krause, 2012; Krause & Bikner-Ahsbahs, 2012; Krause, 2015).

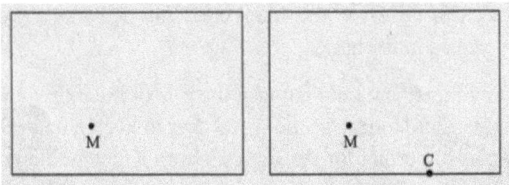

Fig. 6.1: Initial situation given as example for following the folding instructions

With this process, a folding line shall be produced. The next steps consist in (i) drawing a perpendicular to the lower border through the chosen point and (ii) marking the intersection point of this perpendicular and the folding line with a cross. These construction steps, starting with choosing a point on the lower border up to marking the cross, shall be continued until a curve is seen (Concrete formulation: *"Repeat this process until you recognize a curve"*). Fig. 6.2 shows such a possible folding diagram developed from constructing five points.

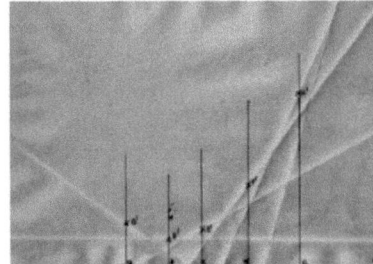

Fig. 6.2: Possible outcome of the construction of the folding sheet following the instructions[72]

The emerging folding lines can be identified as tangent lines touching a parabola in the marked points so that they approach the envelope curve. Based on these folding lines a conjecture shall be stated by the students, concerning the diagram that becomes visible to them (Second part of subtask 1: *"Make a conjecture."*). Subsequently, the students work with the Dynamic Geometry Software (DGS) GeoGebra. Here, an environment is given in which a variable point P can be dragged to the left and to the right on a directrix g. This dragging of P produces a trace of a curve within the environment. In addition, it is possible to vary the distance e between the fixed point B (the focal point of the parabola) and the x-axis. One possible situation is displayed in Fig. 6.3.

---

[72] This folding diagram has been taken from the 'PA7'-data set (see chapter 6.2.2 for the naming of the data) and can be found in the appendix on page 320.

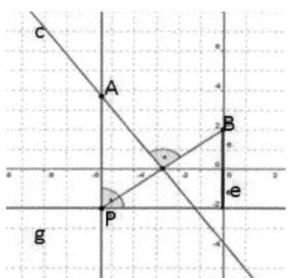 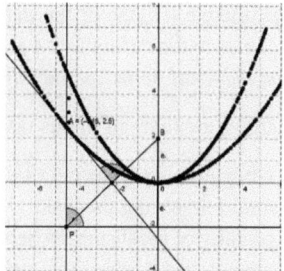

Fig. 6.3: Possible situation from the GeoGebra environment, fixed as printout.
Left side: Without trace;
Right side: Two traces produced for two different values of e

After comparing the folding situation with the GeoGebra-situation (subtask 2a: *"Move the point P. A curve appears. Do you recognize it?"*) and identifying the corresponding components (subtask 2b: *"Which points on the work sheet correspond to the points P, B and A in the file?"*; subtask 2c: *"Which line in the GeoGebra-file corresponds to the lower edge of the work sheet, where you have marked the point C?"*), the displayed situation is provided to the students as a printout on which they can mark anything they like. The task further consists of identifying the curve as a parabola and substantiating this.[73] To do this, the equality of the lengths of the segments $[AB]$ and $[AP]$ shall be noticed (subtasks 3a and 3b: *"Mark the distance between A and B and the distance between A and g. Make a conjecture."*, *"Justify your conjecture."*) and grounded in the folding: The tangent in point A is an axis of reflection of points B and P, the points corresponding to those that have been folded to lie on each other. Starting from the knowledge of $[AB] = [AP]$, the lengths of other segments can be determined by setting them in relation within the diagram.[74] The application of several theorems, for example, the Pythagorean theorem[75] can lead to the general formula $f(x) = \frac{1}{4e}x^2$. Finding this function equation proves the curve to be a parabola, desired to '*convince somebody that it is the curve you recognized in subtasks 1 and 2a.*' (subtasks 4a and 4b). In the last subtask, the students are asked to find a definition of the parabola as a set of points similar to a given definition of a circular path.[76] To solve this task, the equality of $\overline{AB} = \overline{AP}$ becomes important again. Considering an arbitrary point on the parabola and thus

---

[73] More specifically, the students shall find a function equation. The task only asks for justifying the conjecture. While the students preferably start with finding a valid function equation empirically by searching for points, the variation of e makes it necessary to prove that the conjecture is correct in a non-empirical way.
[74] See also the more comprehensive solution given in the appendix.
[75] See the subject-didactic analysis as attached for this proof and others, using the 'Triangle altitude Theorem' and the similarity of triangles $\triangle AGC$ and $\triangle CGP$.
[76] This definition is stated as „A circle is a set of points that all have the same distance from a fixed point (the center)".

ignoring the concrete lengths of the distances from this point to the focal point/the directrix, one comes to the definition '*A parabola is a set of points that each have the same distance from the focal point and the directrix*'.

Several structures are laid out in this task, as for example concerning the following:

- In the beginning, the seeing of a curve as a structure within the folding diagram is central for solving the task.
- The equality of the distances between points A and B and point A and the straight line g needs to be generally justified in the fact that the folding line is an axis of symmetry for points B and P, running through point A (subtask 3b).
- There are various ways to find the function equation, each demanding for another geometric structure to be seen within the diagram. For example, a perpendicular to the y-axis through point A can be taken into consideration so that the Pythagorean Theorem can be applied to the right triangle that emerges.[77]

6.1.3.2 Arithmetic-Analytic: Continued Fraction (CF)[78]

An arithmetic-analytical task concerns the continued fraction

$$f(x) = 1 + \cfrac{2}{1 + \cfrac{2}{1 + \cfrac{2}{\ldots}}}$$

The first four fractions $f(0)$ till $f(3)$ are given, starting with $f(0) = 1$. The first structure laid out in the task design is a compositional one: The value of x that denotes the number of the continued fraction can be recognized as corresponding to the number of +'s (plus-signs), the number of fraction bars, and also to the number of 2s. This structures hint at an explicit rule to compose the continued fraction $f(x)$. After determining the first seven elements of the sequence $(f(x))_{x \in \mathbb{N}_0}$ that builds the continued fraction,[79] a conjecture shall be stated concerning a recursive pattern (subtask 1.2: "*Look at the seven terms f(0) to f(6) and their calculation. Can you find a pattern to get from one term to the next one?*"). While potential construction rules to get from one step to the next

---

[77] See the analysis of the task as given in the appendix for further details and for additional geometrical structures to be applied in order to solve the task.
[78] This task has been designed in the project "Effective knowledge construction in interest-dense situations" with specific contribution of Nava Gilboa and was used in several publications (Behrens, Krause, & Bikner-Ahsbahs, 2014; Bikner-Ahsbahs, Dreyfus, & Kidron, 2010; Bikner-Ahsbahs, Kidron, & Dreyfus, 2011; Cramer, 2011; Janßen, 2010; Kidron, Bikner-Ahsbahs, Cramer, Dreyfus, & Gilboa, 2010; Kidron, Bikner-Ahsbahs, & Dreyfus, 2011; Priwitzer, 2010).
[79] Two perspectives on the relation between sequence and continued fraction can be adopted: "the sequence building the continued fraction" and "the sequence determined by the continued fraction". The perspective taken here takes into account that the students were given the continued fraction and developed the recursive formula for determining the elements of the sequence based on it.

one can concern different ways of substituting or complementing (See the detailed analysis given in the appendix.), the recursive rule can be formally noted as follows:

$$f(x) = 1 + \frac{2}{f(x-1)}, x \in \mathbb{N}_0 \text{ with } f(0) = 1.$$

Furthermore, the students were asked to reason how and why this pattern works (*"Explain the pattern – why does it work?"*). In the next step, the first 20 elements of the sequence shall be added in a table as a fraction and also as decimal numbers.[80] Again, a pattern shall be noticed and conjectured on (subtasks 2.2, 2.3, and 2.4: *"Look at the sequence in the table and make a conjecture.", "Justify your conjecture.", "Why does this justify your conjecture."*). This can concern the determination of a fraction from its ancestor (See the comprehensive analysis of the task in the appendix on pages 297-304.),

$$\text{if } f(x) = \frac{a}{b}, \text{ then } f(x+1) = \frac{2a \pm 1}{a},$$

where the algebraic sign in the numerator alternates from one element of the sequence to the next one, or seeing a structure in the decimal numbers: Beginning with f(3), the integer numbers are alternatingly a "1" or a "2", and the numbers of nines and zeros in the decimal places increase. That the distance to 2 decreases according to a certain pattern can then be seen within the entire sequence, and also for the two subsequences defined by even and odd elements. By combining both, it can be conjectured a two-sided convergence towards two. More in detail, the sequence alternates and converges towards two. That in turn can be substantiated by finding two subsequences and proving their convergence to two by mathematical induction on $x \in \mathbb{N}_0$. Rather than carrying out such a formal proof the students are expected to make this convergence plausible.

6.1.3.3 Logical Reasoning: Induction (IN)[81]

The third task has an argumentative character. It targets on finding and reasoning a strategy, following in its logic the principles of mathematical induction. In a word problem,[82] a situation is described from which a decision on how to react shall be conjectured: An undefined number of persons sits in a circle. Everybody wears a hat and there is at least one hat marked with an 'X'. Everybody sees every hat except their own one. Every five minutes, a bell rings and one needs to decide whether to signal by raising the hand or not, depending on whether one's own hat is marked or not. The task consists about *'thinking about how one can pass this examination'*. More concretely, the strategy shall

---

[80] A calculator was provided. The original subtask 2.1 is given as follows: *"Calculate some more terms of the sequence until you have 20 elements. Use the pattern that you have found. Use a calculator to represent the continued fractions of the sequence as decimal numbers. Note all decimal places that are displayed by the calculator."*
[81] The Induction-task IN has been designed in the context of the project "Effective knowledge construction in interest-dense situations" by Raz Harel and was used in (Cramer, 2010) and (Krause, 2015).
[82] See the full task given in the appendix on page 286.

be described: *"You sit in the circle and see n marked hats. Describe your strategy. Explain why your strategy works"*. In a formal way, this can be solved using mathematical induction on the number $n \in \mathbb{N}_0$ of marked hats visible for me, as the consultant being examined. One necessary condition concerns the knowledge that there is at least one marked hat. Based on this it can be concluded that I wear this marked hat in case I do not see any marked hat ($n = 0$). The induction step to $n = 1$, hence to seeing one marked hat, demands for a distinction of cases concerning the reaction of the consultant wearing the marked hat: If he raises his hand after the first ring of the bell, he does not see any other marked hat, as considered for the base case $n = 0$. If he does not raise his hand, he does see another marked hat that I do not see and consequently, I myself wear a marked hat. Similarly, the induction step can be generally induced from the reaction of the other persons by reconstructing how many marked hats they see, and then concluding whether their own hat is marked.

There are two main types of structures needed to be seen when solving this task: Firstly, the induction step is derived from a distinction of cases. Secondly, the induction step is always the same for an infinite process.

As has been expounded on in this subchapter, the three tasks provide structures of different character, varying from those that are visually accessible in the task design to those that rather concern a logical structure.

6.2 Methods of the Analysis

This section presents the pathway from the raw data to the generation of results, comprising the choice of data and its preparation, as well as the steps of analysis.

6.2.1 Data Preparation

The verbal and non-verbal contributions to the social interaction have been fixed in transcripts, partly in the course of the project the data stems from, partly in the course of this study. To each verbal utterance, a number has been assigned in a linear way. To provide neutrality as good as possible, the transcripts follow a transcription key that considers intonation:

Table 6.1: Transcription key

L:	(Student) L is speaking.
/L:	L starts speaking while the previous speaker is still speaking as well.
exact.	A full stop refers to a dropping of the voice.
exact'	An apostrophe sign represents the raising of the voice.
,exact	A comma directly in front of a word marks a new onset of the voice.

EXACT	Words are written all in capitals when spoken with a loud voice.
Exact	Underlined words are emphasized.
e-x-a-c-t	Dashes mark the stretching of whole words or single parts of them.
(.), (..), (...)	One, two or three seconds pauses are indicated in this way.
(4sec) (or 4s)	Used for pauses longer than three seconds.

The non-verbal activities are described in *italics* and for denoting the temporal relationship of gesture and speech has been developed a simplified transcription code:[83]

- [Squared brackets] indicate beginning and end of a gesture phrase/gesture stroke.
- [[Double square] brackets visualize gesture units that consist of [more than] one [gesture phrase]].

This way, the co-timing of gestures to the verbal expression becomes transparent without disturbing the flow of reading when reading the utterances.

The transcription of the gesture phrases has been added only for the episodes of interest with respect to the analysis of gestures (see section 6.2.3.1). The video data gathered is only available in the German language but the transcripts have been translated partly in the context of the project "Effective knowledge construction in interest-dense situations" and partly in the context of this project.

6.2.2 Choice of Data

The gathered data is taken as a data pool from which specific data is chosen for analysis based on theoretical considerations. First, an analysis across the task that offers the richest and most diverse semiotic environment shall provide a starting point for developing the categories. This is considered to allow the best carrying out of the representational function of gestures in relation to both, speech and inscription. Second, one pair of students is observed working on all three tasks. These choices make it possible to compare the results across different pairs of students and across different tasks in a last step of analysis. These two axes of comparison will be named *horizontal axis* (one task, different pairs) and *vertical axis* (different tasks, one pair of students) (see Fig. 6.4). Along the horizontal axis, the elaborations of the parabola task have been compared. For the vertical comparison a female pair of students was chosen due to their use of gestures.

---

[83] See also chapter 3.2.1 for the definitions of gesture phrase/stroke and gesture unit.

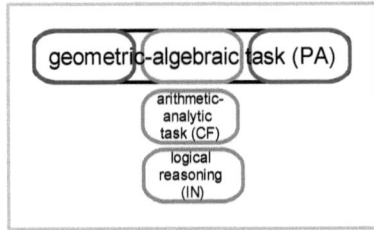

Fig. 6.4: Horizontal and vertical axis of analysis

In more detail, the reasons for this choice are the following: The parabola task provides the widest semiotic variety. That is, it is the task prompting most the usage of highly diverse semiotic resources, so that the largest semiotic bundle is expected to be developed. The three representations – folding sheet, GeoGebra-environment (DGS) and printout – display similar situations but provide different ways of handling them. The folding sheet is associated with the process of folding and the inscriptive products emerging from this. In the GeoGebra-environment, a coordinate system is visible, but it is also possible to make it vanish. Furthermore, this representation allows the distance between the focal point and the x-axis to be varied while observing the impact of this variation on the parabola. The printout makes it possible to produce marks within the diagram, to add components or tags, or to highlight them. Because of these similarities and differences between the representations provided by the task, it can be observed how gestures help to coordinate them.

During a first sighting of the data, the gesture use of the two female students attracted attention. At first sight, both of them seemed to use gestures differently: While one student made many gestures large in extent, the gestures of the other one appeared to be rather small and generally scarce in frequency. Their use of inscriptions seemed to be goal-oriented and beneficial regarding the epistemic process. In addition, they revealed another peculiarity within the social interaction: The number and duration of speech pauses appeared to be remarkable but surprisingly, they needed less time to elaborate the tasks compared to the other two pairs of students. These two reasons led to the decision that the observation of the epistemic processes will become promising for answering the research questions, both from an epistemic, as well as from a semiotic point of view.

The data sets are named according to the task that is elaborated and the pair of students. To the latter, numbers are assigned. The horizontal axis comprises the data sets PA5, PA6 and PA7, and the vertical axis the data sets PA7, CF7 and IN7.

6.2.3 Analysis of the Data

The analysis of the data encompasses six steps. The results from steps one to five are documented within a table that will be presented following step 5 in Fig. 6.8 on page

107. This table only serves an analytical purpose. Its components are explained in the description of the respective step.

6.2.3.1 Steps 1-3: Preparatory Steps

The first three steps ground the basis for the main analysis in which the research questions shall be answered. As result of these three steps, a new data base is produced. This data base allows the simultaneous representation of the epistemic process as reconstructed from speech act analysis and the reconstruction of the meaning of the gestures within the social interaction. It does not concern the entire working process, but episodes that are considered relevant for answering the research questions from a theoretic point of view (see chapter 5.3). The three steps, in brief, consist of

1. reconstructing the epistemic process according to the GCSt-model,
2. reducing the data to relevant episodes based on theoretical considerations, and
3. interpreting the gestures in these episodes within the semiotic bundle.

All three steps have a different input and output of data and will be explained in detail in the following subsections:

Step 1: First Reconstruction of the Epistemic Process: Epistemic Actions Reconstructed from Speech Act Analysis

Based on the transcripts, the epistemic process is reconstructed through identification of the epistemic actions according to the GCSt-model (see chapter 4.1). For this, the speech acts are interpreted respecting the three levels following Austin (see section 5.3.2.1), the non-verbal activities are considered as far as they are described in the transcripts. To get a more comprehensive picture of the social interaction in which the epistemic process takes place, some social actions are considered to influence the working process, e.g. initiating, explicating, or valuing, are reconstructed as well (See Table 6.3 and below.). The working processes of the students are summarized in condensed process diagrams by means of signs (Table 6.2):[84]

Table 6.2: Signs representing the epistemic actions accomplished

•	⊔	⌐	⌐•	⌐⌐
gathering	Connecting	structure-Seeing	concretizing gathering	elaborating connecting
basic action	accumulating action		extending structure-seeing	

---

[84] The signs (and the condensed process diagrams are based on those used by Bikner-Ahsbahs (2005, pp. 200-202). The sign for 'resumptive gathering' has been added.

These signs represent the main epistemic actions. Others have been used to describe other aspects of the social interaction. Although these actions are not explicitly considered in later steps, they are captured at this point in the analysis as part of the social interaction that influences the epistemic process:

Table 6.3: Signs representing other actions accomplished with respect to the working process

⌈●⌉	→←	⌒	e	v	w
resumptive gathering	objecting	initiating	Explicating	understanding (german: *verstehen*)	valuing (german: *werten*)

Resumptive gathering takes place when by gathering, something is repeated that has already been elaborated on within the epistemic process in order to summarize it or to formulate a solution of the task[85] (Priwitzer, 2010, pp. 54-55). *Objecting* is identified when disagreement is expressed, an idea is refused, or an alternative opinion is stated. *Initiating* means that a question is posed or a prompt is made, related to bringing forward the collaboration on the task. *Explicating* is assigned when someone is asked to explicate an idea or a thought. When an effort to understand someone is expressed, the **v** is used. If an idea, thought, hypotheses, but also the task, becomes appreciated or valued (for example saying "nice", "oh my god", "difficult" etc.), the **w** is used. Signs are combined when actions are related, e.g. when entities are gathered while they are connected, or when a connecting action leads into structure-seeing. An example of an excerpt of such a condensed process diagram as used in the original analysis is presented in the following (Fig. 6.5):

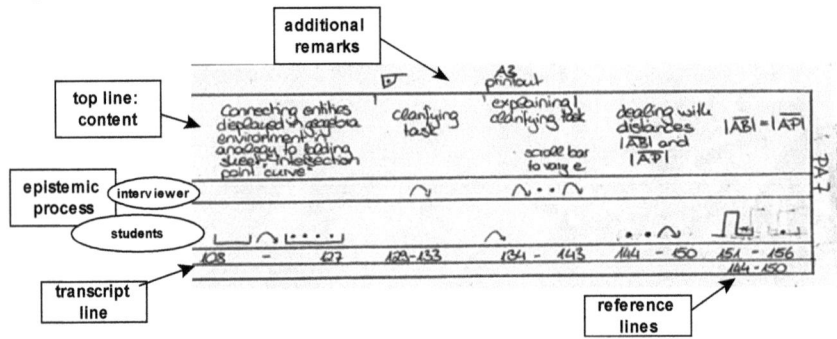

Fig. 6.5: Example of an excerpt of a condensed process diagram

---

[85] This subform of the gathering-action has only been used to get a better understanding of the epistemic processes. In later steps of the analysis, this refinement is subsumed as gathering action.

The first, top line provides a description of the content of the social interaction. In the second and third line, the epistemic process is presented by signs. For this, it is differentiated between the contributions of the interviewer and those of the students to make transparent the role of the interviewer within the students' epistemic process. Beneath it, it is noted which lines of the transcript are summarized. The bottom line displays references identified to recent points within the working process, denoted as lines of the respective utterances.[86] During the process of analysis additional remarks were made above the top line, for example the number of the subtask or the introduction of a new representation. The example displayed above in Fig. 6.5 is taken from an epistemic process of two students working out the parabola task. In line 134, they start working on task A3 (subtask 3) and the printout is introduced as representation. Lines 134-143 concern the introduction of the task, initiated by the students but mainly carried out by the interviewer. Right after, in lines 144-150, dealing with the distances $\overline{AB}$ and $\overline{AP}$[87], the students gather mathematical entities.[88] This leads to seeing the equality of these distances as a structure and justifying it (151-156). The reference to the gathering actions just before is noted in the reference line.

Using condensed process diagrams makes it possible to detach from the concrete verbal utterances but summarizes the epistemic process by the epistemic actions accomplished in social interaction. This allows the reconstruction of the epistemic process to be represented on only a few pages in a condensed way. The references noted in the bottom line provide orientation and a better understanding of the students' epistemic process.

In this step, the *input data* is given by the transcripts, the *output data* is given by the condensed process diagrams. These condensed process diagrams are used to detect substantially interesting episodes in the second step:

Step 2: Selective Sampling Based on the Analysis of the Epistemic Process
Based on the condensed process diagrams, episodes that are considered being substantial regarding the epistemic process are selected, that is, episodes in immediate proximity to actions of structure-seeing. More in detail, they are chosen to be contextually closed in

---

[86] The condensed process diagram has been first used in this form in the collaborative master theses by Janßen (2010) and Priwitzer (2010). They added the top and the bottom line to the form previously used by Bikner-Ahsbahs (2005) and added a contextual dimension to the overview.
[87] The notations used in the condensed process diagram displayed in Fig. 6.5 are also common in German schools and have been used in the first, manual version of representing the analysis of the epistemic process. It differs from the notation used in this book that has been chosen to provide international standards.
[88] In this case, they gathered observations on lengths of segments visible in the printout diagram.

interaction, so that they are understandable as a unit.[89] Fig. 6.6 exemplifies the localization of the structure-seeing as core of such an episode, based on the condensed process diagram as presented in Fig. 6.5.

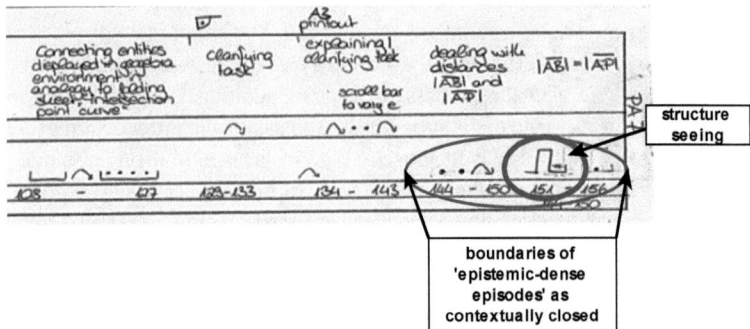

Fig. 6.6:   Example of determining an 'epistemic-dense episode' (PA7e2)

This choice is reasoned as driven by theoretical considerations to ensure "theoretical representativity" (Beck & Jungwirth, 1999, p. 242, my own translation referring to Hermanns, 1992, p. 116): It is expected that gestures appear particularly in situations in which new building blocks of knowledge are constructed. In these situations, approaches and ideas are often not elaborated yet and are therefore expressed in an insecure or incomplete manner. This can become reflected in a fragility of speech. Gestures are considered to be used as supplementing means of expression. Supporting this assumption, Behrens suggests from observations made in her master's thesis (Behrens, 2013) that "gestures are particularly used to carry out connecting actions and to formulate the seeing of a structure" (ibid., p. 86, my own translation). Furthermore, the contextual embedding of the structure-seeing episode allows tracing back to starting points of upcoming ideas within the interaction. The necessity of this backtracking becomes reasonable with respect to the development of the semiotic bundle: The meaning of signs, especially gestures as idiosyncratic signs, and their use in social interaction cannot be analyzed as a momentous event but as developing within the process of constructing knowledge. Following Bikner-Ahsbahs' notion of "interest-dense situations" (see chapter 4.1), I refer to the selected episodes as *epistemic-dense episodes* (EDEs).

Each of these EDEs becomes labelled with a code, based on the name of the data set. For example, the EDE represented in Fig. 6.6 is **e**pisode number **2** in the data set **PA7**, so it is named PA7e2. In addition, a numeration of the scene within this episode may follow.

---

[89] Where these contextually closed meaning units start and end has to be decided individually by means of the social interaction as transcribed.

6.2 Methods of the Analysis

The epistemic-dense episodes are displayed in tables that allow for a more detailed representation of the components that are important for the analysis of the functions of gestures as aimed in this study.[90] As a result of this step, the respective transcriptions are extracted and constitute one column of the table, with one line for each utterance. Furthermore, the use of non-verbal and non-gesture modes of expression is coded next to the transcript by using the following signs:

Table 6.4: Pictograms marking the use of non-verbal and non-gestural semiotic resources

An inscription is produced.	The cursor is used within the DGS-environment.	The situation is varied within the DGS-environment.	An inscription is produced within the DGS-environment.	The calculator is used.	

The main *input data* considered in this step is given by the condensed process diagrams, produced in step 1, the transcripts are used to refine the boundaries. The output data consists of the tables of EDEs, with the corresponding transcripts noted in one column.

Step 3: Analysis of the Gestures within the Semiotic Bundle

In a third step, the gestures appearing in the epistemic-dense episodes are analyzed. These analyses are carried out with respect to the developing semiotic bundle. According to this, the analyses comprise gestures' relation to other semiotic resources such as speech, inscription, and eventually to artefacts like Dynamic Geometry Software, pens, the calculator, etc. Within this semiotic bundle, and assuming co-expressivity of gesture and speech, the verbal utterance provides the context for gesture analysis. In addition, the reference to inscriptions is respected. First, a reconstruction of the meaning of all gestures is carried out for all epistemic-dense episodes of all data sets. This is done respecting the three factors that influence the interpretation of meaning as described in section 5.2.3.2, within the social interaction but without reference to the epistemic process. This proceeding follows a methodological decision: Since gestures have no conventionalized meaning, this explication of their reference as developing within the semiotic bundle shall clarify the analytical basis for answering the research questions. The meanings as reconstructed for the gestures are added to the table of EDEs in the column next to the corresponding verbal utterance (see Fig. 6.8). The co-timing to speech is

---

[90] The complete and detailed representation of the final table and an explanation of the single components and columns is given in Fig. 6.8 on page 107.

noted in the transcript respecting the transcription rules as presented in chapter 6.2.1, starting times of gestures are noted above the respective utterance as time marks. 'Longer' gestures that are held across turn or that are not finished within one turn are noted with a straight arrow next to the transcript column, starting at the beginning of the gesture unit, and ending at the line where the gesture is considered to be released. Assuming the contextual cohesion explained in chapter 3.2.2, *Growth Points* and *catchments* are marked by stars and a keyword of reference (see Fig. 6.8). When gesture-speech-mismatches (see chapter 3.2.2) are identified, they are marked by a dotted ellipse.

The tables of EDEs and the videos have been used side by side as main *input data*. The verbal utterances provided the referential frame to reconstruct the meaning of the gesture from the video data, analyzed frame by frame in a micro-analysis. The students' inscriptions have been considered for reconstructing exact references of pointing gestures. As *output*, the tables of EDEs are refined to capture the reconstruction of the meaning of gestures.

To capture also the epistemic processes as reconstructed from speech within the tables of episodes, they are added next to the transcripts on the left side by using the signs introduced in Tables 6.2 and 6.3.[91]

6.2.3.2 Steps 4 and 5: Main Steps

Based on the prepared tables of EDEs, the research questions are answered.

4. First, the representational and the epistemic functions of gestures are developed and coded (See also chapters 5.4.1 and 5.4.2.).
5. Second, the reconstruction of the epistemic process as carried out in step 1 is refined by respecting also gestures' contribution to the shaping of the immediate mathematical object.

Step 4: Developing the Functions of Gestures

The functions are developed through constant comparison of newly considered data with the data that has already been analyzed. This leads to a refinement of the descriptions, as well as to a cyclic reanalysis of the data to validate new hypotheses (See also chapter 5.4.2.). A main criteria for considering a hypothesis to be relevant, concerning the investigation of the research question, is the empirical grounding: A function is considered to be relevant if it becomes identified frequently and in diverse situations. Furthermore, the categories shall be determined as distinctive as possible without being redundant, and they shall also lie on a similar level of abstraction. This way, the codes assigned to a gesture provide a differentiated description of the functions fulfilled by the gesture.

---

[91] These tables of episodes are inspired by the so-called "time line", introduced by Arzarello, Ferrara, and Robutti in 2011 (Arzarello, Ferrara, & Robutti, 2011). However, these time lines have been modified to fit the specific purpose of this study.

6.2 Methods of the Analysis

Step 4a: Developing Categories of Representational Function of Gestures
In a second run of gesture analysis, categories are generated as grounded in the data in order to approach the first ancillary research question, *"How can gestures contribute to shaping the immediate mathematical object in social epistemic processes?"* For this, the SRA (see chapter 5.2.3) is kept in mind in its multimodal form, answering for every semantic unit 'how the gesture may influence the shaping of the immediate object'. The semantic units just mentioned refer to gesture phrases or gesture units (Kendon, 2004, p. 112; McNeill, 1992, p. 25; see also chapter 3.2.1), and their interplay with speech and inscription. The results gained from this step are noted in column 5 of the tables of the EDEs and are also captured quantitatively.

Step 4b: Developing Categories of Epistemic Functions of Gestures
The epistemic functions are approached first through a *semiotic summary*. In these summaries, parts of the episodes are reconstructed adopting a multimodal approach with special focus on gestures. They include other semiotic resources if this provides a better comprehension of the epistemic process looked at through a multimodal lens.

From this the first hypotheses are derived that answer the questions: *How can gestures contribute to acting epistemically? Which beneficial ways of using gestures can be identified in this? Which possible pitfalls can emerge?*

The gestures are coded according to their epistemic function, where function is understood in the way Bavelas referred to it: *"What does this gesture do and how?"* (Bavelas, 1994, p. 202, italics in the original). Different to the step before, the meaning units used for the identification of epistemic functions of gestures is oriented at the epistemic actions rather than at the segmentive feature of the gesture (see chapter 5.2.3).

For the development of both kinds of functions (representational, as well as epistemic), the analysis is started with a data set concerning the parabola task (PA5). This approach has been chosen for the same reasons for which this task has been considered for the horizontal analysis: To solve this task, many representations come into play and need to be coordinated by the students. I assume this to prompt gestural references to the mathematical objects under investigation. The testing and refinement of the categories is then accomplished by bringing in the PA7-data. This provides the basis for the further elaboration of the categories on the vertical axis, considering gesture use that may be specific for the respective pair of students. The third parabola-related data set (PA6) is used for further testing of the categories as developed from PA5 and PA7 and possibly complementing them if needed. In the same way, the categories are tested and refined on the vertical axis to test their relevance for different mathematical topics and tasks and to refine the descriptions (see also chapter 5.4.2). Since the Parabola-task has been chosen as a starting point because of the semiotic variety provided by different representations, the order of analysis on the vertical axis is chosen to be determined by decreasing variety of representations given: CF7 is the first data-set to elaborate the categories that shall be finally tested and refined by analyzing the IN7-data.

The representational functions are noted in the tables of episodes (see Fig. 6.8). Arrows mark possible relations between them. While first observations on the epistemic benefits of gestures are noted within the tables, the final reconstruction of the epistemic functions of gestures is done separately: Using Atlas.ti[92], a time mark is set in the video data right after utterances in which a gesture is identified to fulfill an epistemic action. This makes it possible to write down the analysis as comment belonging to the time mark. The time mark provides information about where to find the respecting gesture within the video data. The comments contains information about the data set, the number of the epistemic-dense episode and the number of the corresponding utterance in the transcript. This way, a close relationship is kept between video data and transcripts in the process of identifying and documenting epistemic functions of gestures.

Step 5: Extended Reconstruction of the Epistemic Process: Integrating Gesture Analysis

To include how gestures can serve as a means of expression, the epistemic processes are reconstructed once more, complementing the implicit reference to epistemic actions becoming noticed alone from gesture analysis. This complementation becomes visible in diagrams representing the epistemic process in epistemic-dense episodes in a condensed way (Fig. 6.7). In these, the description of the epistemic process within the EDE is reduced to be summarized by the **dominating** epistemic actions. The epistemic actions reconstructed from speech are represented in black, those solely reconstructable from taking gestures into account are added in light blue. Without further reference to the concrete semantic content, Fig. 6.7 exemplifies how such an extended condensed process diagram of an epistemic-dense episode may look like:

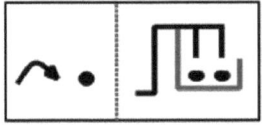

Fig. 6.7: Condensed diagrams of epistemic-dense episodes, episode CF7e10, scenes e10.1 and e10.2

In scene CF7e10.2 (right side of Fig. 6.7), it is verbally referred to two examples to concretize the structure seen just before. Gesture contributes in connecting these two

---

[92] Atlas.ti is a qualitative data analysis and research software.

examples. This connecting action does not become explicit in speech but only apparent in gesture. [93]

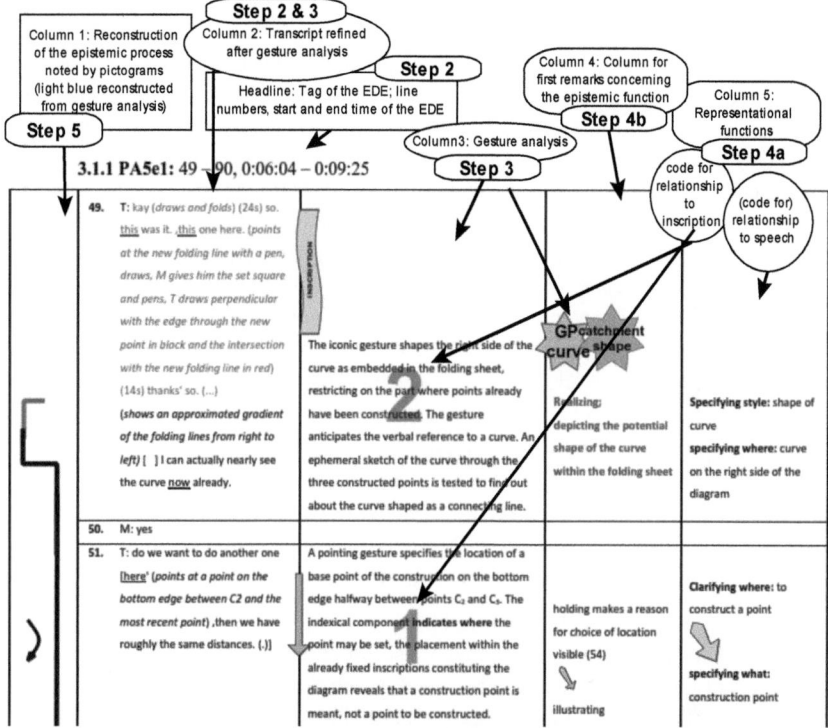

Fig. 6.8: Example of a table of an EDE that served for the simultaneous representation of the outcomes of different steps of the analysis. Column 5, as well as the numbers noted in column 3 concern the results that are described in chapter 7. The green arrow next to utterance 51 as noted in column 2 denotes the holding of a gesture. The arrows in columns 4 and 5 denote possible relations between the codings as they arose in the course of the analysis.

---

[93] In addition to the description given, the curved arrow in the beginning denotes that the social interaction in this scene starts with a students' attempt to keep on working together. This is identified in a substantial question or request.

6.2.3.3 Step 6: Comparative Step

Step 6: Building Hypotheses on the Influence of Gestures on the Epistemic Process along Two Axes
The formation of theoretical building blocks on gestures' contribution to social processes of constructing mathematical knowledge is carried out by comparing the results conducted by steps 4 and 5. In a descriptive analysis, frequencies are compared concerning criteria that are developed in the course of the study. Due to the increasing dynamics of the epistemic process and students growing involvement in the activity (Bikner-Ahsbahs, 2004, section 2), the gestures in epistemic-dense episodes are naturally short-clocked. Because of that, there is no sense in comparing the relative temporal amount of gestures. A *horizontal comparison* leads to hypotheses about gestures' contribution to the epistemic process concerning different pairs of students elaborating the parabola task. A *vertical comparison* aims to make statements about how gestures influence the epistemic processes of one selected pair of students when varying the task.

6.3 Presentation of the Results

Recalling the goal presented in chapter 5.1, *modes of representational and of epistemic functions of gestures* will be carried out in order to make hypotheses on the superordinate issue of *gestures' contribution to processes of constructing mathematical knowledge.*

The research process cannot be portrayed in all its details, including the dead-ends and pitfalls faced while developing the categories. For example, categories considered in early stages of the analysis failed to be found on the same level of abstraction or they subsumed categories that have then been found to be more clear-cut. The presentation of the results restricts on those categories that passed through the testing and revision within the process of analysis.

Two chapters are dedicated to the representational functions of gestures and their epistemic functions, respectively. The functions are introduced by means of illustrative examples in which the main aspects of the function are pointed out. To make the analysis comprehensible for the reader, the transcripts of the excerpts have been complemented by pictures captured from the data. To 'quote the gestures', arrows indicate movement of dynamic gestures. This is done in order to present the gestures as neutral as possible, similar to the transcript of the verbal utterances. Within the transcripts, the pictures of the gestures are positioned beneath the respective verbal utterance.

While the first part of each of the two chapters presents the categories, the second part provides extended findings related to answering the research questions based on the functions. These further findings focused on diachronic aspects of the analysis so the criteria for the theoretical sampling have been adjusted to capture the phenomenon investigated: The categories in the first parts are presented as shorter excerpts, the further findings in the second part embrace longer excerpts, not necessarily being contiguous.

For finally concluding on gestures' contribution to epistemic processes, hypotheses gained from the horizontal and the vertical comparisons described in section 6.2.3.3 are worked out in chapter 9. To make them comprehensible, diagrams and tables are used to summarize the quantitative results with regard to the criteria that are discussed.

Part III
Results

Finally, the groundwork is prepared for carrying out the results of this study. While the following two chapters, 7 and 8, are dedicated to presenting the functions of gestures, I will compare the coding results across the different pairs of students and across the tasks in chapter 9. In conclusion, all results are summarized in chapter 10 in order to state on the overall research aim. In this regard, a theoretical building block is built that integrates the contribution of gestures in the theory of epistemic processes within interest-dense situations.

Overview of Epistemic-Dense Episodes

Since the focus of this study concerns the reconstruction of representational and epistemic functions of gestures, chapters 7 and 8 present the data as analyzed from that perspective. However, these analyses are conducted on a data corpus that is already reduced on the EDEs (see section 6.2.3.1). Although these steps of analysis are not presented in detail, an overview on the input and output data of these two steps shall be given in the following:

The duration of the five data sets that were chosen ranges from 59 minutes (IN7, 607 lines of transcript) to 168 minutes (PA6, 1988 lines of transcript),[92] where the three elaborations of the parabola-task had the longest duration (duration PA7 < duration PA5 < duration PA6 < duration CF7 < duration IN7). The first reconstruction of the epistemic processes led to 83 epistemic-dense episodes across all the five data sets. The epistemic-dense episodes made a total time of 185:54 minutes of gesture analysis. From this time, 10:17 minutes of the PA6-data, during which the reference of the gestures cannot be analyzed due to being covered by body parts or paper sheets, are already excluded. The EDEs have been subdivided in scenes to organize the analysis; each scene is seen as an interaction unit within the larger episode. In the example shown in Fig. 6.7 (page 106), CF7e10 is subdivided in the scenes e10.1 and e10.2. While the data sets PA5, PA7, and CF7 served as main sources for the development of the representational and epistemic functions of gestures (PA6 and IN7 have been coded according to the developed functions, assuming the possibility to complement the codes, if desired), headings have been assigned to the scenes, providing a catchphrase for the epistemic content and/or the gesture analysis.

---

[92] These two extremes concern the two data sets that each stand in the end of the horizontal and vertical comparison; they constitute the data sets to check hypotheses.

7 Representing *Within* the Multimodal Sign

How Gestures can Bridge within the Multimodal Sign to Shape the Immediate Mathematical Object

Just as speech, gesture is a means of expression so it has to be considered in taking part in the forming of the immediate object of a multimodal sign. Furthermore, gesture and speech are co-expressive. That is, the meaning of a gesture is considered to be conceptually related to the idea that is expressed in speech (McNeill 1992, p.1). In addition, gesture and inscription develop together with speech within the semiotic bundle. This suggests reconstructing the immediate object as it is developing in social interaction against the background provided by visual means of expression. The leading question aimed to answer in this chapter thus concerns *how gestures can contribute in shaping the immediate mathematical object in social interaction when mathematical knowledge is constructed.*

This chapter presents the theoretical tool of the *within-functions*. These functions will be defined as consisting of a *specifying-function* of gestures on speech (7.1.1) and a *referential level* (7.1.2) on which gestures are related to inscription. The within-functions describe the representational functions of gestures within the multimodal sign that may influence the immediate mathematical object interpreted from the entire semiotic composition (see chapter 4.4).

These within-functions of gestures provide a more profound understanding of the social interaction. They allow the analysis of the students' epistemic process to be extended by gesture analysis in two ways: First, the development of meaning as embodied in gestures can be reconstructed within the social interaction by means of the within-functions. Second, the within-functions can reveal information that is only expressed in gestures, but that needs to be considered as framing the interpretation of the utterance and with that, the shaping of the social immediate mathematical object.

7.1 Within-Functions of Gestures: A Tool to Integrate How Gestures May Shape Mathematical Meaning

The within-function of a gesture is considered the combination of its specifying-features and its referential level, the former concerning gestures' relations to speech, the latter its spatial reference to inscription. The first two subsections of chapter 7.1 concern the semantic surplus of gesture use with a focus on the semantic content of speech: the *specifying-features*. The third subsection deals with different referential levels on which gestures can refer to mathematical objects, respecting their relation to an inscription.

7.1.1 Specifying the Immediate Object: Adding What is not Revealed in Speech

Speech is considered to be linear and segmented (McNeill, 1992, p. 19). This property causes difficulties in simultaneously representing different aspects of one object only by means of speech. However, as McNeill points out, "language is unidimensional while meanings are multidimensional" (McNeill, 1992, p. 19, with reference to Saussure). The simultaneous use of gesture can thus enrich the forming of the immediate object in different ways.

7.1.1.1 Clarifying Imprecise Terms: Gesture can Complement the Immediate Object Presented in Speech by Filling a Semantic Gap

Missing words or deictic phrases that simplify the verbal expression make the representation in speech semantically incomplete. This requires a clarification that is often accomplished in gesture.

Excerpt 1.1a, PA5-51 (PA5e1, 0:06:22)[94]

Mike and Tim are working on the parabola task. They have constructed three points corresponding to the points C_1, C_2 and C_3 on the lower border (Fig. 7.1) when Tim mentions that he "can actually nearly see the curve now already." (49).

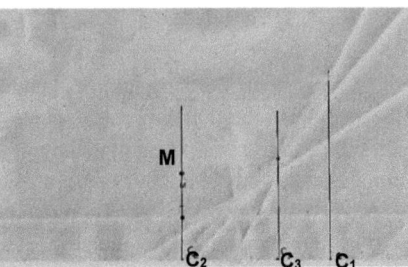

Fig. 7.1: Folding sheet of Mike and Tim as visible in the beginning of PA5e5-49 (represented with higher contrast to provide better visibility to the folding lines; tagging of points enlarged)

---

[94] An analysis of this gesture within the semiotic bundle has already been given on pages 69-70, related to example 1.1.

According to the task instructions, the students shall keep on constructing until they recognize a curve. Although Tim's utterance suggests that this part of the task is completed now, he proposes to construct one more point:

51 T: do we want to do another one [here'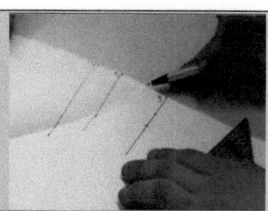
(*points at the bottom edge between C_2 and C_3*)
,then we have roughly the same distances. (.)]

The concrete positioning of this point can only be reconstructed by looking at the indexical gesture. From this, the location is presented as being more or less in the middle between C_2 and C_3, on the bottom edge of the sheet.

An obvious function of this gesture is to *clarify* what is meant with "here". But in addition to this, the gesture frames the interpretation space even more:

i) The precise location identifies the "another one" referred to in speech as being 'point on the lower edge'. This makes the reference more precise since it *specifies a quality* of the point chosen on the lower edge in distinction to the point that is the *result* of the construction (the point on the parabola).

By adding ",then we have roughly the same distance" it is revealed that the choice of this point is not arbitrary.

ii) Holding the pointing gesture until the end of the utterance provides visual access to Tim's reason of choosing this concrete point, giving an idea to which "same distances" he refers.

Respecting this additional information given by using gesture, the imprecise immediate object "another one here" is stated precisely as the 'starting point of the construction, located in the middle between C_2 and C_3'.

When clarifying imprecise speech, gestures can transport more information than has been requested for filling the semantic gap.

In addition to a *clarification where* of deictical words like "here" or "there", other imprecise terms require clarification of the aspects "what" and "how", as illustrated in the following excerpts.

Meanwhile, the fourth point, corresponding to construction point $C4$, has been constructed, and the right hand side of the curve is fixed within the folding diagram as connecting the points that have been constructed following the instruction (Fig. 7.2).

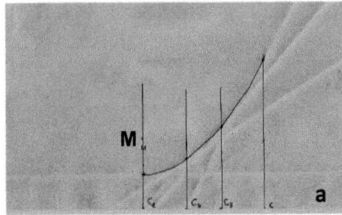 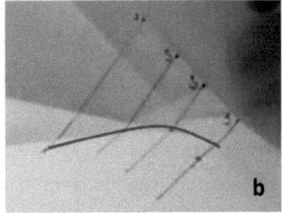

Fig. 7.2: Folding diagram in the beginning of scene 2 in PA5e1, 66-68
(a) the students' original diagram represented with tagging of point M enlarged
(b) perspective used in the quoting of the gestures in the transcript (inscribed curve highlighted to provide better visibility)

Subsequently, Tim offers a conjecture concerning the shape of the curve on the left hand side, where no points have been constructed.

Excerpt 1.1b: PA5-66 (PA5e1, 0:07:58)

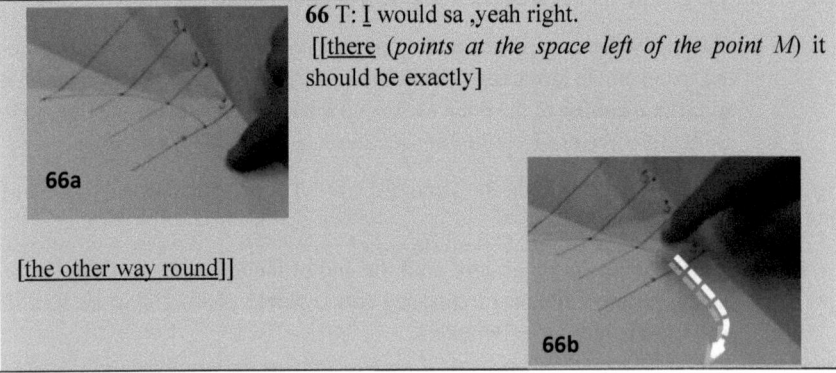

66 T: I would sa ,yeah right.
[[there (*points at the space left of the point M*) it should be exactly]

[the other way round]]

The pointing on the left hand side (66a) **clarifies** the **where** of the verbal reference that is left imprecise in the deictic verbal expression. Then, Tim **clarifies** the imprecisely verbalized clause of manner, "the other way round" (66b): Simultaneous to speech, he ephemerally shapes the left hand side, beginning beneath point M (see Fig. 7.2) and moving to the left in an upwards arc. Considering the fixed representation of the right hand side of the curve, the **clarification how** is seen as being accomplished as a *comparison in similarity*. The shape is the same, but the orientation of shaping is 'the other way round', from right to left. This also **clarifies what** has been referred to by "it". Not being clear before whether speech represents the immediate object 'the construction of points' or 'the shape of the curve', the interpretation is made less ambiguous through gesture.

But aside from clarifying speech, gesture also **specifies relations** characteristic for the curve to contribute in shaping the immediate object: The pointing left beneath the line through points M and C_2 divides the diagram in two sides. The shaping hints at a relation between both sides in an iconic way. The curve as immediate object is hence represented as having two parts that are similar in shape but different in orientation.

This separation is made explicit right after:

Excerpt 1.1c: PA5-68 (PA5e1, 0:08:02)

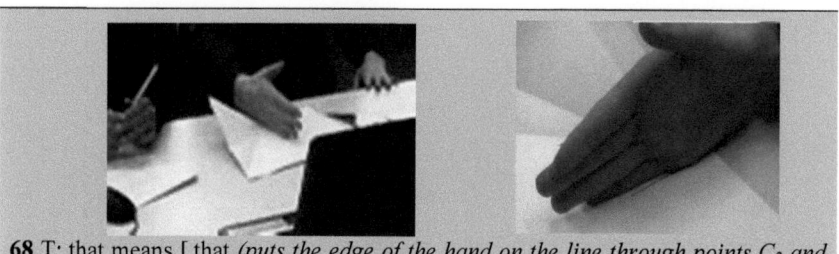

68 T: that means [.that *(puts the edge of the hand on the line through points C_2 and M)* theoretically is our axis] and there is symmetry

The placing of the hand **clarifies what** is referred to with "that": the extended perpendicular through point M and the lower edge (See Fig. 7.2 for the diagram at that stage in the working process.). This visualizes the idea of separating both sides of the curve as two parts, as was just suggested before. Interpreted within the entire utterance, it represents an axis of reflection in anticipation to verbally referring to symmetry.

In these excerpts, the use of gestures provides more information about the immediate object than is given from speech alone. That is, the gesture *specifies* aspects of the *what* and the *how* of the immediate object and furthermore also *specifies* the *relation* between the two sides of the curve.

Gestures' contribution in shaping the immediate object can now be described by a *specifying-function of gesture within the multimodal sign*:

Gestures can specify the where, the what, the how and relational aspects of the immediate objects by adding information that is not provided in the verbal utterance. They clarify imprecise terms but also add information about aspects not mentioned in speech.

Notabene: *If* a gesture specifies and furthermore, which aspect it specifies, crucially depends on its interpretation. This in turn is carried out within the social interaction in which it is embedded and with respect to the development of meaning within the semiotic bundle. The importance of this background becomes apparent in situations when speech suggests that there are several objects of the same kind of which a specific one is identified. Such a case is illustrated by the next excerpt:

Excerpt 1.1d: IN7-73 (IN7e1.2-73, 0:12:38)

This example is taken from the induction task, elaborated by the two female students. In scene IN7e1.2, Lisa and Rosa figure out the knowledge of the consultants in the specific case of '*two marked hats existent among five consultants*', assuming that nobody reacts at the first ring ("then they all know that there is more than one" (68/70)). Earlier, they have drawn a simplified representation of this situation (Fig. 7.3). The meaning of the single components of the stylized diagram has been clarified before when the first situation was captured in inscription (19).

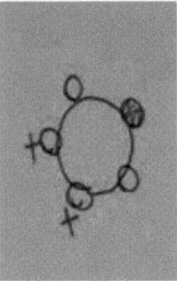

Fig. 7.3: Idealized representation of the case "five consultants, two have a marked hat"

Lisa then states more precisely the knowledge of one specific consultant who sees only one cross and uses gesture to represent the single entities within the fixed representation:[95]

73 /L: yes of course when when- the first ring was- then knows- of and-	
(*points at the upper "X in the second diagram with her little finger*) [[„that one looks now] and	
(*points at components of the diagram without looking at it*) [sees only one cross] then [he knows] that there is a second [then he]] must have it [because all others] don't have one more cross.	

The pointing to the circle with the 'X' indicates a consultant who has a marked hat within the simplified representation of the case. Speech only refers to "that one" and leaves imprecise also the kind of object to which is referred. The components are treated to stand for the respective objects, that is, the diagram represents the situation, the small circles represent the consultants, and the 'X' denotes consultants that wear a marked hat. This knowledge is important to identify information that is added by the use of gesture

---

[95] The example is given to illustrate the importance of taking into account the concrete formulation of the verbal utterance when interpreting the additional information given by the use of gesture. To concentrate on this, the other pointing gestures in this utterance are not made topic.

to the one provided by the verbal utterance: The *what* is not specified, but the gesture distinguished the consultant from the others by *specifying* the consultants *location* within the circle of consultants. Furthermore, the following pointing specifies the location of the consultant referred to in speech ("he"). A 'specification who' does not make sense when defining specifying-functions as those functions of gestures that provide additional information about the immediate mathematical object. The aspect that is specified to identify one consultant in distinction to the others concerns its position in the circle and is thus a spatial one (*where*).

This example points out that it is not enough to consider the specific words and ask questions like "Is there a 'what', a 'where' or a 'how' missing?". The function of the gesture does not necessarily need to concern a clarification of seemingly imprecise words that may derive their meaning within the concrete situation from context. Moreover, gesture may specify another aspect that may not be necessarily clarified to overcome imprecise speech. This will be topic of the following section:

7.1.1.2 From Clarifying to Specifying: Providing a Real Surplus of Meaning

Within the last four excerpts, it became visible how gestures can **clarify** imprecise terms to complement the semantic content of the whole utterance. Words like "here", "there", "that", "this", "like that", clauses of manner etc. demand for further information about the **where**, the **what**, or the **how** of the reference. With this clarification a **specification** can come along of the immediate object in some respect. That is, a real surplus of meaning is added to the aspects that are already considered in speech. The importance of the precise verbal formulation thus becomes salient: While the identification of a clarifying-function is triggered by a term that already hints at the respecting aspect in an imprecise way, the specifying-function is considered to offer collateral information.

So far, specification has been considered as a byproduct of clarifying-gestures. Hence, the clarification can be seen as an occasion to gesture that may lead to give more information than is referred to in speech.

Actually, gestures have also been identified to specify without compensating imprecise speech. That is, they may not have been considered important within the reconstruction of the epistemic process as only conducted from an analysis of the *verbal* discourse. However, they may crucially influence the interpretation of the verbal utterance by making its interpretation less ambiguous.

The following gives an example for such a gesture that others may deem redundant:

Excerpt 1.1e: PA6-23 (PA6e1, 0:04:07)

This time, the students Kris and Simon are working on the parabola task. They have constructed three points so far and are in the middle of constructing a fourth one. Kris is folding the sheet when Simon proposes the location of one more point to construct:

23 S: *(mumbles)* let us better make one
[on the <u>other</u> (*points at the left side of the sheet*) side.]

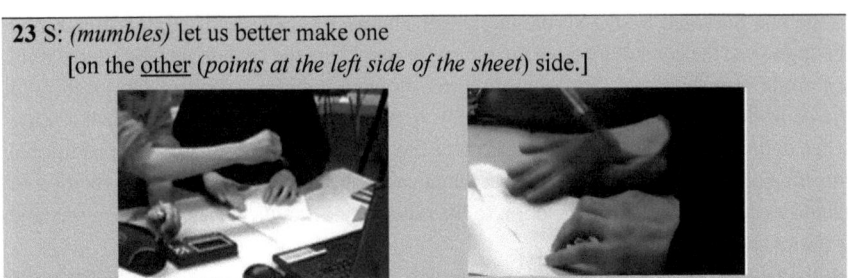

Within the situation of constructing points according to the instructions as given by the task, the term "one" does not need to be clarified. It is of no importance whether Simon refers to a base point or a point as result of construction. He could even mean a folding line but that does not seem to be of any importance. Furthermore, the reference of the verbal "the <u>other</u> side" is natural from having all points constructed so far on the right side of point M. Hence, "on the <u>other</u> side" may mean 'on the other side of point M' or 'on the other side of the sheet' without making any difference. Although there is no need for a specification since there is no obvious ambiguity concerning the area Simon refers to, he points towards the left side of the sheet, co-timed to "on the <u>other</u> side". With this gesture, he **specifies the location**[96] as actual spatial information. This provides a visual representational approach to his proposal.

The analysis of the representational function of gestures of the preceding excerpt leads to the following:

*Specifying-gestures can **explicate** a visual approach to the immediate object.*

As already suggested in excerpt 1.1d, their status of clarifying- or specifying-gesture hardly depends on what is in the common ground of the students that develops within the social interaction (Clark & Brennan, 1991). In the following excerpt, the importance of this consideration becomes even more salient. It illustrates not only how a gesture **specifies** the **how** and the **where** of the verbal reference, but also how the visual approach to the immediate object is established between both students.[97]

Excerpt 1.1f: PA5-289-291 (PA5e3.2, 0:22:26 – 0:22:53)

Mike and Tim try to find a function equation that describes the curve as visible on the screen. They consider the curve to represent an exponential function but change their mind after getting hints by the interviewer. They recognize that the shape of the curve on the left hand side does not fit the shape of an exponential function. Now, Tim states a new conjecture about the form of the function equation:

---

[96] Hence, the ‚where'.
[97] An in-depth epistemic analysis of the preceding excerpt will be given in section 8.1.2.3 to point out how reasoning is benefitted by the interplay of speech, gesture and inscription in this case.

289 /T: right then it is yet a uh ,then it is either a to the power of ,some function with uh ,with an even-numbered exponent. *(looks at Mike, Mike nods)* (.) because to the power of three would look differently.
290 M: yes (..) or (.) wait. (.) ah right **must** be.

(0:22:46)
(.) [elsewise that would go down.] (*sketches the form in the air with a pen*)

291 /T: yes (..)

Tim reasons his conjecture about the equation being a function "with an even-numbered exponent." and uses as a counter-example that x^3 "would look differently" (289). This suggests a reference to the shape of the curve as a valid argument and provides the background in which Mike's utterance in 290 has to be interpreted. He validates Tim's argument and sharpens it by making explicit the *specific style* of a difference: After he verbally mentions the existence of this difference in style verbally, he sketches the left branch of an x^3-graph in an iconic gesture from right to left. Co-timed to "[elsewise that would go down]", it is **specified how** it "would go down" (shape) and from **where** to where, the direction of the fictive motion (Talmy, 2000; see also chapter 4.2.1) of the curve.

7.1.1.3 Specifying-Gestures in *Mismatch*-Situations: Gestures Take the Lead

Thus far, gesture and speech have been considered to represent different aspects of the same immediate object. When speech and gesture represent different mathematical objects or conflicting aspects of one mathematical object, the assumption that gesture and speech are co-expressive evokes a conflict. That is, gesture does not add information, but presents information that stands in concurrence with the one presented in speech. Following Goldin-Meadow's notion (Goldin-Meadow, 2003, p. 25), a *mismatch* appears.[98] As will be presented now, the study showed that there are two kinds of specification of gestures in mismatch-situations:

7.1.1.3.1 Correcting Speech: Gestures can Provide Alternative Information

When gesture and speech refer to different immediate objects, one of them may not 'fit' the context established in social interaction. In this regard, the interpretation of the utterance - and the immediate object it represents - differs depending on the means of expression from which it is derived. Gestures thus can specify the immediate object in

---

[98] Note that Goldin-Meadow highlights that a mismatch does not need to be a *conflict* between speech and gesture (Goldin-Meadow, 2003, p. 26). Here, this term is used to contrast the conflicting situation to the 'adding information to speech'-situation.

some respect by providing an alternative interpretation frame, such as in the following excerpt:

Excerpt 1.1g: PA7-849 (PA7e10, 1:14:01)

Lisa and Rosa work on the parabola task. They already concluded the general function equation from empirical results derived for the cases $e = 1, e = 2$, and $e = 0.5$. To demonstrate the validity of the equation, they investigate general relations within the diagram such that a function equation can be derived from the knowledge about the general point $A(x/y)$ on the parabola. The idea is to identify segments within the printout diagram for which the length is known. This enables them to set up an equation in x and y that can be transformed to an equation of y in x.

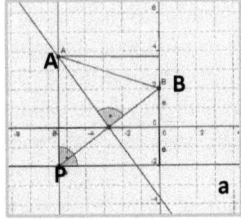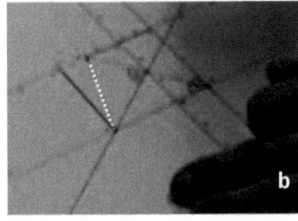

Fig. 7.4: Printout with auxiliary line fixed perpendicular to the y-axis through point A, in the beginning of PA7-849
(a) Students' original represented with higher contrast and tagging of points A, B, and P enlarged. Auxiliary line is red, perpendicular to the y-axis through point A.
(b) Perspective used in the quoting of the gestures in the transcript (color added to optimize visibility: red auxiliary line: white, dotted line segment $[AB]$)

Just before, they made a fresh start with a new printout (Fig. 7.4). Now, they negotiate which segments are known:

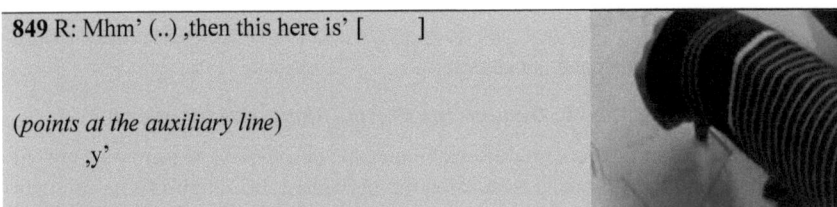

849 R: Mhm' (..) ,then this here is' [   ]

(points at the auxiliary line)
 ,y'

(L sets the pen to write, then takes the pen up again) (..) ,no [,this here.] *(Rosa points at the segment $[AB]$)*[99]

---

[99] This pointing gesture is covered by the arm but its reference can be re-constructed from the relations within the diagram.

Rosa identifies the auxiliary line as the segment having length y, although she revealed before that she is aware of the horizontal distance corresponding to the x-value. Her gesture indicates a segment having length x, while verbally referring to the value of y. This is a mismatch respecting the notational conventions that have been negotiated before within the social interaction. Rosa's gesture *specifies the location* of the considered segment within the printout as an offer for Lisa, who notes the lengths of the segments next to them. This concrete identification represents information about the location that does not fit for Lisa: She hesitates in noting the length and does not accept Rosa's implicit offer. This in turn makes Rosa rethink and she explicitly corrects her mistake: Verbally, she rejects the former identification ("no") and offers another segment ("[‚this here.]"), spatially specified through gestural indication.

In fact, it has been confirmed within the data that gestures take the lead in such situations. If gesture and speech referred to an aspect in a contradictory way, the students relied on gesture rather than on speech.

7.1.1.3.2 Linking Two Objects in One Aspect: Gestures Can Single Out the Overlap in Aspect

Gesture and speech can also refer to two different immediate objects so that they have an informational overlap. Although the objects are different, the aspect that is specified can link both in some regard. The way this may look is presented in the following:

Excerpt 1.1h, PA6-1940 (PA6e24, 2:44:20)

Kris and Simon try to define the parabola as geometrical locus. They propose a definition very similar to the given definition of a circle, using that all points of the parabola should have the same distance to a fixed point B (transcript line 1936). Their approach does not satisfy them since this may suggest that this distance is the same for all points on the parabola (transcript line 1939). This is when Kris tries a fresh start and goes back to the comparison of the points on the circle and the points on the parabola:

1940 /S: from a- ,yes [this fixed-] *(briefly points at the circle in the task instruction with the pen)* (.) ,this fixed point actually always is [an- *(points at different points on the circle path)* arbitrary point] on the- parabola that

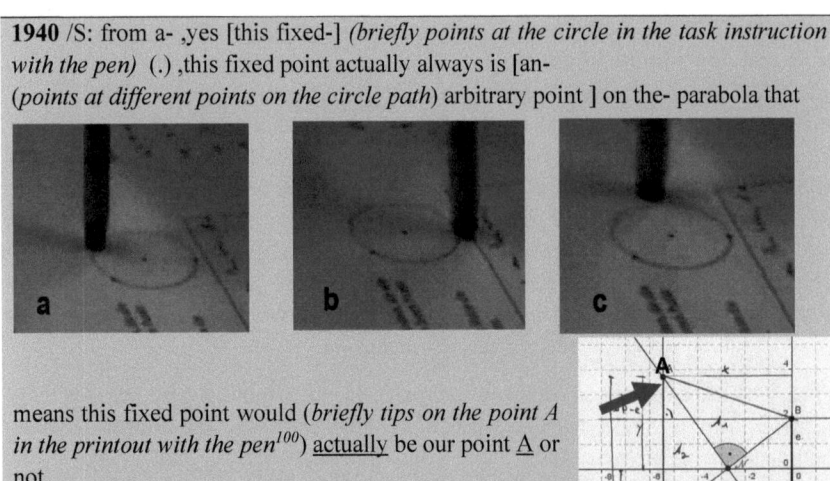

means this fixed point would *(briefly tips on the point A in the printout with the pen[100])* <u>actually</u> be our point A or not.

The first pointing precedes the more precise reference to the circle that follows in turn: The indication of different points on the circle path is co-timed to "an- arbitrary point". Different to what can be suggested from speech, referring to "our point A", the points do not specify the *where* as **points on the *parabola***, but as **points on the *circle path***. This clearly is a *mismatch* since gesture represents another immediate object than speech does (points on circle path vs points on parabola). Nevertheless, both components of the semiotic composition overlap in that they represent points that constitute a shape –parabola in speech, circle path in gesture - and with that rather *specify* a *defining aspect* (the what) than a spatial one (the where). Furthermore, the first reference to the circle, accomplished in gesture co-timed to "[this fixed-]", may mix up to 'fixed points constituting a shape'. To this 'fixed points constituting a shape' is referred in speech then as concluded correspondence, introduced by the verbal "that means": "that means this fixed point would <u>actually</u> be our point A". The indication in gesture has been perfectly integrated in the verbal utterance to constitute a common meaning although it is not mathematically correct: Point *A* on the parabola is not fixed.

From the mismatch-situations can now be concluded:

Gestures do not only complement what is given in speech but also shape a powerful expression themselves. They superimpose the verbal reference or link both immediate

---

[100] The indication does not become visible in the video data since the camera focusses on the task instruction. That it is actually indicated point *A* and not point *B* is reconstructed from the video recorded from the front perspective. Just before, in line 1936, the point *B* has been indicated, as becomes visible with the camera focusing on the printout. At the same time, its position as visible in the front perspective can be identified. The pointing in the end of 1940 is clearly left spatially above the pointing in 1936, at point *A*.

objects in one aspect. Through this, the semiotic composition can represent another, perhaps more abstract, immediate object.

7.1.1.4 Summarizing Overview about the Specifying-Function of Gestures on Speech

Specifying-features of gestures describe how gesture can take part in framing the interpretation of the verbal utterance. Without any claim for exhaustiveness, the specifying-features investigated here are derived from the data at hand as a means to describe how the representation of mathematical objects can differ in gesture and speech. It turned out that there are four aspects that are added through gesture that potentially represent further information about the immediate object:

Spatial Aspect – Where: Gestures can determine *static* and *dynamic spatial aspects*. That is, they can localize objects within an environment,[101] distinguish them from others of the same kind or add information about the direction of fictive movement.

Defining Aspect – What: By adding information about the *kind of the object* referred to, gestures can take part in defining it. That is, it refers to a characterizing property.

Describing Aspect – How: Gestures can refer to the *style* of an object or an action. This concerns *how* something is shaped or accomplished.

Relational Aspect: When gesture visualizes a *relation* between two or more aspects and this relation is not revealed in speech, it can specify this relation. This concerns the establishment of a diagram or an iconic representation within a diagram, but also relations to which is referred by the use of gesture in a metaphoric way.[102]

The potential of this representational function of gestures becomes especially salient in mismatch-situations. Co-expressiveness of speech and gesture leads to a coordination of the information provided in both modalities. If there is a linking aspect, the overlap of its peculiarities can be extracted; if both immediate objects contradict each other, it has been observed that gestures take the lead. That is, the aspects specified in gesture actually strongly influence the understanding of the immediate object.

---

[101] I avoid the term ‚representation' here since ‚environment' comprises 'real-world deictics' as well. These have been observed in particular in the induction-task (IN7) to refer to consultants in the circle as pointing forward. I hypothesize that such pointing gestures are especially initiated by tasks in which the reference to a "real-world" problem is more obvious than its mathematical reference.
[102] This has been primarily observed in IN7 and CF7, connected to reasoning actions. There, the relation between two different conjectures or two different cases has been specified as opposing by metaphorically referring to them on two opposing sides of the gesture space. An example for such a relation will be given in chapter 9.2, excerpt 5.1b.

7.1.2 Specification and Non-Specification: Further Details

The specifying-features are empirically grounded and developed from the observation of the data to investigate how gesture contributes to shaping the immediate mathematical object. For this, 908 meaning units have been interpreted within the epistemic-dense episodes, including 71 non-mathematical gestures. Setting the focus on the gestures that directly concern the elaboration of the task[103] led to coding 825 mathematical gestures[104] that have been analyzed according to the leading question of this chapter. It turned out that almost 91.4% of the illustrators specified the immediate object in some way, representing information about it that has not become explicit in speech (Fig. 7.5).

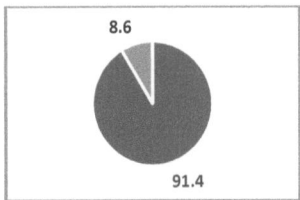

Fig. 7.5: Relation between specifying (dark) and non-specifying (light) illustrators (in %)

Furthermore, 323 out of these 754 specifying-gestures (42.8%) have been identified to add information in more than one aspect (Fig. 7.6).

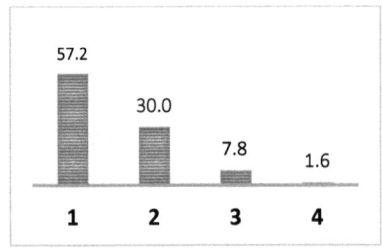

Fig. 7.6: Amount of specifying-gestures specifying a certain number of aspects (in %)[105]

---

[103] The 71 non-mathematical gestures concerned, for example, gestures that have been related to the organization of the working process.
[104] I refer to gestures and meaning units synonymously in this chapter. Most of the meaning units embrace one gesture phrase, but in some cases the meaning of the gesture unit revealed another aspect of the immediate object. Note that this does **not** mean to assign different meaning to the gesture unit and to its gesture phrases. This would contradict the global-synthetic property of gestures (McNeill, 1992, p. 19, see also footnote 20 on page 29). The gesture as a whole has a meaning but different aspects may be revealed in different strokes as they may also refer to different entities.
[105] Note that values do not add up to 100% because gestures specifying compound aspects are not included. The compound aspect has been identified when the gesture phrases did not specify an aspect but the gesture unit did.

7.1 Within-Functions of Gestures: How Gestures May Shape Mathematical Meaning

These results confirm that gestures do not only illustrate mathematical objects as described in speech anyway, but that additional information is often implicitly given by the use of gesture. Considering gestures' ability to represent, their representational function (Hoffmann & Roth, 2007) provides a more comprehensive picture of how the immediate object is shaped than is given by considering speech alone.

To make the coding of the gestures as transparent as possible, it will be described more in detail in the following section.

7.1.2.1 Coding Guideline for Illustrators[106]

The epistemic-dense episodes are chosen around structure seeing situations, but they still contain short scenes or single utterances which rather concern meta-discourse on the task. These utterances can be ignored for the coding of the gestures. Furthermore, gestures that are "understood as making a reference to the interlocutor rather than to the topic of the discourse" (Bavelas et al., 1992, p. 471) are excluded as not representing a mathematical object. Bavelas et al. term these 'interactive gestures' and distinguish them from 'topic gestures' as two subclasses of illustrators. For illustrators they state:

> They specify and often clarify verbal references, and they can denote meanings that may not be in the accompanying words […].If this is true than we can explicate or paraphrase the meaning of an illustrator in the context in which it occurs (i.e., the preceding conversation and the co-occurring words, intonation, and facial displays)- *just as we do with words.* (ibid., pp. 470-471, italics in the original)

Encouraged by this confirmation of the specifying-functions as developed from the data, and inspired by Bavelas' coding tree presented in a manual for identifying interactive gestures,[107] the following coding guideline (see Fig. 7.7) has been developed to make transparent how can be decided about a gestures' property to specify the immediate object. This guideline shall not be seen as giving a static, predetermined algorithm that unambiguously leads to 'the' specifying features of a gesture. It rather provides a way to make the coding more comprehensible, taking into account that the meaning of the gesture is reconstructed within an interpretative approach.

---

[106] See also sections 3.2.1 and 3.2.3.
[107] This manual, including the coding tree for identifying interactive gestures, has been provided by courtesy of J. Bavelas (personal communication, May 2013).

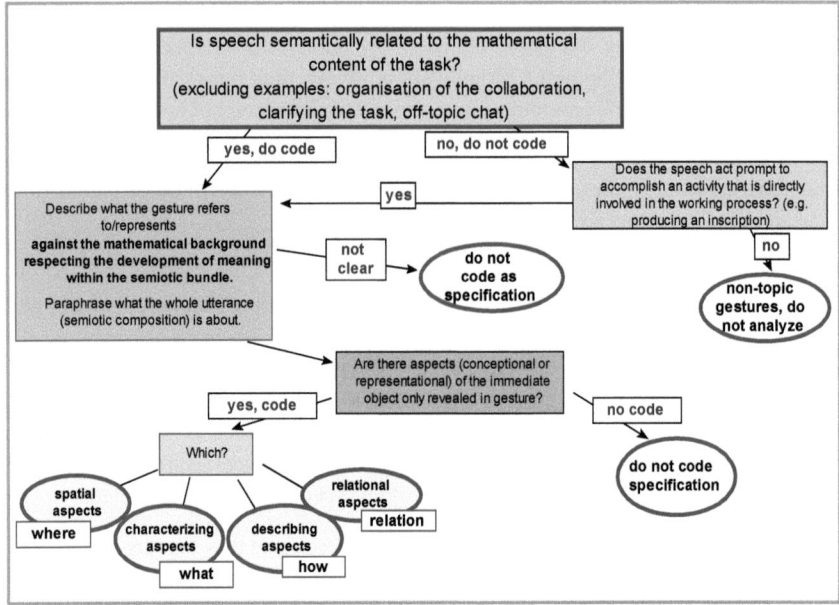

Fig. 7.7: Guideline for the coding of specifying-functions of gestures

The non-topic gestures (path directed to the right side) are skipped. They are related to the organization of the working process in general rather than to the mathematical topic. In this study, these gestures have been explicitly excluded from analysis and it has been focused on the left path of the guideline.

In chapter 7.1.1 it has been illustrated how the 'positive' outcome of this manual may look like and the four aspects that may be specified (bottom left) are summarized in section.7.1.1.4. However, the 'negative' outcome, i.e. when the gesture is not coded to specify the verbal utterance, shall be considered as well. It will be addressed in the following section.

7.1.2.2 Non-Specifying Illustrators

In the course of analysis, three different main kinds of non-specifying illustrators have been identified. There is no claim that these three kinds cover all cases of non-specification. They have been identified as the most prominent and recurring kinds of non-specifying gestures.

Type I: The meaning of the gesture does not become clear at that point in time, gestures seemingly have no reference to the semantic content of the verbal utterance. It may however happen that the reference of the gesture can be reconstructed within the semiotic bundle in retrospect: A diachronic analysis may identify gestures of type I as possible

7.1 Within-Functions of Gestures: How Gestures May Shape Mathematical Meaning

Growth Points of upcoming ideas, as it has been described in chapter 3.2.2. That is, they may indicate the starting point of an idea, expressed consciously or unconsciously. While the epistemic impact of these gestures will be explicated in chapter 8.2.2, an illustrative example is already given at this point in the framework describing its representational function:

Excerpt 1.2a, PA7-601-603 (PA7e7, 0:50:30 – 0:50:37)

Lisa and Rosa are working on the parabola task and try to justify their conjecture on the curve representing a parabola. Right before the start of episode 7, the students consider the slope of the curve as approach for their justification when the interviewer prompts them to think about the form of the function equation that describes a parabola (575-579). Guided by the interviewers' advice and his explanation of the function of the scroll bar, the students vary the value of e and trace the curves for $e = 1.25$ and $e = 0.5$. The diagram visible on the screen is represented in Fig. 7.8 and Fig. 7.9.

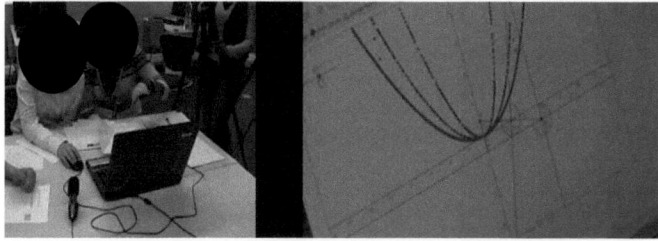

Fig. 7.8: The situation after having produced the traces in the GeoGebra-diagram represented in a part of the split-screen.
Left side: left front perspective. Right side: screen

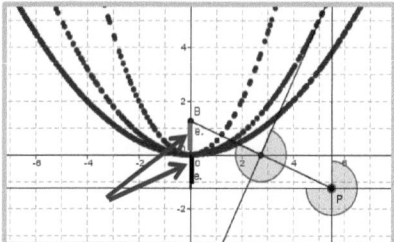

Fig. 7.9: Cutout of GeoGebra-diagram as visible on the screen. The arrows indicate the segments with length e

Rosa and Lisa conclude that "this f of x has something to do with e." (597) and observe that the distance between the lowest point and point B always equals the distance between the lowest point and the straight line g (598-600).

598. L: Hm' (7sec) (*points at the apex of the parabolas with the cursor*) ,but the ,lowest po-int' (..)
599. R: (*Lisa multiply goes along the segment between the origin and the apex with the cursor*) No the segment (*looks at the first print, then at the screen again*) uh
600. L: Is always- the same. (..) (*moves the slide control to e=1*) uhou. (..)

Their observation concerning the distances does not become immediately clear from their verbal utterance but can be reconstructed from the following substantiation they give:

601 R: Yes because
[[e uhm]
(*lifts her left hand towards the screen, moves her hands in opposed directions*)

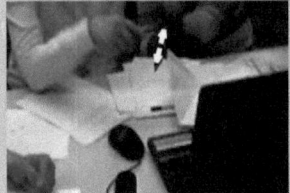

602 /L: e [ ]
(*moves her right index up and down*)

603 R: [because (*continues the movement of the hands*) ,we have <u>two</u> e]] quasi
(*takes down her hands, folds her arms, looks at the Interviewer*)

In line 601, the gesture is referring to GeoGebra, as can be concluded from its indexical feature. The co-timing to speech, furthermore, reveals that the reference is linked to *e* in some way. In the preceding social interaction, *e* has only been considered as value related to the change of the curve. This warrants the indexical reference to the screen since the curve is displayed there. However, it does not explain the alternated up-and-down-movement of the gesture, which is left unclear. Almost synchronically, Lisa performs a similar reference (602) but reduces it to one hand. With an up-and-down-movement the gesture has an iconic feature directed towards the screen directly after verbally mentioning "*e*". The simplicity of the iconic feature combined with the representation of the segments having length *e* as orientated vertically within the diagram allows an interpretation of the gesture as reference to a segment having length *e*. The vertical movement does not need to refer directly to the representation visible on the screen, but may rather

7.1 Within-Functions of Gestures: How Gestures May Shape Mathematical Meaning 131

be grounded in her tracing of the segment with the cursor in line 599. The indexical feature though is weak enough to leave imprecise to which of the two segments on the screen Lisa refers. Considering this interpretation of the up-and-down-movement, Rosa's ongoing two-hand-gesture in 603 becomes interpretable: With her gesture in front of the screen, she visualizes the idea of a 'midpoint', repeatedly moving both hands a bit up and down so that they meet in one point. Each hand iconically refers to one of the two segments with length e, set in relation as one below and one above the meeting point. The "because"-part of the utterance verbally refers to the following illocutionary act of reasoning the aforementioned general observation (598-600). The *Growth Point* of the reasoning idea is already recognizable in Rosa's gesture in 601, *before* reference of speech and gesture matches in 603. In 603, speech provides the context to interpret the two-hand gesture that at the same time specifies the relation between the segments that is used for Rosa's justification.

Type II: In these cases, the meaning of the gesture can be clearly interpreted, but it does not add any *semantic* surplus to what speech represents. Although these gestures have no specifying-function as defined in this chapter, they provide a visual access to what is expressed verbally. This visual access does not necessarily need to be related to the *semantic content* of speech (case ii), but it can be (case i). Without any claim for exhaustiveness, two kinds of gestures of this type II can be described:

i) Gestures do not specify but represent the immediate mathematical object when representing in a *perfect match* what is mentioned in speech. Vice versa, a perfect match is considered when there is no aspect added in gesture to speech. Not representing any specificity of the object, the gestural reference has a rather general character.[108]

Excerpt 1.2.b, PA5-319 (PA6e4, 0:24:51)

This example has already been introduced as excerpt 1.2e. Tim and Mike justify the suggested structure of the function equation and Tim excludes linear functions as follows:

319. /T: if we had a to the power of <u>one</u> function *(looks at Mike)* ,so <u>just</u> x *(Mike nods)* it would be a [straight line]. *(draws a straight line in the air)*

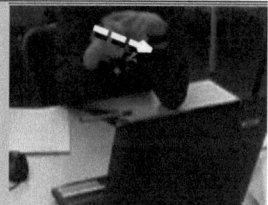

---

[108] Note that this general reference is a necessary, but not a sufficient condition. Whether gesture specifies or not crucially depends on the concrete formulation of speech.

The graph representing a "to the power of <u>one</u> function", hence a linear function, is identified to be a "straight line" both in speech and in gesture. While speech explicates this conclusion, gesture illustrates it by sketching a prototype of a straight line. The kind (graph) and the style (straight) are already given in context and speech,[109] and a relation to another entity is not presented. The movement has a direction and the graph a certain slope, but both aspects are not relevant for the representation of "a straight line". The use of an indefinite article points at a general object and gesture provides a concrete example without loss of generality.

The concrete example to which is referred may also be given by pointing to a concrete inscription. To decide whether a reference specifies or not needs to be interpreted in the context given by speech.

ii) Gestures do not specify but represent the immediate mathematical object when referring to the mathematical structure of discourse rather than to its semantic content. This has been particularly observed in argumentation; the use of gesture space metaphorically illustrates, for example, the step from case to conclusion, or the relation between two cases in comparison or distinction. Similarly, such a reference to the structure of discourse has been observed in the metaphorical reference to the structure of the induction. Gestures of that kind are considered to represent the mathematics 'behind' the semantic content as some kind of mathematical 'meta-structure' of discourse. The gesture can represent this meta-structure without specifying the semantic content. The epistemic value of these gestures will be considered in section 8.1.2.6. At this point, it is illustrated why they do not specify the immediate mathematical object, using the example of what I will call the 'conclusion-gesture':

Excerpt 1.2c, IN7-297 (IN7Ee, 0:29:33)

Lisa and Rosa are working on the task in which it is aimed to carry out an inductive argumentation in order to justify a strategy to pass the 'king's consultants-test' (IN7). They already hypothesized how to behave after the nth ring of the bell, dependent on the number of marked hats they see and the behavior of the other consultants. After checking their conjecture for some specific cases, they conclude their strategy to be valid. Nevertheless, the question arises whether there is another method that reveals their own status already at an earlier point. Prompted by the interviewer, the students start to investigate this possibility by assuming the case that all consultants have a marked hat. For this, they consider the concrete example of five consultants:

---

[109] As can be seen in excerpt 1.3e, Tim's preceding utterance specifies that graphs are considered. There he refers to the graph as displayed on the screen, specifying 'what' in gesture.

7.1 Within-Functions of Gestures: How Gestures May Shape Mathematical Meaning   133

297 R: *(takes the black pen)* the at the fourth ring they <u>still</u> see- *(writes: "4ᵗʰ ring: no")* (4sec) none ‚oh- *(scratches out "no")* ‚there I had always written nobody *(writes "nobody")* (...) *(writes "⇒" in a new line)* (0:29:44)
‚that means there have to ‚be-

(briefly raises her hand and turns it forward in one fast movement)

[more than four]
because they are only five everybody has one. *(writes: "all")*

In the first part of the utterance, Rosa informs Lisa about the situation that all consultants see nobody raising his hand ("at the fourth ring they <u>still</u> see none").[110] In turn, she records this observation of the behavior of the consultants on note sheet 4 (see Fig. 7.10). This inscription is structured similarly to the ones noted before (See also page 336.):

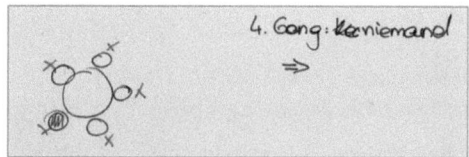

Fig. 7.10: **Rosa's inscription representing the case of five consultants when everybody wears a marked hat (inscription on the right side: "4ᵗʰ ring: nobody")**

On the left side of the inscription displayed in Fig. 7.10, a diagram is recorded that represents the considered case of the five consultants, all of them wearing a marked hat. On the right hand side, the behavior of the consultants is written down as "4ᵗʰ ring: nobody" with an arrow added beneath. In the previous cases, the conclusion about how many consultants wear a marked hat has been written down next to the arrow as "more than 1", "more than 2" and so forth (see note sheet 4 on page 338). This is stated in the same manner for the present case by

---

[110] It is suggested that she refers to "no one signals" with "none" from considering the utterances in line 293 and 294: In line 293, Rosa claims that in case everybody wears a marked hat, "they STILL all don't report" after the ring given by the number of persons. In line 294, Lisa already expresses the hypothesis "if at the <u>fourth</u> stroke of the ring nobody reports- ‚at the fifth then everybody reports. " that is investigated more in detail now in the present excerpt.

Rosa right in turn as "more than four". The gesture co-timed to this consists of raising the hand a bit and quickly turning it forward arc-shaped. The verbally mentioned "that means" already hints at a forthcoming conclusion. Gesture represents its *status* as an argumentative step in the chain of conclusion while speech formulates its *concrete manifestation* regarding this case.

The non-specifying-gestures are considered to **not** influence the shaping of the immediate mathematical object by specifying aspects of it. Nevertheless, they provide visual access to what is also expressed in the verbal utterance.

As has been described in the coding guideline, the interpretation of a gesture has to be carried out within the semiotic bundle as developed in social interaction against the mathematical background. In the following sections the interplay with inscription is integrated in the interpretation of meaning shaped by the use of gesture.

7.1.3 Gestures can Refer to Mathematical Objects on Three Spatial Levels

To describe how gestures frame the interpretation space of an utterance, it is also important to consider how they provide non-verbal access to an object. Each aspect has been observed to be specified on different spatial levels, more or less linked to a representation at hand. These different referential levels will be presented in the following examples.

7.1.3.1 Presentation of the Three Levels of Gestural Reference

On a **first, concrete level**, they refer to actually present inscriptions, such as in the following utterance:

Excerpt 1.3a, PA5-77 (PA5e1, 0:08:28)

Mike and Tim want to write down their conjectures about the curve they got in the folding diagram. They have already fixed the right hand side of the curve (see Fig. 7.2, p. 104) and discussed its symmetry when trying to formulate the conjecture:

77 T: u-m ,I would say *(points at the two outside red points with thumb and middle finger)* we describe [that as a graph ,or']

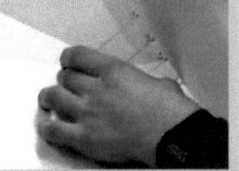

The two-finger pointing gesture indicates the fixed part of the graph and thereby clarifies to which extent it is considered being described as a graph. This shapes the semiotic composition by restricting on what is actually inscribed. *On level 1, the meaning of the gesture as an index to concrete inscription is given by the meaning of the inscription.*

In contrast to this, the shift to the **second referential level, the level of the hypothetical**, becomes visible subsequently when Tim revises his proposal:

Excerpt 1.3b, PA5-79 (PA5e1, 0:08:32)

79 T: [that is actually (*spins the hand above the sheet*) a graph.]
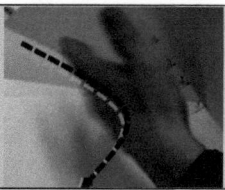

Different to before, the gesture does not merely refer to an iconic representation, but provides one itself. It shapes the curve and complements the fixed part of the curve by its left hand side. This further **specifies what** is considered as a graph, ephemerally extending the concrete inscription indicated before on level 1. *On level 2, the meaning of the gesture is dependent on its embedding into a diagram.*

In the following excerpt it is exemplified how gestures **specifies** a '**what**' on referential **level 3**:

Excerpt 1.3c, PA5-629 (PA5e8, 0:48:32)

Tim and Mike aim to find the function equation that describes the curve. They already determined the function equation to have an even-numbered exponent (289) and aim to reason this conjecture by referring to the symmetric shape of the curve:

629 T: well I know this for sure like this ,because anyway with an odd exponent. ,uhm it is that way in any case ,when we now have an x to the power of three or x to the power of five ,there is at least uhm a saddle point. (*looks at Mike*) (...)

and then there would ,it would look (*draws a parabola in the air*)

,the whole thing wouldn't be axially symmetric but (*looks at Mike*) point symmetric. (...)

The shape of the parabola is sketched in the air as anticipated **specification what** is meant by "the whole thing". It may be directed towards the screen, but its *interpretation does not require to consider any reference to it. The meaning of gestures on level 3 is independent from co-occurring inscription.*

Another example for a referential level is given in

Excerpt 1.3d, PA7-692 (PA7e8, 0:58:14)

Lisa and Rosa are working on the parabola task. They have already found the function equation $f(x) = \left(\frac{x}{2}\right)^2$ for the case that $e = 2$. Just before, they adjusted the GeoGebra representation such that the point on the parabola has x-coordinate 3 and the corresponding y-coordinate becomes visible. In the following, Rosa checks the function equation by substituting x by 3, calculating the y-value.

692 /R: (*at the same time*) that's about right (*sits up, looks upwards*) if ,because three

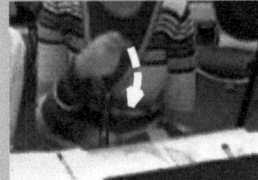 

[[divided by two'] *(lifts up the right index and then moves it downwards)* equals one point five'] and *(Lisa takes a pen and turns the note sheet towards herself)* ,fifteen square equals two hundred and twentyfive. *(Lisa writes at the end of the list for e=1 "3,2.25", in the meantime she looks at the screen)* (7sec)

Detached from the concrete example mentioned in speech, Rosa's gesture refers to the general top-bottom-relation of a fraction as notation in the gesture space: The downwards movement co-timed to "divided by two'" **specifies the relation** of top-bottom and with that **specifies what** is referred to as symbolic notation of a fraction.[111]

On the three presented levels, the degree to which the interpretation of the gesture depends on a fixed representation decreases. On the first level, a concrete fixed representation of the object exists and gesture refers to it. Gesture derives its meaning from the meaning of the inscription. On a second level, gesture embeds an ephemeral representation within a diagram that provides relations considered for interpreting the gesture. The reference of a gesture performed in the gesture space is independent from an existing fixed representation. Recalling the excerpts presented in chapter 7.1.1, excerpts 1.1a (pp. 102-103), 1.1b (p. 104) and 1.1c (p. 105) showed gestures with level 2-reference,[112] 1.1d (pp. 105-106), 1.1g (p. 110) and 1.1h (p. 111-112) specification on level 1, and 1.1f (pp.

---

[111] Edwards (2009, p. 138) refers to this as iconic-symbolic gesture. In chapter 3.2.3 it has already been discussed why the term 'iconic-notational' is considered to be more suitable.

[112] In 1.1a, the mentioned point is embedded in the diagram, in 1.1b, the same is done with the non-visible part of the parabola. The gesture in excerpt 1.1c extends the perpendicular and highlights it by incorporation. It is not concretely referred to the fixed line as perpendicular through M and C_4 (the entity actually constructed according to the instructions), but to a representation of an axis.

108-109) a gesture with level 3-reference. An even different case becomes visible in **excerpt 1.1e:**

23 S: *(mumbles)* let us better make one [on the <u>other</u> *(points at the left side of the sheet)* side.]

The gesture does not refer to something concretely fixed and therefore is not a level 1-reference. On the other hand, it does not represent itself in an iconic manner but refers to a **location** respecting the diagram as underlying representation of the situation. This provides reference to a larger area but no concrete reference within this area. Gestures that refer in that manner are suggested to have representational flexibility between 1 and 2. They point to an existing representation of something, but not referring concretely to an entity. Thus they are coded with '1/2'.

Furthermore, there are gestures that carry meaning themselves, but for which cannot be decided whether they are interpreted against an inscribed diagram or not. The reference to the inscribed diagram is not explicit but possible, which makes the interpretation ambiguous. The consideration of a level between 2 and 3 was needed especially in cases where a fixed diagram was given in a vertical oriented plane:

Excerpt 1.3e, PA5-317-319 (PA5e4.1, 0:24:58)

Tim and Mike aim to write down their conjecture of the style of the function equation as being even-numbered. Mike has just asked whether the function has to have an exponent and with that has implicitly requested whether it can also be a linear function. With this, he expressed doubts in the negotiated solutions that the exponent is even-numbered. Tim now argues for the function being not linear:

(0:22:49)
317 T: *(shapes the displayed curve on the screen in the air)*
     [elsewise it is not <u>curvy</u>] or is it'

318 /M: yes ,right.

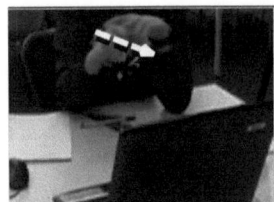
(0:24:51)
319 /T: if we had a to the power of one function (*looks at Mike*) ,so just x *(Mike nods)* it would be a [straight line]. *(draws a straight line in the air)*

In line 317, the reference to the curve as displayed on the screen is identifiable without a doubt (level 2). The case is different in line 319: Tim shapes the straight line in the gesture space that crosses his line of sight. It could be possible that he refers to the screen and shapes the straight line to highlight that both curves are different. But it may also be the case that he shapes a straight line in the gesture space to show the difference in the motion of shaping something straight. This cannot be decided so it has been considered a level 2/3 in such cases.

7.1.3.2 Summarizing Overview about the Referential Levels of Gestures

Gestures have been observed to refer to the representation of an object on different levels. The higher the level, the less the interpretation of the gesture depends on a fixed representation and the more detached it is from a concrete situation.

Level one is considered the *level of the concrete*. Gesture refers to something actually represented in a fixed diagram. It does not represent itself but works as an index to hint at something already represented.

On a **second level,** gesture is embedded in a fixed representation but does not merely refer to an already fixed concrete component. It represents itself but needs to be interpreted against the background of an inscription that provides relations. Such gestures may shape and highlight already fixed entities but also depict new entities in an established diagram as ephemerally embedded in it. A gesture that specifies on level 2 represents a 'hypothetical something' not there but *potentially 'thought into' a present diagram*.

The **third level** concerns what Müller (1998, p. 102) calls "free gestures" performed in the gesture space. On this level, meaning is transported by the gesture itself such that it can be regarded as autonomous. Without being dependent on a present referential frame, the interpretation of the gesture is detached from the concrete. That is what makes it 'free'. Furthermore, its iconic rather than indexical feature allows for establishing 'body diagrams'[113] in social interaction. These gestures may reveal a more conceptual than contextual idea of a mathematical situation or object.

---

[113] This term has been borrowed from the section 'Body Semiotics' of the DGS (German Association for Semiotics) that announced the topic of the 2014 conference "Body Diagrams: On the Epistemic Kinetics of Gestures". I considered gestures to provide a 'body diagram' when an icon of relation is represented by

When gesture indexically refers to a representation as a whole rather than to a concrete represented aspect, it refers on **level 1/2**. When it is not clear whether the interpretation of a self-representing gesture is framed against the background of a fixed diagram or not, it is considered to refer on **level 2/3**.[114]

The three main levels, 1, 2, and 3, are related to a shift of an idea from a representation of the concrete to a general representation (Fig. 7.11). Level 1 is anchored in the concrete case visible, gestures on level 2 use a concrete diagram to visualize the potential (e.g. dynamics) and a gesture on level 3 is considered to depict the nucleus of the idea, that is it provides a view onto the core aspects that can be represented in generality.

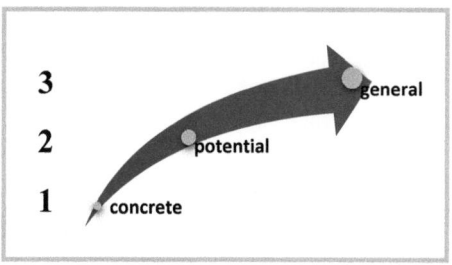

Fig. 7.11: **Shift of reference from the concrete inscription into the gesture space**

7.1.4 Closing Remarks on the Within-Functions of Gestures

In this chapter the two representational features of gestures have been presented that together constitute gestures' influence on shaping the immediate mathematical object of the multimodal sign. These are the specifying-function of gestures and the three levels of reference to inscription within the semiotic composition. While there is always exactly one referential level assignable to a gesture, the coding revealed that only 57.2% of the specifying-gestures identified throughout the data specified only one aspect (see Fig. 7.6 on page 126). This is actually considered as one of the core benefits of the specification in gesture: Their multi-specifying character makes it possible to provide information about more than one aspect at a time, enriching speech in a multifarious way.

---

using gesture as an embodied means of expression. As it conforms to the Peircean understanding of diagrams (see also footnote 15 on page 23) the term 'body diagram' is used as descriptive term in this book.
[114] Only level 1, 2, and 3 are considered here as levels of reference. Gestures on these levels have been clearly identified to represent a certain object.

7.2 Reconstructing Meaning in Social Interaction: The Development of Associated Signs and Information Bundles

Meaning is assigned to gesture within the social interaction. It develops in its use and is "always contextually embedded, the context deriving from speech" (Parrill & Sweetser, 2004, p. 200). Yoon et al. (2011) already describe how 'free gestures' can derive their meaning from blending the real space of physical movement of the hands in the gesture space and the mathematical concept as presented in speech. However, how can a shared meaning of gestures - and more generally, of signs - be established when they are used to represent *newly* constructed mathematical objects and concepts that are *not yet* accessible in speech? Following Hoffmann and Roth (2007), this concerns the shared *collateral knowledge* that is needed to interpret a sign (e.g. a gesture) as well as to use it in order to intentionally represent something (ibid, pp.107-108, pp. 118-119, see also chapter 3.1.3). This consideration is essential for the integration of gestures into the reconstruction of the epistemic process: The shared collateral knowledge, not only about the mathematical object, but also about the meaning of signs used, makes understandable how a shared interpretation of gestures in social interaction becomes possible when new knowledge is constructed.

A diachronic analysis of the semiotic bundle developing within social interactions revealed how signs can be endowed with meaning during the social process of constructing mathematical knowledge. I observed how the meaning of 'mathematical' gestures'[115] used in the social interaction develops in its use together with speech and inscription. For this the catchments produced by the students have been identified. As has already been described in chapter 3.2.2, a catchment is identified "when one or more gesture features occur in at least two (not necessarily consecutive) gestures" (McNeill, 2005, p. 116). The idea behind this approach of identifying catchments is that gestures similar in shape, style or dynamics are considered to refer to similar or related references. By identifying catchment-gestures produced by the students and considering the contexts provided by the verbal utterance, it can, thus, be reconstructed *which* meanings are linked for them. Furthermore, the recurring feature makes interpretable *what* this link may consist of. This means that using a sign in a certain sense or context may establish a link so that a renewed use of this sign may refer to the context as well. The analysis within the semiotic bundle thus allows the observer to reconstruct the development of shared meaning within the working process.

---

[115] With this notion I refer to gestures that are considered to represent a mathematical idea, i.e. object or concept, in an iconic or metaphoric way.

7.2.1 Detailed Reconstruction of Two Illustrative Examples

Rather than symbolically 'standing' for some mathematical idea or object, a sign can become *associated* with this idea or object in the course of social interaction. That is, the mathematical meaning of the sign is grounded in its use during the social construction of the mathematical object represented by the sign for the students. This meaning does not need to be conventionalized and does not even need to be 'mathematically true'. It is moreover *situationally conventionalized* and represents the current stadium of an idea associated with this sign. Non-visual examples for such associated signs are linguistic terms invented in social interaction, expressing a mathematical idea before an appropriate, culturally conventionalized notion is available. Most likely, such situationally conventionalized notions are chosen to verbally capture a significant aspect of the mathematical idea it refers to. This idea can be transferred to non-verbal associated signs in which a significant aspect is captured. For example, a gesture that iconically refers to an inscribed representation of a mathematical object may not refer to it in all the aspects that are fixed, but resemble those aspects that seem important at the moment the gesture is used. Gestures of this kind are consequently called *associated gestures*. In the next section, associated signs concerning the immediate object 'shape of the curve' will be reconstructed:

Example 2.1a, PA5-49ff. (PA5e1, 0:06:04)

Tim and Mike have just started to work on the parabola task. They have constructed three points (see Fig.7.1 on page 114) when Tim mentions that he sees a curve:

49. T:
   [ ]
   (makes an arc-shaped movement above the sheet, connecting the constructed points)
   I can actually nearly see the curve <u>now</u> already.

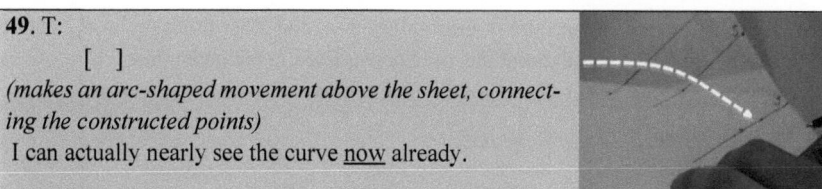

Tim shapes the curve on *level 2* above the folding sheet and **specifies** its considered *location* within the diagram as well as its *style*. This style now becomes closely linked to the curve as iconically representing its shape. This link between his gesture and the mathematical object 'curve shaped within the folding sheet' is reconstructed from his verbal utterance in which he mentions that he 'nearly sees the curve already'. Shortly after, he explicates the connection of the constructed points to shape the curve within the folding sheet, repeating the gesture in the same style as before while concretely indicating the constructed points one after the other:

(0:06:29)
53 /T: [

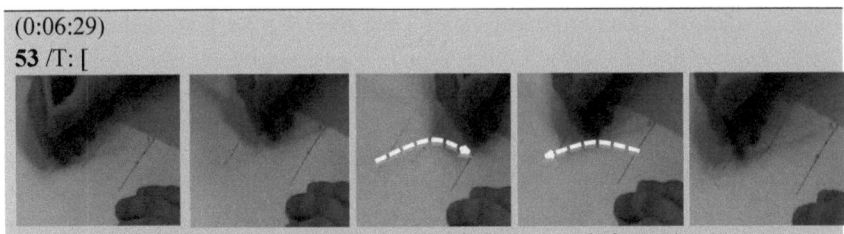

(traces the red dots on the folding sheet with the pen)
well no actually when one] connects these three points by a good curve one can theoretically predict it already. (.) ,approximately.

On *level 1*, the points are **specified spatially** within the printout. The movement conducted by their successive indication leads to shaping the curve as connection of the points on *level 2* and again **specifies** its *location* and *style*.

The gesture now represents the curve not only in its style and located within the printout: Its accomplishment while concretely specifying the points mentioned in speech also explicates the curve to be represented as connection of the constructed points. In its use, the gesture becomes more and more associated with meaning and through specification, more and more *information* is *bundled* within the gesture. So far, the *information bundle* associated with the gesture referring to the 'shape of the curve' is thus *packed* to consist of 'the style of curve', 'the location of the curve', and its being the connection of constructed points.

In the following scene, the gesture is accomplished several other times on level 2 above the diagram in which a fourth point has been constructed in the mean time:

(0:07:14)
55 T: *(traces the red dots with the pen)*
[shall I now try]

to draw this curve approximately'

Right after, he sketchily repeats this gesture less accurately:

(0:07:21)
58 T: yes we [can just there-]
(sketches the curve from [MC_4] to the right)

we simply draw a curve in a different color.

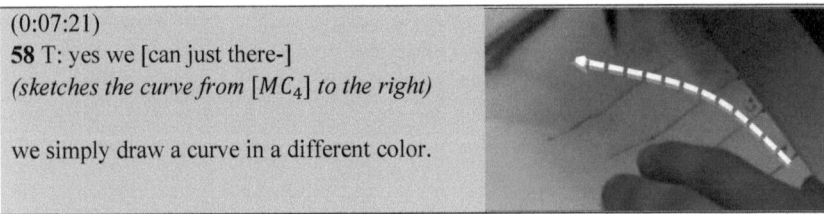

7.2 Reconstructing Meaning in Social Interaction: Associated Signs and Information Bundles

Both utterances have in common that they illocutionary propose to fix the curve within the printout, accompanied by the ***specification where*** and ***how*** of the shape that shall be fixed. In turn, Tim produces the following inscription, fixing the curve as formerly established by using gesture:

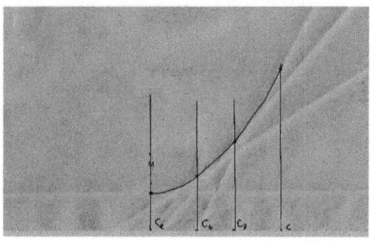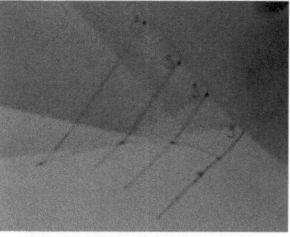

Fig. 7.12: Folding diagram as elaborated in line 65

Following Arzarello (2006, p. 281), the semiotic bundle is now enriched by *genetic conversion* of a gesture to an inscription. The shape of the curve that has been established in gesture over time - tried out on *level 2* as a connection of components that are indicated on *level 1* - is fixed as new, lasting sign representing exactly what has been associated with the gesture before: The connection of the four points in a certain style shaping a curve. This is the representation of the 'shape of the curve' as social immediate mathematical object developed in gesture through specification of aspects. Its development can be traced in the following Table 7.1:

Table 7.1: Development of the associated sign referring to 'shape of the curve' in PA5

Line	Associated sign	Current Information Bundle	Immediate Mathematical Object
49	Gesture on level 2	- style of the curve - location of the curve within the folding sheet	Shape of the curve constructed by folding
53	Indication on level 1/movement on level 2	- style of the curve - location of the curve within the folding sheet - curve is a connection of constructed points	Shape of the curve as a connection of points constructed by folding

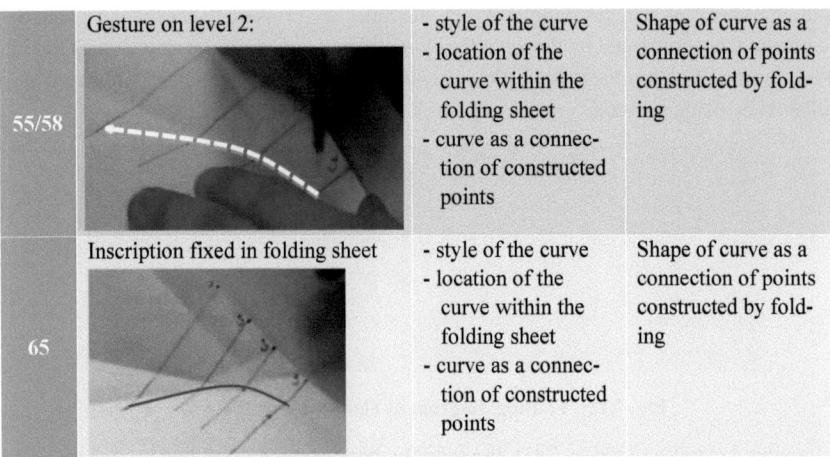

Reconstructing these aspects by tracing the development of the associated sign within the social interaction means to reconstruct the social immediate mathematical object represented in the gestural sign. Moreover, it is this immediate mathematical object that is represented also in the inscriptive sign as fixation of the gestural sign.

This fixation of this concrete social immediate mathematical object 'shape of the curve on the right hand side as connection of the four constructed points' enhances a further elaboration of the social immediate object 'shape of the curve':

After fixing the right hand side as a connection of the points constructed so far, Tim uses a variation of the gesture to express how the curve shall look like on the other side:

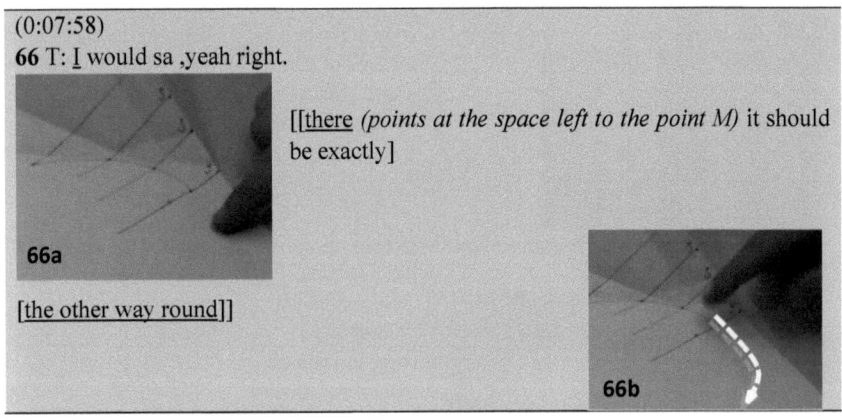

The style of the gesture is similar to the one that has been established and fixed. By that, it can be used to refer to the 'shape of the curve' as it has been socially conventionalized

7.2 Reconstructing Meaning in Social Interaction: Associated Signs and Information Bundles 145

before. The gesture is not equal but differs in some aspects: First, the *location* is *specified* on the left hand side of point *M* as potentiality on *level 2*, explicitly clarifying speech ("there"). Second, the *style* is not the same but "exactly the other way round", specified as shaped by the same dynamic movement but performed in another direction, from right to left. The similarity of both gestures thus marks them as related; the differences between them *specify* this *relation* as '*relation between both sides of the curve consists in being shaped the same but the other way round*' (See also excerpt 1.1b on page 116.). Complementing Table 7.1, the following line is added:

Table 7.2:   Use of the associated sign to elaborate the immediate mathematical object

Line	Associated sign	Current Information Bundle	Immediate Mathematical Object
66	Gesture on level 2	- style of the curve - location of the curve within the folding sheet - curve as a connection of points - dynamics of the curve (from right to left)	Potential shape of the curve on the left hand side as **symmetric** complementation to the shape of the curve as a connection of constructed points

This relation in turn enriches the social immediate mathematical object 'shape of the curve', now consisting of two sides that look 'the same but the other way round'. Mathematically, this can be condensed to 'the shape of the curve is symmetric' as Tim does right in turn:

68 T: that means [,that theoretically is our axis] and there is symmetry	

Tim represents the axis of reflection as embedded within the folding sheet on *level 2*. This highlights the vertical line separating left and right side of the curve, being shaped 'the same but the other way round' as developed before.

The representational connection made by bundling meaning within an associated sign to represent the social immediate mathematical object is condensed in the verbal explication of the mathematical structure of 'symmetry'.

Another example of an associated sign has been identified in Rosa's and Lisa's elaboration of the parabola-task. They also establish a variation of an associated sign 'shape of the curve' but furthermore, it develops to be linked to a sign representing the 'axis of reflection' for the students:

Example 2.1b, PA7[116]

After constructing three points on the right hand side of M (corresponding to points E, F and C on the lower edge), Lisa and Rosa construct a fourth point, located on the left hand side of point M (corresponding to point G on the lower edge):

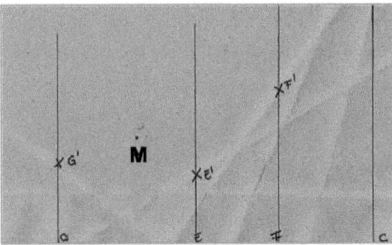

Fig. 7.13: **Folding sheet as constructed just before the curve is first suggested to be seen in line 58 (tag for point M enlarged)**

Sketching the curve on *level 2* above the folding sheet, Lisa and Rosa also established a gesture that represents the shape of the curve:

(0:06:32) 58 R: ‚there it goes [up again] doesn't it' (*moves her little finger from the bottom edge beneath point M to point G')*	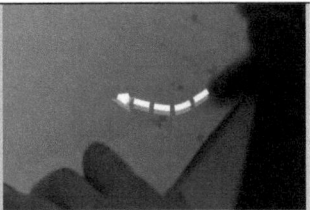
59 L: M-m (*puts the pen on the table, moves her chair towards the table)* (.) 60 R: Do we recognize a curve' yes. 61 L: yes. (incomprehensible)	

In line 58, Rosa *specifies* the *where* (",there": on the left side of M from right to left), the *what* ("it": shape of the curve) and the *how* (the arc-shaped style in which it "goes up again") by the gesture performed simultaneous to "up again" (58). Implicitly, her formulation ",there it goes up again" suggests that 'it went down before', presuming that the shape of the curve is already shared collateral knowledge although it has not been explicated yet. In line 60 she explicates the reference to 'a curve' and with that makes sure the mathematical object specified for Lisa is actually the one she considers. This

---

[116] In this example it is referred to various excerpts throughout the data. Representing results from a diachronic analysis concerning the whole working process, it cannot be restricted to a certain episode, scene or utterance.

makes her immediate mathematical object represented through gesture the social immediate mathematical object 'shape of the curve'. *The associated sign thus represents this object in style and location within the folding sheet situation.*

When Rosa refers to the immediate object 'shape of the curve' the next time, she already uses gesture detached from the concrete folding sheet representation. Lisa has just explicated a general structure she has seen respecting the co-variation of the points on the curve: She specifies her verbal utterance ",the further the point lies away" (73) "the higher is-" (75) by a horizontal (73) and a vertical (75) movement of the hand on level 2 above the folding sheet. Rosa complements this general consideration of a curve as a set of points by adding a visual reference to the shape:

(0:07:49)
76 R: *(holds the left hand on chest height and makes a swift gesture to the upper right)*
[ ]

the curve. ,is a parabola probably or so. ,but- (.)

By shaping the curve in the gesture space on *level 3*, Rosa expresses a more general understanding of the curve, probably inspired by Lisa's general consideration she aims to complete. This may also cause Rosa's restriction on shaping only the right side of the curve, while the left hand is held as reference to the axis. This *specifies two aspects* of the mathematical object: First, the *'what'* is *specified* as a 'curve' associated with the gesture by similarity of style, that is, iconic reference to the gesture that was associated with the mathematical object in line 58. This object is explicated by Rosa as a mathematical object referred to right after in speech. Second, the restriction to one side and the representation of the axis *specifies* a *relation* to this axis: In line 58, a similarity between the two sides has already implicitly been suggested in the verbal utterance. Leaving aside the left hand side of the curve but adding the left hand as reference within the body diagram on *level 3* (see footnote 113 on page 138) specifies the shape of the curve as being symmetric and furthermore, potentially blends the meaning of 'axis of reflection' as a relational reference to the left hand held within the gesture space.

That is, the information bundled in the gesture is a more general one now: The gesture corresponds in style, so that its iconic feature carries the association to the immediate mathematical object 'curve'. A spatial aspect is not specified as a concrete location in the gesture space, but it is bound to the specification of the relation 'symmetry of the shape of the curve': 'where' as 'on one side of the axis'. Embedded in a larger body diagram, the *specification* of the *relation* as symmetric is understood in connection with the axis of reflection represented by another component of the body diagram. Although the reference is not explicitly shared yet, this associated gesture can be interpreted as referring to the axis of reflection and with this to the symmetric shape of the curve in social interaction.

Nevertheless, this general connection is not made explicit earlier than in line 528. In line 500, Rosa and Lisa state that the seen curve is a parabola as solution on subtask 4a. Subtask 4b requires to justify this conjecture and makes them resume some properties of the curve. Lisa recalls the symmetry of the curve with reference to the concrete situation fixed on the printout:[117]

(0:44:36)
521 L: a-nd (…) (*lets the hand off the mouse, takes a pen*) ‚then basically- (*puts the pen on the y-axis on the printout*) {[t-h-i-s] ‚is also an axis of reflection. (*Rosa looks at the printout, then at the screen*) (4sec) **522** R: yes. ‚yes yes. (incomprehensible) **523** /L: *(takes the mouse)* because A always moves a-w-a-y (.) (*drags point P to the right*) **524** R: yes. **525** L: (*moves point P to the left*) by the- (.) same distance. (.)

By placing a pen, Lisa *specifies the orientation* of the axis of reflection (521). The temporary fixation does not specify the immediate mathematical object 'axis of reflection' concretely within the printout but rather in a more general way on *level 2*. Although the displayed situation is concrete, the specification does not refer to the concrete component of the diagram but to the pen representing a vertical axis more generally on level 2. The shape of the curve is not displayed and Lisa uses the dynamic representation in GeoGebra to make visible that "A always moves a-w-a-y (.)" "by the- (.) same distance." (523/525). Shortly after, Rosa takes up this symmetric property made visible in GeoGebra and shifts its representation into the gesture space. With this, she explicitly introduces the associated gestures to represent the social immediate mathematical objects 'shape of the curve' and 'axis of reflection' on *level 3*:

| **528.** R: well ‚we have [just *(takes up her hands and lays the palms of her hands against each other on chest height, the fingertips pointing up and forward)* ‚said' ‚because of | 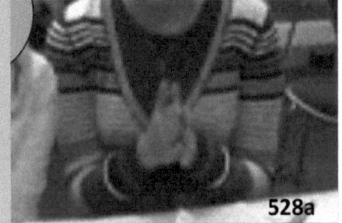 528a |

---

[117] The curved bracket in line 521 denotes that the pen has been laid down.

7.2 Reconstructing Meaning in Social Interaction: Associated Signs and Information Bundles   149

[[this axis of reflection] *(moves the right hand to the right, then indicates the shape of a parabola by moving both indexes up and down)*

[both sides are the same.]]

529. L: mmh'
530. R: which is already clearly indicative <u>for</u> *(repeats the movement of the indexes from 528 once)* [a parabola.]
531. L Mhm'

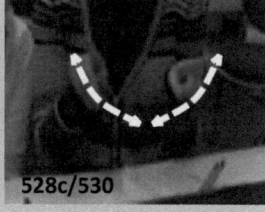

Rosa specifies the verbally mentioned "axis of reflection" as being represented by the hand held at chest height (528b). The shaping of both sides of the curve is reminiscent of the shape and *specifies how* they "are the same" (528c). This explication connects the two associated gestures on a semiotic level so that the gesture 'axis of reflection' is associated with the information bundle that also contains 'symmetric shape of the curve'. The representation of the 'axis of reflection' and its link to the shape of the curve can thus be considered to be collateral knowledge on these representations and their situationally conventionalized meaning.

The following Table 7.3 gives a succinct overview about the development and linking of the signs associated with 'shape of the curve' and 'axis of reflection'.

Table 7.3: **Development of the associated sign referring to 'shape of the curve' and linking it to the associated sign 'axis of reflection' in PA7**

Line	Associated Sign	Current Information Bundle	Immediate Mathematical Object
58/60	Gesture on level 2	- style of the curve - location of the curve within the folding sheet (-relation to the other side of the curve: implicit!)	Shape of the curve constructed through folding

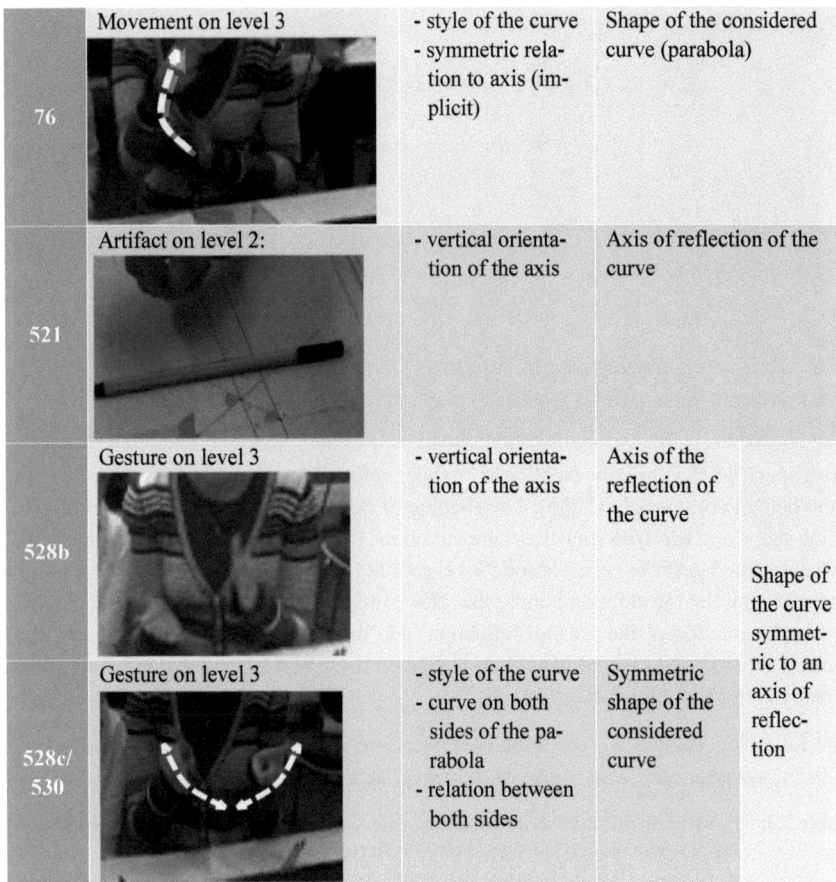

7.2.2 Evidence of the Use of Associated Signs

So far, it is described how shared meaning of gestures *can* arise from the concrete representation to a general one in social interaction without giving evidence that the participants *actually do* associate similar immediate mathematical objects to these gestures. The specifying-functions as well as the associated signs provide *a frame* for the interpretation of the semiotic composition in order to reconstruct the immediate mathematical object. Following the semiotic reconstructive analysis, it can be decided on whether the meaning of the gesture 'fits' for the other participants by considering the interpretant as a reaction on this sign (see chapter 5.2.3).This way, two possibilities of deciding whether the same meaning is associated with a gesture have been identified:

7.2.2.1 Evidence through Irritation when the Use of a Gesture does Not Fit the Context

In Subsection 7.1.1.3, I presented how gestures take the lead in mismatching situations. This at least holds true as long as the immediate object expressed in speech and the one expressed in gesture do not contradict each other. In social interaction, this can be used as an indicator for identifying whether the immediate mathematical objects expressed in gesture and expressed in speech fit together for the partner as well. To illustrate this, an example is taken from Rosa's use of the associated sign 'axis of reflection' (see example 2.1b). In this example, the gesture is used to represent 'a perpendicular bisector' as expressed in speech:

Example 2.1c, PA7-1029-1039 (PA7e13.3, 1:32:34 – 1:33:08))

Lisa and Rosa work on subtask 5, aiming to define the parabola as a set of points similar to the definition of a circle given in the task description. Since that definition deals with a distance from a fixed point, Rosa starts to offer an approach that also considers distances:

(1:32:34)
1029 R: this parabola-
(moves her left hand on chest height and holds it straight, edge down, in front of her chest) [has a perpendicular bisector of the sides]

[where *(holds her right hand against her left hand and then moves it away to the right, once horizontally, once in an upwards curve)*

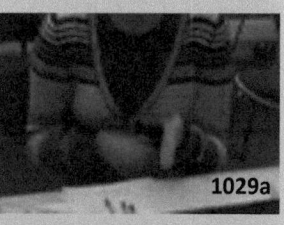
1029a

the points on the one side-]
(spins her left hand and step by step moves her right hand to the right)

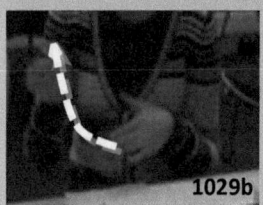
1029b

[,where the -]] *(still holds her left hand in the same position, moves the right hand to the right again, this time in one go)*

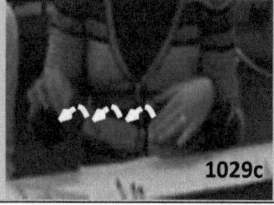
1029c

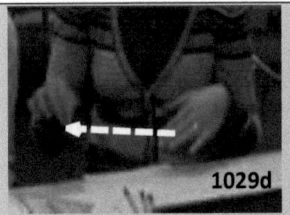

[points x-value-]

if [we *(lets her right hand sink, moves her left hand top down and then again holds it at chest height)* see it like this'

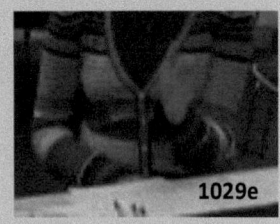

1030 L: mmh-

1031 R: [Then-]
(multiply goes with her left hand to the right, Lisa takes the mouse and goes along the straight line c with the cursor) [(..) wides-]
(low) ,uh no uh. ,increases-
[,the *(repeats the gesture a few times , looks at Lisa, Lisa lets the mouse go and briefly looks at Rosa, then at the screen again)* x-value']

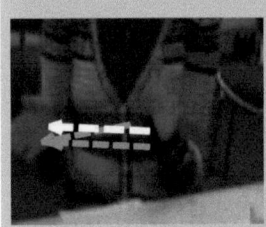

[from the] *(moves her right hand to the left hand)*
[a a]] axis of- *(moves her hand apart)* (low) r r reflection- to- *(takes the blue pen and opens it)* ,wait a moment.

(draws a vertical line on the second note sheet, draws an arrow next to it and labels it "axis of reflection") (9sec) *(puts the pen on the table and takes the black one, sets it left of the axis of reflection)* ,I can't paint parabolas.

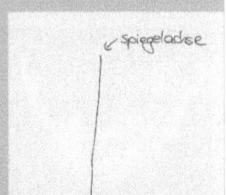

1032 L: what is for you the axis ,axis of reflection now'

1033 R: *(moves her hand top down on chest height, edge downwards)* the [this-]

Shortly after, Rosa continues to utter her idea and completes her inscripton :

7.2 Reconstructing Meaning in Social Interaction: Associated Signs and Information Bundles   153

1037. R: *(still draws)* That theoretically becomes a parabola now' *(looks at the diagram, laughs)* ,great- *(closes the black pen, puts it on the table, takes the red pen and opens it)* (..) ,

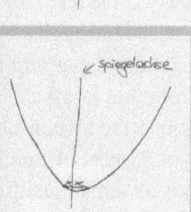

that here- *(draws horizontal arrows from the axis of reflection to the parabola, close to the point of intersection between the parabola and the axis of reflection)* (.)

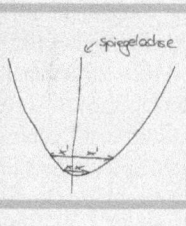

is theoretically the same. *(labels both arrows "x", draws two more arrows a bit further up and labels them "x'")* (6sec)

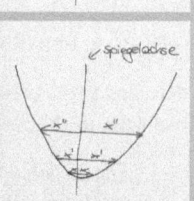

1038. L: Mhm'

1039. R: And that becomes larger and <u>larger</u>- *(draws even further up two more arrows and labels them "x''")* (.) ,somehow.

Rosa starts with resuming that "this parabola [has a perpendicular bisector of the sides]" (1029). Co-timed to this, she positions the left hand straight with edge down on chest height in the gesture space (1029a). Since gesture and speech are co-expressive and this gesture has been formerly established to represent the axis of reflection (see example 2.1b), this can be considered to be a slip of the tongue, implicitly corrected later on when she uses the words 'axis of reflection' first in speech and then also in inscription (1031). The verbal mistake is probably initiated by two concurring occurrences of the mathematical object "axis of reflection" earlier in the working process:

- First, the folding line/tangent line has been interpreted as axis of reflection of points P and B, but also as perpendicular bisector of $[PB]$ (455-459).
- Second, Rosa and Lisa mention the existence of two different objects that both have been termed 'axis of reflection'; one of them being the perpendicular bisector of $[PB]$, the other one the axis of reflection of the parabola (563-567).

Not focusing on the representation in gesture but the idea aimed to communicate, Rosa mixes up both mathematical objects in her verbal and gestural expression. Immediately following (1031), she proposes an approach to a solution based on symmetry. This idea becomes more apparent when she starts again and makes use of inscription (1037-1039). For her idea, she combines two considerations: First, there are always two points on the parabola having the same distance from the axis of reflection. Second, this distance increases upwards. In the beginning, Rosa visually supports the verbal expression of this newly considered idea by using gesture, but changes to source her idea out to inscription when she struggles with too many components to be represented.[118] Her speech becomes fragile and insecure ("[Then-] [(..) wides-] (low) ,uh no uh. ,increases- [,the x-value'] [from the] [a a]] axis of- (low) r r reflection- to-") and her gestures become less precise so that the switch to fixing the components one after the other lightens the cognitive load. Instead of simultaneously coordinating two ephemeral modalities, speech and gesture, as semantically complementing each other, the inscription makes it possible to add one component after the other, functioning as inscriptive memory (1031, 1037-1039). This fixation of her idea includes tagging a vertical line with "axis of reflection". After this reconstruction, it becomes clear why the double reference to the gesture as "perpendicular bisector" (1029a) and then as "axis of reflection" (1031) seems hard to combine for Lisa in the context of this newly considered idea: She associates Rosa's gesture used in 1029 with the situationally conventionalized meaning 'axis of reflection'. To clarify the meaning of Rosa's gesture, Lisa explicitly asks her which axis of reflection she refers to (1032). In turn, Rosa' confirms the associated character of the gesture (1033).

There are two possible reasons why Lisa does not claim for clarification right after Rosa's use of the gesture in 1029:

- The possible interpretations of the utterance can coexist side by side in a state of validation, as long as Rosa's idea is sketched rather than sharply outlined. As long as the idea does not become too definite, Lisa still has a chance to figure out what Rosa refers to by co-constructing her idea.
- Lisa recognizes that Rosa's idea just comes to existence while speaking and is still fragile. An interruption may disturb her train of thought.

In fact, both reasons may take a part in Lisa's decision to let Rosa keep on explaining her idea. When Rosa starts to capture her idea inscriptively, Lisa is forced to decide on an interpretation to understand the meaning Rosa assigns to the parts of the inscription. It is also this fixation that allows for an interruption of Rosa's train of thought since the initial part of the idea is fixed as a lasting reminder on the note sheet.

Résumé: Lisa's request for clarification is seen as an indicator for her irritation concerning the social mathematical objects represented in the co-timed gesture in the beginning of 1029. From the use of the gesture and the co-occurring speech, several possibilities

---

[118] See also section 8.1.1.1, Excerpt 3.1b for problems with representing complex relations by the use of gesture.

for interpreting the utterance are offered to her. To make sure to follow the same idea, she needs to have definitely clarified to which axis of reflection Rosa refers. From the pure fact that this need for clarification arises for Lisa, we may conclude that she sees two concurring objects in the gesture-speech-reference to the "perpendicular bisector of the sides" (1029). Hence, the meaning of the associated gesture representing the 'axis of reflection' can be considered to be socially conventionalized.

7.2.2.2 Evidence through Students' Joint Packing and Shared Use of the Associated Sign

Another indicator for a gesture being associated with a social immediate mathematical object – that is, the mathematical meaning of the gesture being shared in social interaction – can simply be found in how it is used by different participants of the social interaction. In the following, this will be exemplified by the shared use of the 'increasing numbers'-gesture:

Example 2.1d, CF7e15.1

Lisa and Rosa are working out the task dealing with the continued fraction. They found the recursive pattern that allows the successor of an arbitrary fraction $f(x) = \frac{a_1}{b_1}$ to be written down as $f(x+1) = \frac{2a_1 \pm 1}{a_1}$. Based on this, they now try to justify the conjecture that the sequence approaches 2 more and more:

(0:46:12)
542 L: ehm- the- (4sec)
[,fractions always become [more <u>precise</u>.]
(*slight moves her hands up and down above the table*) ,so only [with more <u>numbers</u>.]] (*the hands are horizontally moved apart above the table*)
(.) you understand'
543 R: with more digits- ,more.

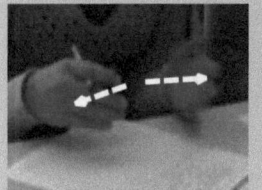

544 L: [with more digits
(*horizontally moves the hands apart and together above the table*) ,yes]
,well with <u>larger</u> numbers.

545 R: yes.

Lisa and Rosa establish the associated sign that refers to the numerator and denominator becoming larger. Lisa's gesture co-timed to the verbal "with more <u>numbers</u>" (542) is performed on *level 2* in the horizontal interaction space above the table. Speech explicates the reference first to the "fractions" and then, co-timed to the horizontal movement, to the "numbers". This suggests a reference to a growing extent in the sense of the characters, *specifying what* she refers to (number of digits) by *specifying how* they become more (not numerically but in extent of the inscription). Being aware of using a non-conventionalized sign to express this newly considered idea, Lisa requests a confirmation that Rosa actually understood ("you understand' ", 542). Rosa in turn understands this request as an invitation to refine Lisa's expression and explicates the meaning of the gesture as reference to "more digits" (543). Lisa repeats the gesture and, at the same time, agrees to Rosa's proposal of the mathematical object referred to by this gesture ("with more digits ‚yes", 544). In addition, she also offers the reference to "larger numbers" without making precise whether the "larger" means numerically larger or larger in a semiotic extent for her. In line 544, the gesture is already at a stage of detaching from level 2 as the virtual interaction space that needs the inscriptive reference to be understood. Nevertheless, it is still not considered a free gesture within the gesture space (level 3), but at a transitional stage in which it becomes 'transportable' between the students.[119] The packing of information associated with the 'increasing numbers'-gesture is accomplished by both students together, mutually elaborating the social immediate mathematical object 'number enlarging in digits and numerically'. The following meaning of the associated gesture thus develops socially while performed by Lisa. Table 7.4 summarizes this beginning elaboration:

Table 7.4: Associated sign referring to 'increasing numbers' in CF7

Line	Associated Sign	Current Information Bundle	Immediate Mathematical Object
542	Gesture on level 2	- numbers grow in extent	Increasing numerator and denominator of fractions representing elements of the sequence
544	Movement on level 2/3	- numbers grow in extent - numbers get more digits	Increasing numerators and denominators of fractions representing elements of the sequence

---

[119] I thank Sotaro Kita for this pinpointing metaphor of gestures becoming *transportable* between the students in social interaction when they reach level 3 (personal communication, July 2014).

7.2 Reconstructing Meaning in Social Interaction: Associated Signs and Information Bundles

That this sign and its meaning is shared by both students becomes even more explicit shortly afterwards, when Rosa performs it as well:[120]

547 R: but- ehm ,what I actually wanted to say' ,eh [[because these numbers]
(slightly moves both hands apart horizontally twice)
[always become larger-]]
,this always becomes less significant] well ,in comparison.

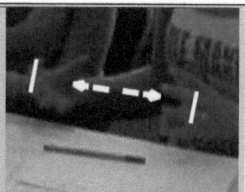

Rosa uses a gesture similar to the one performed by Lisa before in lines 542 and 544. The ± 1 "always becomes less significant" when the "numbers always become larger-". With this, Rosa may equally refer to the growing number of digits and the growing numerical value, the latter resulting from the former. Considering that she referred to the term $\frac{2a_1 \pm 1}{a_1}$ just before, the thing suggested to always becoming "less significant" may be the ± 1 in the numerator. The gesture *specifies* the *what* of the verbal utterance by recalling the information bundle associated with it and with that referring to the numerator and the denominator becoming larger. At this stage, the gesture potentially becomes a *basic sign* (see chapter 4.3.3) referring to the idea of 'numbers becoming larger'. Different to before, when Lisa's gesture specified the how of the verbal utterance, Rosa's gesture rather represents the 'increasing numbers' in a perfect match to speech. She gives visual access to the observation that 'the numbers become larger' from which the decreasing significance of the ± 1 is concluded.[121]

The students jointly negotiate the meaning of the gesture in the first excerpt: Lisa performs and Rosa proposes a possible meaning that in turn is confirmed by Lisa. The performance of the gesture within the interaction space above the table (*level 2*) allows Rosa to interpret Lisa's utterance with respect to the inscribed representation of numbers and to refine her verbal formulation "with more numbers." to "with more integers-". Confirmed by Lisa, the gesture is situationally conventionalized so that Rosa can use it to refer more generally on *level 3* (547).

7.2.3 Closing Remarks on Associated Signs

The meaning of associated signs develops within the social interaction by packing information bundles to which can be referred by using gestures. These bundles are packed over time by enriching meaning through specification by gesture and show which aspects may be stored within a sign. Their consideration can illuminate the current interpretation frame for reconstructing the immediate mathematical object, assuming that the shared

---

[120] To exclusively highlight the reference to the 'increasing numbers' it is restricted to part of the larger utterance 547.
[121] See the section on non-specifying gestures, Type II (i) on pp. 118-119.

meaning of the associated sign allows the students to shape a social immediate mathematical object. Furthermore, the aspects specified by means of gesture and stored within it reveal which aspects are considered important to represent the social immediate mathematical object. By this, it can also be traced how a social immediate mathematical object is grounded in inscriptive reference.

The examples presented in this subsection illustrate how mathematical meaning of gestures can develop, so that a shared meaning becomes established in social interaction: In example 2.1a, the curve is shifted from the concrete reference indicating points fixed within the folding sheet, to a more general idea of the 'shape of the curve'. This allows the potential shape of the curve to be partly represented as embedded on level 2 in the folding diagram, making it possible to consider new perspectives to elaborate the mathematical object in social interaction.

As we have seen in examples 2.1b and 2.1d, a sign can also be shifted on level 3 to represent a mathematical object within the gesture space. Here, the representation is reduced to its essential features. It is not dependent on the concrete representations and becomes 'transportable'. These associated gestures are closely related to basic signs (see chapter 4.3.3). In a Peircean sense, the latter ones are used as symbols while the former ones can also evoke their reference as icons that look similar to an inscription.

7.3 Representing Within the Multimodal Sign: Within-Functions as a Methodological Tool

This chapter illuminates how gestures can shape immediate mathematical objects in social interaction and how this can be traced when reconstructing the epistemic process from observing the signs used by the students. Including that and how gestures shape meaning of the object as represented by the semiotic composition, the multimodal sign triad can be visualized as follows:

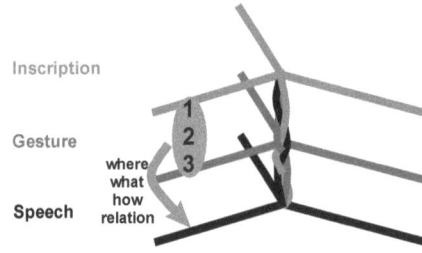

Fig. 7.14: Integration of the within-functions as influencing the shaping of the immediate object

7.3 Representing Within the Multimodal Sign: Within-Functions as a Methodological Tool

The basic constitution of this model of a multimodal sign has been described in chapter 4.4. Each triad represents one modality of the multimodal sign, distinguishing speech (red/lower triad), gesture (blue/middle triad) and inscription (green/upper triad) as semiotic resources possibly constituting a semiotic composition.[122] Each component may represent other aspects of the mathematical object. How gestures can bridge between the components is displayed in the object-dimension: They function as verbal specifiers with respect to the four aspects 'where', 'what', 'how', and 'relations' and refer to mathematical objects on three spatial levels that are more or less related to an inscriptive representation. The representation of an aspect of the immediate object is not bound to one mode of expression: Speech may seize aspects that have been specified in gesture before, or gesture may lead to the production of an inscription.[123] These considerations allow for a more detailed description of the development of the semiotic bundle.

Not necessarily made explicit, the four specification features and the levels of reference are always kept in mind to reconstruct how gestures contribute to the accomplishment of epistemic actions (see chapter 5.2.2).

7.4 Summary

This chapter describes how gesture, similar to speech, may (implicitly) transport information that is involved in shaping the immediate mathematical object. Gestures can *specify* the *where*, the *what* and the *how* of the object as represented in speech and add information about *relations*. This is done on three referential levels, on which the relationship between gesture and inscription differs: Indicating concrete inscribed entities (level 1) gesture is used as a means to refer to the meaning assigned to the inscription. As an index, it points out an aspect already represented in this inscription. The ephemeral embedding of a gesture into an inscription (level 2) prompts an interpretation against the fixed background. With this, the gesture itself represents something and is interpreted as an ephemeral component of a diagram. Gestures performed in the gesture space (level 3) are independent and detached from the concrete. The combination of these two features, *specification-function* and *referential level*, defines the *within-functions* of a gesture. While to each gesture, exactly one referential level can be assigned, a gesture can specify more than one aspect of the immediate object. In this case, it accomplishes a *multi-specifying-function*. These within-functions are supposed to provide the basis to

---

[122] The use of artifacts is analyzed depending on how they are used: The temporarily placed pen in example 2.1b, PA7-521, is rather seen as a nonverbal reference to the axis on level 2, even though it is not a gesture. However, the same placing of the pen would be considered to refer on level 1 if the verbal utterance was to the y-axis instead to the axis of reflection. The main difference hence consists of whether the artifact refers in its use to something concretely represented or to something potentially represented.
[123] This is what Arzarello refers to as "genetic conversion" (see also chapter 4.3).

understand how gestures bring to existence collateral aspects of the immediate objects and with this "create ideas" (Goldin-Meadow, 2003, p. 178) in social interaction.

Whether a gesture has the potential to specify or not can be decided following a coding guideline (see section 7.1.2.1). The interpretation of the gesture, and with that the identification of specified aspects, needs to be carried out against the mathematical background with regard to the development of the semiotic bundle. Three types of non-specifying illustrators have been distinguished within the analysis: First, those gestures whose meaning does not become clear (at that point in time). Second, redundant illustrators that depict the mathematical object in the same way as speech refers to it. And third, gestures that metaphorically represent mathematical ideas that rather have a mathematical meta-structuring function. Not having a specifying-function, these gestures still have a representational function and provide visual access to the mathematical idea. However, almost 91.4% of the gestures specified the verbal utterance, 42.8% had a multi-specifying-function.

Tracing the within-functions fulfilled by gestures with recurring features allows the reconstruction of the meaning of these gestures within the social interaction. These gestures indicate situations that are linked in catchments, that is, situations that are somehow contextually related (See also the explanation of 'catchments' given in the glossary on pages 265-266.). The immediate mathematical objects can then be reconstructed within these situations, taking into account the *information bundles* that have been *associated* with these gestures in their former use. Within this process, the meaning of gestures may become *situationally conventionalized* as shared among the participants and can be traced back to reconstruct where it is grounded in.

8 The Epistemic Process in Progress

How Gesture can Shape the Semiotic Composition to Act Epistemically

The previous chapter dealt with the *representational functions* of gestures within mathematical epistemic processes; how they can represent a mathematical object and with that, can contribute to the shaping of its actual formation within the social interaction. The gesture analysis was carried out within the semiotic bundle to have access to the meaning that may have been assigned within the interplay of the semiotic resources. This chapter will deal with the ability of gestures to be used as means to constitute knowledge, that is with their *epistemic* functions (Hoffmann & Roth, 2007).[123]

Within this study, gestures are considered to fulfill an epistemic function when they contribute to the social construction of mathematical knowledge; that is, when they influence the accomplishment of one of the three epistemic actions gathering, connecting, or structure-seeing. The epistemic-dense episodes have been reconstructed once more, this time setting a focus on the contribution of gestures to the epistemic process. The resulting findings lead to the conclusion that gestures are not only requisites in the construction of mathematical knowledge, but that they can 'play their own parts'. As we will see, gestures can not only illustrate, complement and clarify the verbal utterance, they can even actively promote epistemic action.

The first part of this chapter will present epistemic functions in two categories, the forming- and the performing-functions. While the forming-functions concern different ways in which gestures can make mathematical objects 'graspable', the performing-functions reveal how and why the use of gesture may enhance the accomplishment of epistemic actions. The second part of this chapter will then provide a detailed analysis of the use of gestures and their epistemic functions within an entire scene. This will seize the contribution of gestures by giving an idea how the epistemic functions of gestures actually merge to benefit the epistemic process.

---

[123] See chapter 3.1.3 for 'epistemic functions of signs'.

8.1 Epistemic Functions of Gestures: Detailed Descriptions of the Categories as Reconstructed within the Data

The gestures considered in this study are 'illustrators'.[124] Per definition, they are suggested to refer to the semantic content of speech. More specifically, they refer to a mathematical entity co-expressively to what is uttered verbally. Investigating *how* and *why* the illustrating gestures can enhance the epistemic process, two types of epistemic functions will be distinguished. On the one hand, there are *forming-functions* of gestures. These functions relate to how a mathematical entity is literally made *graspable* by giving visual access to it through using gesture. The performance of these gestures provides a non-verbal representation of mathematical objects that may lead to epistemic actions. On the other hand, there are *performing-functions*. These influence the epistemic process in a more direct way than only offering a visual representation to the mathematical object: A gesture fulfilling a performing-function takes active part in the accomplishment of an epistemic function. What this means concretely is presented in this chapter.

The different forming- and performing-functions will be described and represented by means of illustrative examples. For each epistemic function, two examples have been chosen to show how diverse their accomplishment may be in concrete cases.

8.1.1 Forming-Functions: How Gestures Can Provide Visual Access to a Mathematical Object that is Involved in an Epistemic Action

Three kinds of representing mathematical objects through using gestures have been identified to be beneficial for the epistemic process.

8.1.1.1 Sourcing Out

Sourcing out a reference means to link the *verbal* reference to a *visual* reference by means of gesture. The gestural reference can be clearly assigned to a lexical affiliate that is to a mathematical entity mentioned in speech. The following excerpt illustrates how sourcing out is accomplished to trace the 'line of speech' by indicating corresponding components fixed as entities within a diagram.

Excerpt 3.1a, PA5-494-500 (PA5e7.3, 0:37:19 – 0:37:58)
Working out the parabola-task, Mike and Tim aim to justify their conjecture about the distance between the points A and B being equal to the distance between the points A and P ($\overline{AB} = \overline{AP}$). Mike has just stated the general rule that each point on the tangent line has same distance to B and to P (episode PA5e7.2, 475-479). Tim proposes to check this generality by looking at the situation as fixed in the printout diagram (Fig. 8.1).

---

[124] See chapter 3.2.1.

8.1 Epistemic Functions of Gestures: Detailed Descriptions

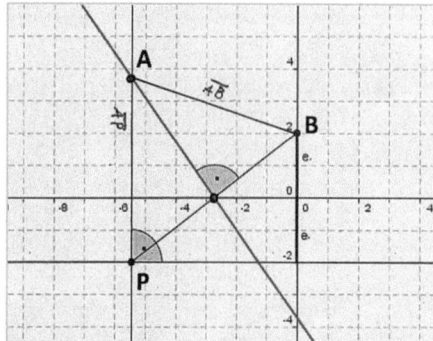

Fig. 8.1: Printout diagram in the beginning of scene PA5e7.3, line 494 (tags of points enlarged)

He identifies the tangent line by tracing it from top left to bottom right in the printout diagram and highlights it by redrawing it red so that "one can distinguish it better" (first part of line 494). This outsourcing by identification and subsequent redrawing is accomplished in the following for the points *B*, *P* and *A*, first by using gesture, then by producing inscription:

494 T: and they are equally far away from [P'] (*points at point P*)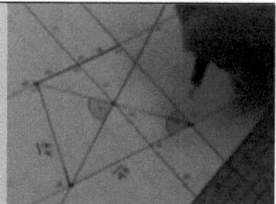

 **495** /M: well (*points at point P*) [ ] ,yes exactly that one now it is correct

496 /T: from P ,and B. (*thickens up the dots for points P and B, points at the connecting line*)
(0:37:40)
(4s) [that is already very clear because of <u>that.</u>] (*points at the intersection of the connecting line and the red perpendicular*)

164　　　　　　　　　　　　　　　　　　　　　　　8 The Epistemic Process in Progress

(0:37:42)
497 M: these are simply mirrored' and the point A <u>is [lo-cated]</u> (*points at point A*) on the tangent. ,and then they should move away from each other by the <u>same distance</u>. (.) uh have the same distance from each other.

(0:37:50)
498 /T: yes exactly. that means if I draw in any point now' (*Mike traces a part of the red line*)

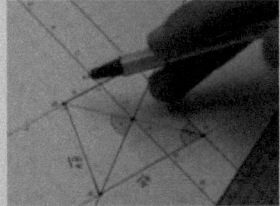 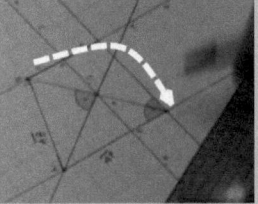

[B and P] (*points at points B and P*) are equally far away there and is

(0:37:55)
499 /M: yes [one that lays (*points at the red line at the y-axis*) on the tangent] of course

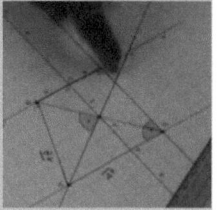

500 T: yes (.) (*draws in a point on the red line at y=-5*) here ,let's take this one'

In lines 494/496, Tim continues Mike's idea stated in lines 475 to 479 and refers to the points on the tangent line ("they") having equal distance to points P and to B. The pointing in line 494 **specifies** *the location* of point P within the printout (**level 1**). By this, Tim *sources out* the verbal reference to the point P to the concretely visual shared interaction space. He makes this reference visible for himself as well as for Mike and links a representation to the point **gathered** verbally. Mike repeats this identification in gesture by also pointing to point P before he confirms in speech as well (495: ",yes exactly that one now it is correct"). In line 497, Mike **specifies where** point A becomes visible in the printout (**level 1**) by pointing at it and with that *sources out* its location on the tangent line from speech ("the point A is [located] on the tangent") to also become visible in the printout diagram. The **representational gathering** of the entity through gestural indication illustrates one part of the **connection** between 'point A is a point on the tangent line' and 'points B and P have the same distance to this point'. In line 498, Tim detaches from the concrete point A and instead comes back to a general conjecture stated by Mike before; that 'all points on the tangent line have equal distance to points B and P' (477).

8.1 Epistemic Functions of Gestures: Detailed Descriptions

Referring to "any point" (498) instead of the concrete point A, he connects back to this structure. He consecutively indicates the *locations* of points B and P in one movement and with that, *sources out* a reference to them once more, making it not only audibly but also visually accessible. Mike considers this "any point" to be too imprecise and ***specifies*** a *location of a possible example* (*level 2*) in speech and in gesture in 499: He adds the feature "that lays on the tangent" (499) and sources out the reference by indicating the tangent in the lower right part, close to the y-axis. This extends the **connecting action** as started by Tim in 498 by refining a condition that needs to hold: The "any point" mentioned by Tim (498) needs to be a point that "lays on the tangent" (499).

In this example, gesture sources out entities that are mentioned in speech to also become visible within the printout. Through this, the epistemic actions **gathering** (494, 499) and **connecting** (497, 498/499) are made graspable by providing a concrete reference to the mathematical entities gathered or connected. The *specification of concrete and possible locations* is a remarkable feature becoming visible in this case of sourcing out in gesture. This prompts the following hypothesis about the benefit of outsourcing regarding the epistemic process:

Sourcing out helps to adjust common knowledge about where mathematical entities are, or can be, located within a diagram. Gathering is not only accomplished in speech, but simultaneously in gesture as well. The non-verbal gathering can **support** *the verbal* **gathering** *action by giving visual access to representations of the gathered entities. Furthermore, a gesture can also support a connecting action through indicating the connected entities as represented on level 1. The verbal* **connecting** *action is then* **supported** *by a non-verbal-**gathering** action* **realized** *through gesture.*

However, sourcing out the representation of mathematical entities is not restricted to the specification of locations within already existing, fixed diagrams. As we will see in the following excerpt, sourcing out does not need to be indexical. The gesture itself can represent the mathematical object that is sourced out 'to the eyes'. The next excerpt is part of example 2.1b, which has been introduced in chapter 7.2.1 as an illustration of the concept of associated signs, related to the representational function of gestures. This time, how this representation sourced out into the gesture space can benefit the epistemic process will be focused on.

Excerpt 3.1b, PA7-528-537(PA7e6.4, 0:45:25 – 0:45:45)

Lisa and Rosa are working on subtask 4 of the parabola task. In order to justify their conjecture that the curve represents a parabola, they search for properties characterizing a parabola. The idea behind that is to find a property that can be proven for the curve displayed in the representations worked on. Rosa proposes to consider the style of the slope of the curve as a characterizing property. By sourcing out the axis of reflection and the shape of the parabola into the gesture space, she visually accompanies the resumptive gathering of the seen symmetry of the curve as secure knowledge and provides the representation of relations within the gesture space. With this, Rosa sets the stage for using situationally conventionalized signs to illustrate an idea on level 3, aiming to complement insufficient verbal expression:

528. R: well ‚we have [just
(takes up her hands and lays the palms of her hands against each other to chest height, the fingertips pointing up and forward)
‚said' ‚because of

[[this axis of reflection]
(moves the right hand to the right, then indicates the shape of a curve by moving both indexes up and down)

[both sides are the same.]]

529. L: mmh'
530. R: which is already clearly indicative <u>for</u> *(repeats the movement of the indexes from 528 once)* [a parabola.]
531. L Mhm'

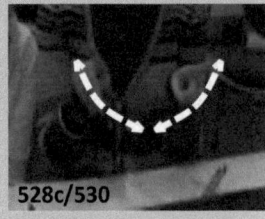

(0:45:33)
532. R: Means we now] *(takes down the hands)*

[only have to'
(multiply makes a curvy gesture from her lower left to her upper right with her right finger)
prove that it is this ‚right' *(looks at Lisa, repeats the gesture)*

533. L: *(moves point P to the right in GeoGebra)* that it is'*(looks at Rosa)*]
(0:45:38)
534. R: *(repeats the gesture three more times)* [uhm] [this-]

8.1 Epistemic Functions of Gestures: Detailed Descriptions   167

(takes the hands up to chest height, moves the palm of her hand from the right to the left)

 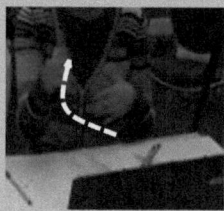

[„more and more upwards but not so much any more- *(makes a gesture similar to the one in 532 but with her hand bent a bit)* ‚well how can one describe]] that- *(looks at the screen)* (..)
535. L: *(briefly leans backwards, smiles)* well-
536. R: *(looks at Lisa)* you know what I mean' roughly'
537. /L: yes.

In lines 528 and 530, the **resumptive gathering** recalls the already made connection between the symmetry of the curve and the axis of reflection.[125] Sourcing out the components 'axis of symmetry' (528b) and 'two sides of the parabola' (528b) also prompts the *specification* of the *style* of the shape. This makes it possible for Rosa to make a special remark on this aspect in the following. In 532, she only shapes the right hand side of the curve. The symmetrical *relation* between the two sides is not mathematical knowledge in the focus but was needed to be recalled to reason the restriction on one side at this point in time. Both sides are the same and have a certain shape, and this shape is *sourced out* by a gesture to become visually accessible as a meaningful aspect of the entity. In line 532, Rosa uses gesture to overcome the insufficient verbal formulation "that it is this ‚right'": She modifies the former outsourcing of one side of the curve to compensate speech and suggests "this" to refer to the specific slope. The concrete performance of the gesture is different to the shaping of each side in 528c/530: The movement of the gesture always starts below, where the imagined axis of reflection has been located within the gesture space before. This does not merely refer to the shape of the curve, but also to a suggested development, starting from the axis and directed to the outside. Lisa's illocutionary *request to specify* 'what is referred to' (533) makes explicit that the reference to the changing slope has not become visible. In turn, Rosa refines her idea in speech and in gesture: In the verbal utterance, she further describes a development ("‚more and more upwards", 534) and co-timed to this, the curve is shaped by a slower movement of the hand, tracing this development. Simultaneous to speech, she *sources out* the development of the slope of the curve and **supports the connecting action** between the shape of the curve and the development of the slope into the gesture

---

[125] See chapter 7.2.1 (example 2.1b) for the information bundle associated to this gesture.

space, performed on *level 3*. Mathematically, her idea concerns the change of the instantaneous rate of change; a relation between relations. Representing this idea in the ephemeral mode of gesture seems to be a hard task for Rosa. However, the gesture fulfills its function to give Lisa an idea about what Rosa is not able to express in words either so that she 'roughly knows what Rosa means' (536/537)

In this excerpt, sourcing out the reference **supports** the **resumptive gathering** in line 528 as well as the **connecting action** in lines 532/534 by illustrating its subjects. Furthermore, sourcing out representations of the gathered entities (528/530) **prepares** the upcoming **connecting action** by recalling and specifying the aspects that are needed for the illustration of the connecting action. This allows Rosa to reduce the gestural representation to what is considered as important: the development of one side of the curve.

Possible Limits of Outsourcing become visible in the representation of the changing slope. This idea concerning a relation between relations may demand for too many aspects to be represented at the same time. Gesture, as an ephemeral means of expression, may not be useful in this case.

The analysis of this example allows for a modification of the hypothesis concerning the beneficial effect of sourcing out the reference as has been proposed before. Respecting the use of the gesture space and gestures contribution to the epistemic process, it is stated as follows:[126]

Sourcing out helps to adjust common knowledge about meaningful aspects of mathematical entities by specifying them. This may provide a common ground for the collaborative elaboration of ideas. Together with speech, it gives a more complete picture of a mathematical idea. Both modes of expressions are spontaneous and ephemeral and their combination allows several 'dimensions of meaning' to be grasped simultaneously. **Representational gathering** *can* **support gathering** *by giving visual access or can* **realize gathering** *itself. This may* **support connecting** *actions by visualizing the connected entities, but it may also* **prepare connecting** *actions by setting the ground for interpreting gestures that clarify imprecise speech.*

However, the analysis of the latter scene also shows that outsourcing into the gesture space has its limits when too many components are represented at the same time.

8.1.1.2 Depicting

In the latter excerpt (3.1b), the reference to "both sides" (528) of the curve was sourced out into the gesture space by *depicting* its shape. In this special case, depicting is one kind of sourcing out but it may also stand for itself. In general, a gesture is considered to have a depictive function when it establishes an ephemeral 'body diagram', that is when it represents an icon of relations as embodied in gesture (see footnote 113 on pages

---

[126] Compared to the hypothesis gained from the first example, extensions are highlighted with bold letters.

138-139). As will be seen, this body diagram may be performed on level 3 in the gesture space or on level 2 as embedded into a fixed diagram. Its main feature consists in shaping relations. These may concern graphical as well as notational relations:

Except 3.1c, PA5-629 (PA5E8, 0:48:32)

This excerpt has already been considered in chapter 7.1.3 (excerpt 1.3c) as an example for a specification on level 3. Mike and Tim work on subtask 4b of the parabola task and search for arguments that may convince somebody that the curve is "a function with even-numbered exponent", as proposed in subtask 4a. Their strategy is to eliminate other types of functions as possibilities to describe the curve (622, T: "yes. and we can do that in two possibilities. either we exclude all others' (.) or."). Just before the start of this episode, the interviewer prompts at the invalidity of this method (I: "how do you take all others in that case"…" all other kinds of functions" (623/626)). Tim concludes "right ‚so we theoretically have to prove that it is exactly that one." (627) and Mike confirms non-verbally by nodding his head. Even though this reveals that they understood that it is not sufficient to exclude functions to prove their conjecture, Tim recalls the symmetry as argument why the function cannot have an uneven exponent:

---

629 T: well I know this for <u>sure</u> like this ‚because anyway with an <u>uneven</u> exponent. ‚uhm it is that way in any case ‚when we now have an x to the power of <u>three</u> or x to the power of <u>five</u> ‚there is at least uhm a <u>saddle</u> point. (*looks at Mike*) (...)

and <u>then</u> there would ‚it would look (*draws a parabola in the air*) [ ]

‚the whole thing wouldn't be <u>axially</u> symmetric but (*looks at Mike*) <u>point</u> symmetric. (...)

Tim starts a new approach by stepping back to secure knowledge ("well I know this for sure"). This knowledge concerns the **connection** between the symmetry of the curve and the properties of functions with an uneven exponent (",there is at least uhm a saddle point.", "point symmetric."). The main argument is *recalled* by depicting the parabola and with that its symmetry as has been associated to the shape before (See excerpts 1.1a, 1.1b, 1.1c in chapter 7.1.1 and example 2.1a in chapter 7.2.1.). Tim struggles with verbal expression, not being secure whether to refer to the missing saddle point or to the shape of the curve as a whole ("and then there would ‚it would look") to point out the difference. The depiction of the shape *unburdens speech* and *specifies what* is referred to as "the whole thing" in the gesture space. It releases him from the pressure of finding a suitable word such that *it becomes easier to express the connection* between the symmetry and the function equation as an excluding argument. This **supports** the verbal **connecting** action.

This excerpt shows how an epistemic action is benefitted by the depiction of a mathematical object in two ways: The gestural depiction terminates the search for words by offering a non-verbal reference to the mathematical object without making precise on which aspect Tim focuses. That is, it *simplifies the expression* by being used to represent meaning that has been associated with the gesture within the epistemic process. This in turn may *recall* aspects that give visual access to the idea expressed in speech (the symmetry of the curve, in this case).

*Depicting gestures can represent a mathematical object by a (situational) convention established in the epistemic process. This can ease expression by substituting verbal reference and may recall ideas that are related to the mathematical object. With that, the depiction can **support** a **connecting** action.*

In the following excerpt, the relations within notation are depicted on level 2. This reveals a more general idea than is expressed in speech.

Excerpt 3.1d, PA7-667-671 (PA7e8, 0:56:21 – 0:56:48)

Lisa and Rosa work on subtask 4b of the parabola task. Their current aim is to determine the function equation in order to prove the curve to represent a parabola. The students have gathered and written down x- and y-values of points laying on the curve from the GeoGebra environment, first for $e = 1$, then for $e = 2$ and finally for $e = 0.5$ (See Fig. 8.2 and note sheet 1 attached in the appendix on page 324.[127]).

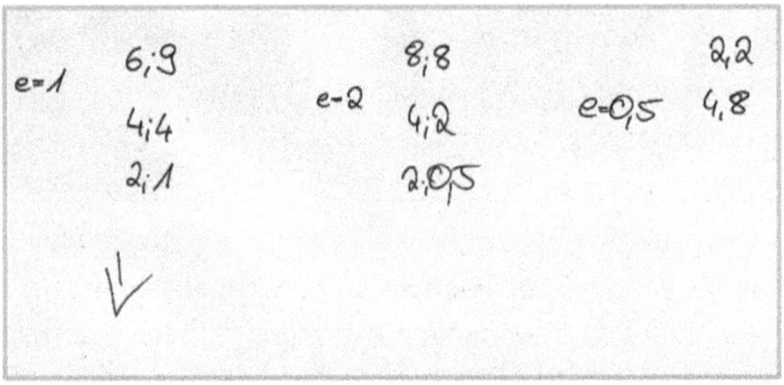

Fig. 8.2: List of x- and y-coordinates of points on the parabola, gathered for different values for e (PA7e8-667)

Comparing the values for $e = 1$, they recognize that the y-values "are square numbers" (649). They transfer this regularity to the case of $e = 2$, where they find: y-values "are

---

[127] In the students' inscriptions represented in the appendix it can be seen that as a conclusion from these lists, they find the formulas $\left(\frac{x}{2}\right)^2 = f(x)$ and $2\left(\frac{x}{4}\right)^2 = f(x)$ for $e = 1$ and $e = 2$, respectively. From these two concrete examples they empirically conclude $e\left(\frac{x}{2e}\right)^2 = f(x)$ as the general function equation describing the parabola.

twice the square numbers." (657) and similarly for the case $e = 0.5$: "that are zero point five times some square numbers", 666). Having realized this regularity of 'the y-value is some factor times a square number', Rosa integrates the x-value and proposes a more general way to relate the y-value to it:[128]

667 R: *(takes the cap off the pen and briefly points at the first list of numbers with it)* [(.)] (.)
I once need this one here *(draws an arrow from the list downwards, Lisa puts the first print in front of herself and looks at it)* (.)
(first points at "2,1", then on a location further down, moves the hand on an upwards-arc to the right, finally points to the "1" in "2,1". The gesture phrases accompany the following verbal expression)

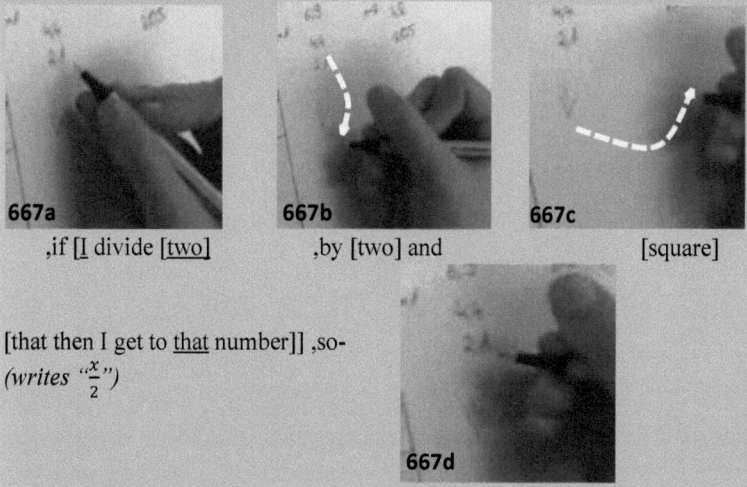

| 667a | 667b | 667c |
,if [I divide [two]   ,by [two] and   [square]

[that then I get to <u>that</u> number]] ,so-
(writes 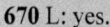*")*

668 L: if you divide <u>two</u> by two' (..) *(bends forward towards the note sheet)* ,if you divide <u>what</u> by two'

(0:56:36)
669 R: if I' *(puts the cap on the pen)* (.) divide the [,x-value here *(points at the left "4" in "4,4")* by ,two-]

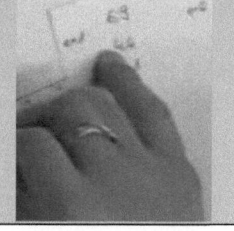

670 L: yes.

---

[128] Note that in this analysis it is focused exclusively on the depicting-function. To avoid confusion, pictures of the gestures that source out but do not depict are omitted. For better comprehension of the movements, the pictures are turned -90° such that they correspond with Rosa's perspective on the sheet.

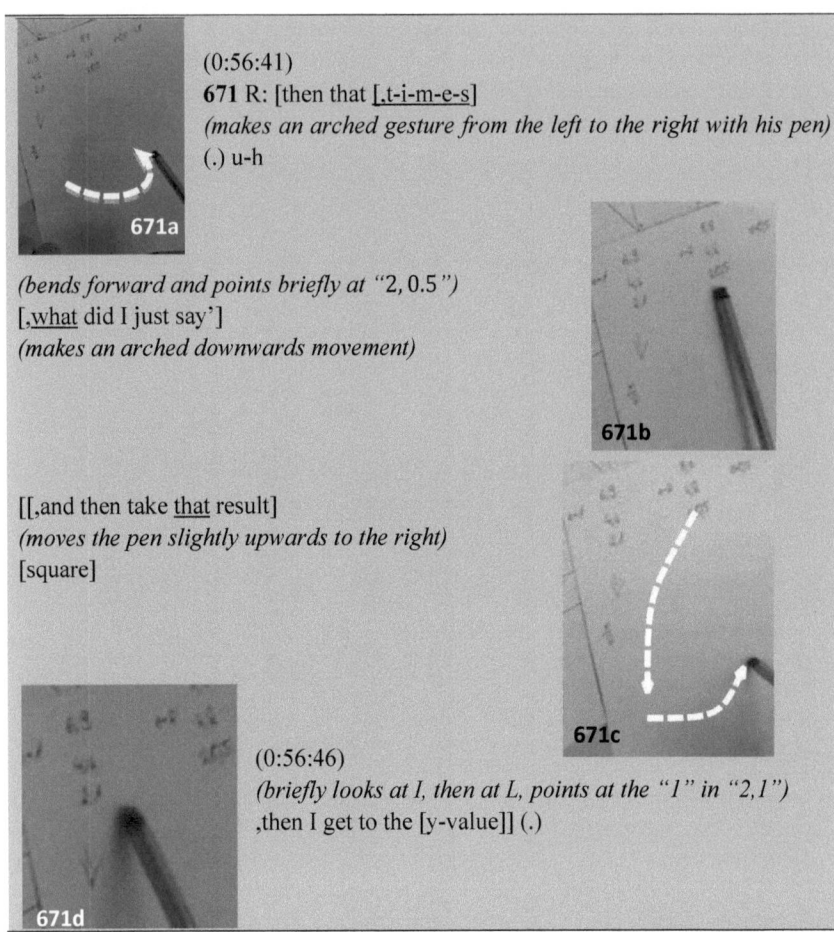

(0:56:41)
671 R: [then that [.t-i-m-e-s]
(makes an arched gesture from the left to the right with his pen)
(.) u-h

(bends forward and points briefly at "2, 0.5")
[.what did I just say']
(makes an arched downwards movement)

[[,and then take that result]
(moves the pen slightly upwards to the right)
[square]

(0:56:46)
(briefly looks at I, then at L, points at the "1" in "2,1")
,then I get to the [y-value]] (.)

In speech, Rosa describes a calculation with which the concrete y-value "1" from its corresponding x-value "2" can be determined. The regularity therein, dividing by two and squaring the result is already revealed in gesture: It depicts the relations of a **general formula** by referring to 'dividing by' (667b) and 'squaring' (667c), while ignoring the concrete values: The downwards-movement (667b, "if [I divide [two] ,by [two]") depicts the spatial relation of numerator and denominator (in fractions) and with this refers to the notation of a fraction as an operational symbol for division. The arc-shaped upwards movement co-timed to "square" (667c) metaphorically depicts squaring by referring to the position of the exponent. The gesture shifts from the concrete reference on *level 1* to *specifying relations* within a non-existent, but hypothetically fixed formula on *level 2*. With this, the top-bottom relation of the division as fraction is depicted as well as the position of the exponent 2 for squaring. This is seen as the Growth Point of the general

8.1 Epistemic Functions of Gestures: Detailed Descriptions 173

formula for the case $e = 1$ that Rosa starts to immediately write down. She is interrupted by Lisa claiming for explication (668): In her utterance "if you divide two by two' (..) ‚if you divide <u>what</u> by two' " she expresses confusion about whether the reference shall be concrete or general. Answering this claim, Rosa repeats her rule, this time generally described in speech as a rule to get from an x-value to a y-value ("divide the ‚x-value by two" "and then take that result square", 669/671). In a first approach, she confused multiplication with squaring: The verbal "t-i-m-e-s" is co-timed to a gesture similar to the one that accompanied "square" in the depiction of the idea expressed in 667c (upwards arc referring to the position of the exponent, 671a). Her hesitation in speech (667) signals insecurity while perceiving the co-timed verbal "t-i-m-e-s" and the square-gesture as a mismatch in this situation. For Rosa, meaning was assigned to her 'squaring'-gesture by using it in 667c and this meaning is recalled by using it once more in 671a. To correct her mismatch, she reformulates her idea, this time co-timing the squaring-upwards-arc to the verbal "square" (671c). Moreover, she recalls what is squared (the result of division) exclusively in gesture: The symbolic-notational reference to numerator and denominator is depicted as in 667b, and recalls what the verbal "that result" refers to; the result gotten from division. The gesture unit is closed with a concrete reference to a y-value in the list for $e = 1$ (671d). The same y-value has been considered in the first exemplification in 667a, the starting point of her idea. Her renewed indication in 671d refers back to this starting point.

The depiction of the general relations within a term expresses a more general understanding of the idea than speech does. Rosa reduces the references in gesture to "dividing" and "squaring", the core points of the general idea. This **prepares** and **realizes structure seeing** by visually **connecting** the components in the movements of the gesture and establishes an iconic-symbolic reference to the structure of the term in a body-diagram on level 2. This benefits the verbal formulation of the structure by storing the significant aspects and recalling them when the gesture is performed.

The abovementioned hypothesis can thus be extended to:

*Depicting gestures can represent and specify mathematical objects **and relations** by a (situational) convention established within the epistemic process. This can ease expression by substituting verbal reference by gesture and may recall ideas that are **stored in the gesture**. This enables depicting-gestures to support **and realize** connecting, **as well as preparing and realizing structure-seeing actions**.*

However, this statement on the benefit of depicting ideas also reveals the necessary condition that is required:

To provide an epistemic function through depicting, a gesture needs to be (situationally) conventionalized so that it can represent a mathematical object or relation.

8.1.1.3 Extracting

Gesture can also single out one aspect of what is represented within a larger diagram, *extracting* it while leaving aside other aspects. Its main feature consists in limiting to one aspect or entity and emphasizing it as core reference considered at this moment.

To illustrate this rather theoretic description, two examples are picked from the data.

Excerpt 3.1e, PA6-268-270 (PA6e7, 0:26:29 – 0:27:09)

In this first example, Kris and Simon work on subtask 3a of the parabola-task, checking whether the distance between points A and B equals the distance between points A and P. They have just expressed this idea, based on comparing and measuring these distances as represented in the printout diagram (263). Varying the situation by moving point P in the GeoGebra-environment makes Simon doubt the equality. These doubts lead to concentrating on the comparison of segments, visually benefitted by extracting them from the diagram (See Fig. 8.3 for the diagram as visible in the beginning of 268.):

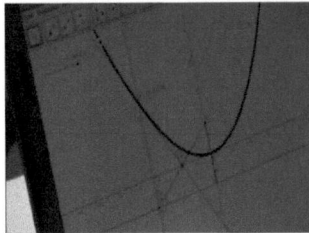

Fig. 8.3:   Geogebra-diagram as visible on the screen in the beginning of PA6e7-268

268 /S: [(incomprehensible) square' (..)] *(points at the screen with the left hand, thumb at point B, other fingers at point A)* ,here	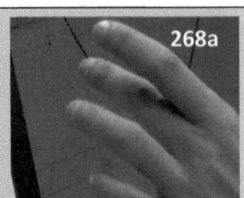
(points once more at points A and B in the same way, moves point P a bit and accompanies the increasing distance between points A and B with the hand) [this distance of course becomes a <u>lot</u> larger and]-	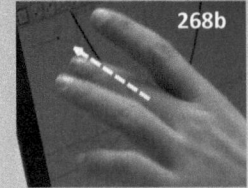

I <u>don't</u> think *(takes back the hand and moves point P to and fro at the left side)* that this one becomes a lot larger. *(adjusts the zoom on the situation such that only the part above the x-axis is visible, the trace disappears, Simon tries to click on point P)* (12sec)

(0:26:50)
269 K: ah okay now you have to again- *(points towards the screen)* [this.] *(Simon moves point P a bit to and fro at the left side and produces a trace)* (7sec)
[*(Simon successively indicates [AB] and [AP] with thumb and middle finger)*

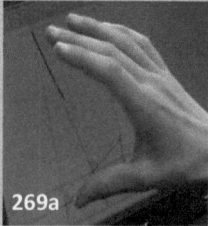
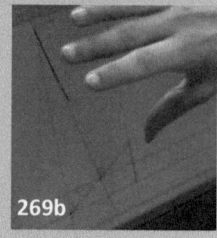
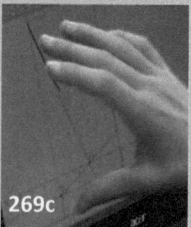

‚y-e-s right that is a-b-o-u-t right. ‚ehm-]
(Simon laughs),can we (incomprehensible) this

(0:27:05)
270 /S: ah yes right
[[that is
(points at the screen first at the angle between points B and P, than traces the triangle APB)
here yes]
a right angle that is ‚ehm]] just like a <u>triangle</u>.
271 K: ah <u>yes</u>.

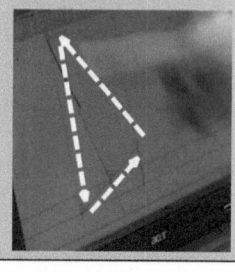

The variation of the situation adjusted in the GeoGebra-environment makes Simon doubt the equality of lengths of the segments $[AB]$ and $[AP]$. Not refusing the proposed equality concerning the specific situation as displayed in the printout, he argues that the distances do not increase equally (268: "this distance of course becomes a <u>lot</u> larger and]- I <u>don't</u> think that this one becomes a lot larger."). Simultaneously, he suggests the increasing distance visually (268b) and by that, introduces the 'distance-gesture' shaped by the thumb and index finger. Simon checks his conjecture on the equal distances by adjusting another situation in the GeoGebra environment in which point P is more distant from the x-axis than in the situation displayed in the printout diagram. Using the 'index-gesture' again, Simon successively indicates both segments in the diagram on the screen (269a-c) and ***specifies*** their ***locations*** on ***level 2***. These static gestures indicate each segment as one entity limited by two points and singles them out as entities of importance. The successive indication of both segments accomplishes a connecting action of comparing them. Kris interprets this as visualization of equality of the distances and agrees (269). Simon extends the idea to not only connecting two points, but all three of them so that they form a triangle (270). He traces this triangle on the screen and through this, the entity 'triangle APB' becomes available as a representation within the diagram, **extracted as connection** of points. Kris again confirms understanding (271) from which can be concluded that the triangle APB has become a visible entity extracted from the diagram for him as well.

The extraction of entities by the use of gestures influences the epistemic processes at several points in this excerpt: First, the distance between points B and A is marked as one entity within the diagram (268). By this, the connection of the two entities is condensed in one gesture that situationally conventionalized refers to the distance as new single entity. Second, a similar gesture is used to successively extract the two segments $[AB]$ and $[AP]$, each as one entity (269a-c). Extracting them as parts of the diagram, represented on referential *level 2*, eases the comparison of these two entities by zooming in on them and fading the coordinate system into the background. Rather than considering the concrete distances, their comparison becomes important as a connection. Third, the triangle APB is extracted as one entity in line 270. This **representational connection** of points establishes the existence of a new entity within the diagram: the triangle.

Concerning the epistemic benefit provided by extracting gestures can thus be stated:

Extracting can single out one entity as represented within a larger diagram. This zooms in on this entity while fading down the rest of the diagram. Concerning the epistemic process, extracting gestures can give visual access to a **connection** *and by this realizes a* **connecting** *action. Furthermore, the gesture can extract parts of the diagram to form the representation of a new entity so that its representation becomes shared knowledge.*

This hypothesis suggests that extracting presupposes a (geometric) shape that is extracted. The next excerpt illustrates that a *relation* can be extracted as well. It is taken from Rosa and Lisa's elaboration of the arithmetic-analytical task that deals with the continued fraction $1 + \frac{2}{1+\frac{2}{...}}$.

Excerpt 3.1f, CF7-299-301 (CF7e9.2, 0:23:59 – 0:24:15)

Lisa and Rosa started working on subtask 2.1, which requires the insertion of the first twenty elements of the sequence defined by the continued fraction in a table, both as a fraction and as a decimal number (See page 331 for the complete table filled out by Rosa and Lisa.).

8.1 Epistemic Functions of Gestures: Detailed Descriptions

$f(3)$	$\frac{11}{5}$	2,2
$f(4)$	$\frac{21}{11}$	1.909090909
$f(5)$	$\frac{43}{21}$	2.047619048
$f(6)$	$\frac{85}{43}$	

Fig. 8.4: Cutout of the table in subtask 2.1 as filled out by the students in CF7e9-296

Lisa, who fills in the values, sees the rule of the numerator being the denominator of the next fraction after inscribing the fraction for $f(6)$: "just say ‚something keeps appearing to me-" (296) (see Fig. 8.4). She explicates her observation right after:

299 L: [[one always has this one]

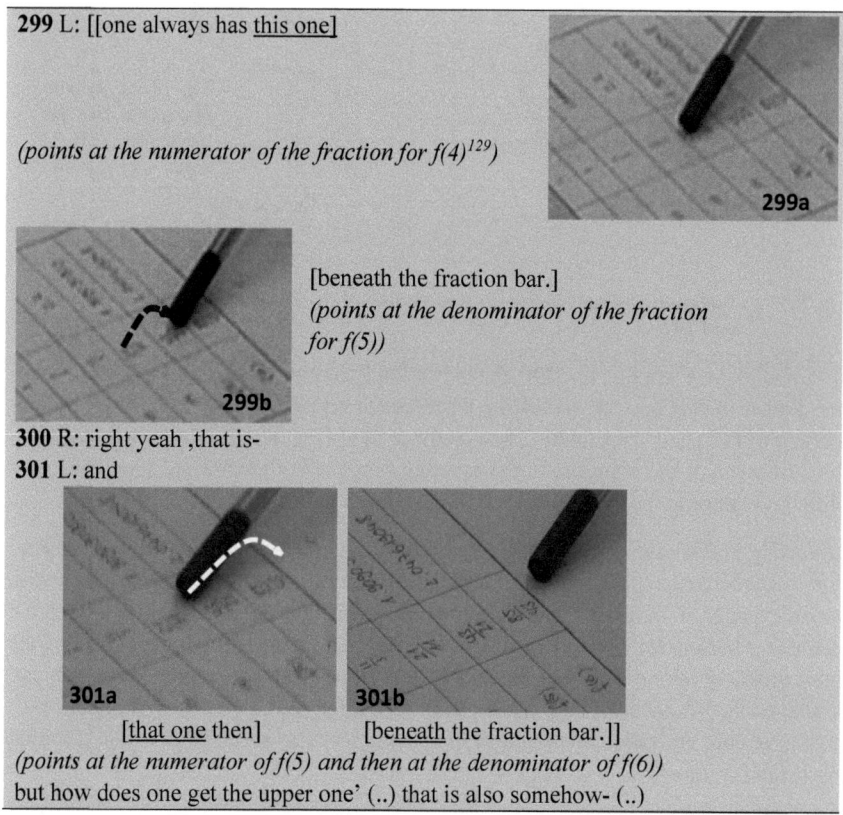

(points at the numerator of the fraction for f(4))[129]

299a

[beneath the fraction bar.]
(points at the denominator of the fraction for f(5))

299b

300 R: right yeah ‚that is-
301 L: and

301a   **301b**
[that one then]   [beneath the fraction bar.]]
(points at the numerator of f(5) and then at the denominator of f(6))
but how does one get the upper one' (..) that is also somehow- (..)

---

[129] See Fig. 8.5a and 8.1b to see what is pointed at exactly.

Lisa offers the **structure** of '*the numerator of an element always equals the denominator of the next element*'. The information 'of the next element' is only given by the indication in gesture that concretizes the structure by *specification where*. Her pointing from the numerator of $f(4)$ to the denominator of $f(5)$ in line 299 is precise (Fig. 8.5a). This clarifies that she refers to the numerator as "this one" and *specifies* the *where* of "beneath the fraction bar" as 'the denominator of the following fraction'. Speech suggests the validity for this observation as a general rule (299: "always"), while gesture indicates its single aspects by means of the concrete example ('numerator of $f(4)$ – denominator of $f(5)$'). Although Rosa expresses understanding (300: "right yeah"), Lisa repeats the connection of equality between numerator of an element and denominator of the next one (301).

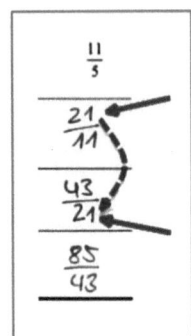
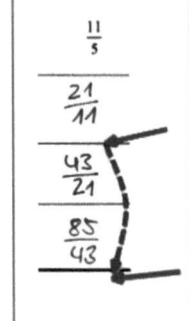

Fig. 8.5a: Pointing gesture in line 299: "one always has this one] [beneath the fraction bar.]"

Fig. 8.5b: Pointing gesture in line 301: "[that one then] [beneath the fraction bar.]]"

Fig. 8.5: Location and movement (dotted arrow) of the pointing gestures in lines 299 and 301

Different to the gesture in line 299, the indication is accomplished in a less precise manner. Rather than concretely indicating what speech refers to, Lisa points above the numerator of $f(5)$ and beneath the denominator of $f(6)$ (Fig. 8.5b). With this she picks up the movement of the reference to the concrete example in line 299 and suggests a simplified reference to the relation 'top of first – bottom of next'.

Extracting in this case is linked to a *shift of the referential level*: In line 299, the indication is a concrete one, accomplished on **level 1** to *specify* the reference of the imprecise verbal expression. In line 301, the reference is not concrete anymore but still interpreted against the background of the inscription. It detaches from the concrete by extracting the *main aspect of arc-shaped movement*: The gesture suggests a reference of the general structure that is reduced to the essential idea of relating 'top of first' to 'bottom of next'. Rosa's gesture embodies this *relation* in a catchment and *specifies* it within a virtual inscription on **level 2**.

A comparison of excerpts 3.1e and 3.1f shows the main differences of the two extracting gestures which both accomplish a connecting action. First and foremost, the connecting

8.1 Epistemic Functions of Gestures: Detailed Descriptions

actions differ from an epistemic point of view in that they do not have the same status within the epistemic process: The first example concerns a connecting action *before* a general structure is mentioned[130] and is thus directed *forward* towards structure-seeing. The extraction in the second example appears while *extending* a seen structure; the connecting action is directed *backwards* with respect to this structure. This difference is marked in bold for excerpt 3.1f in Table 8.1, as well as two other distinctive differences:

The form of *semiotic connection* (physical form that becomes visible as connection) differs between the two excerpts. In excerpt 3.1e, the gestures refer to the immediate mathematical object *in similarity*, being reminiscent of what is commonly conceptualized as *static geometric form* of 'segment' or 'triangle'. In excerpt 3.1f, the immediate mathematical object as a relation cannot be referred to in an iconic manner. The reference thus is a *metaphoric* one: The arc-shaped *dynamic movement* embodies the relation as a feature extracted from the preceding concrete pointing.

Table 8.1: Comparison of extracting-gestures in both excerpts

	Excerpt 3.1e	Excerpt 3.1f
References that are Extracted	Segments [AB] and [AP]; triangle APB	Relation between numerator and denominator
Semiotic Connections	Connection between concrete points	Connection between entities in a concrete example
Forms of Semiotic Connections	Segment/triangle as geometric entity	**Arc-shaped movement as simplified reference to a relation**
Resulting Idealized Gesture Associated to the Semiotic Connection	Iconic static representation	Metaphoric dynamic representation
Connection as Epistemic Action	Pre-structure status (accumulating)	Post-structure status (elaborating)

Considering these differences, the hypothesis for excerpt 3.1e is modified as follows:

Extracting can single out one aspect from a more complex diagram. This zooms in on this aspect while fading down the rest of the diagram. Furthermore, the gesture can extract parts of the diagram to form the representation of a new entity. Concerning the epistemic process, extracting-gestures have two main benefits:

(1) They make it possible to capture a newly considered relation as a new entity by extracting the essential aspects of this relation in a gestural sign.
(2) They make it possible to represent an accumulating as well as an elaborating connection action as embodied in gesture.

---

[130] The conjecture of general equality of the two segments, $\overline{AB} = \overline{AP}$, is first made explicit shortly after the end of this excerpt in PA6e7-274.

The first three forming-functions are closely related to inscriptions: This relationship either concerns the direct reference to inscriptions or the representations of aspects lifted out from inscription, for example as associated sign. While these gestures are iconic-symbolic (see chapter 3.2.3), metaphoric gestures can also have a forming-function, as shown in the following excerpt.

8.1.1.4 Illustrating Generality

Gestures can also express an idea of generality by referring in a metaphorical way to concepts like infinity or continuity, often in the context of processes. Not having any relation to inscription, they are free by definition and reveal a more general view on ideas that may have a concrete reference in the verbal utterance.

Excerpt 3.1g, IN7-558-561 (IN7e8.1, 0:57:36)

Lisa and Rosa have just started to work on the extra task related to the task dealing with logical reasoning. They have to examine whether the strategy they found to decide when they have to signal still works without the condition of at least one marked hat existing. Rosa expresses her considerations on this specific situation:

558. L: ‚if one sees n-o-n-e
559. R: *(moves the right index finger in steps from the right to the left on the edge of the table)*[131] one can not be sure that one has one. ‚that means

[then one can in *(shakes her head, rotates both arms)* ‚there one can also tell nothing from ‚nothing from

fingers turning around each other

560. /L: *(at the same time)* right. yes.
561. /R: the reactions of the others. *(looks at the Interviewer)* ‚I think.]

Lisa starts considering the basic case, seeing no marked hat (558). In turn, Rosa concludes that one does not have secure knowledge on their own status anymore ("one can not be sure that one has one.", 559). Following this, she mentions that "‚there one can also tell nothing from ‚nothing from the reactions of the others." (559/561). While her verbal "‚there" indicates the reference to the concrete case (including the changed condition) of seeing no marked hat, the gesture refers to the general process constituted of induction steps. Both are **connected** by the expression "one can tell nothing from ‚nothing from the reactions of the others.", indicating that the basic case is central for starting an induction. The gesture *illustrates the generality* within the uttered idea and, with that negates the general validity of the strategy.

---

[131] This is considered being a beat-like gesture, accompanying the prosody of speech.

8.1.1.5 Summarizing Overview on the Forming-Functions of Gestures

Forming-functions of gestures concern ways in which gestures 'make mathematical ideas visible' to benefit an epistemic action. The previous section presented four different forming-functions: *Sourcing out, depicting, extracting mathematical entities, and illustrating generality by the use of gestures 'brings mathematical ideas to the eyes' and adds them to the shared pool of ideas in social interaction.* It turned out that this does not only **support** the epistemic process by illustrating the epistemic action reconstructed from speech, but may also **prepare** an upcoming epistemic action. Furthermore, gestures can **realize** actions of gathering, connecting and structure seeing themselves.

Sourcing out in gesture provides a possibility to spontaneously represent an idea visually while it is formulated verbally. Within the social interaction, this may lead to coordinating or adjusting shared knowledge on non-verbal representations of mathematical objects and on their relations. However, some ideas appear difficult to be represented in gesture, as to mention, for example, relations between relations.

Depicting means to represent an iconic representation of relations by the use of gesture. These depictions have been identified to arise from iconic reference to an inscription. Being integrated in a fixed diagram (level 2) or performed in the gesture space (level 3), the gesture refers to the mathematical object that has been associated with the inscription. Depicting-gestures as ephemeral modes of expression only refer to certain aspects of the inscription and not to all its specificities. They can store some aspects and leave out others. By this, they can provide a representation of the essentials. While similar is done by extracting-gestures, the latter are more closely linked to inscription:

Extracting-gestures bring to the fore certain aspects or components represented while sending others to the back. Extraction may concern singling out one entity within a larger diagram but also a relation, extracted in a movement, for example.

Illustrating generality: Gestures purely metaphorical in nature that is, without any reference to the inscriptions used within the social interaction, can give visual access to the generality underlying an idea.

8.1.2 Performing-Functions: How Gestures can Take Part in the Accomplishment of an Epistemic Action

The specific ways in which gestures can enrich the epistemic process are manifold. Gestures fulfilling a performing-function are considered to *catalyze* the epistemic process, influencing it in a directive way. Based on the data, six main performing-functions of gestures have been characterized. Their concrete shaping crucially depends on the specific situation; the main features that cause their epistemic benefits are presented as reconstructed for illustrative examples:

8.1.2.1 Focusing

Focusing means that gesture helps to concentrate on a certain aspect or entity so that the accomplishment of an epistemic action is enhanced through this. The gesture *focuses* attention on what is considered important at that very instance.

For example, this may concern the (pre- or poststroke-)holding of a gesture across turn to keep on an idea linked to the reference of the gesture. As will be seen in the next excerpt, this allows the students to go on within the social interaction while not losing sight of the reference aimed to be considered further.

Excerpt 3.2a, PA7-399-409 (P7e5, 0:34:56 – 0:35:41)[132]

Lisa and Rosa work on the parabola task and have already stated the conjecture of $\overline{AB} = \overline{AP}$ (lines 149-156). In order to substantiate this conjecture, they have compared angles and segments within the printout diagram. They have suggested a similarity of the sides of the triangles PM_iA and AM_iB (173-176).[133] Justifying this lemma means to justify their conjecture as well, so Lisa and Rosa consider finding an argument proving that point M_i is exactly in the middle between points P and B (380-385). This is taken up by the interviewer and he prompts them to consider how the 'process of folding was implemented geometrically' in the GeoGebra diagram (396). Fig. 8.6a and 8.2b show the representations as they are present at that point in time (See page 320 for the final version of Rosa and Lisa's folding diagram.).

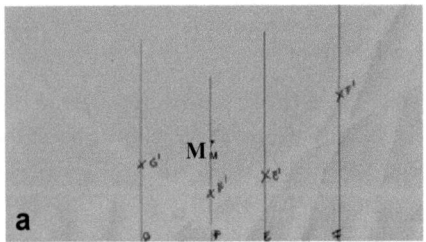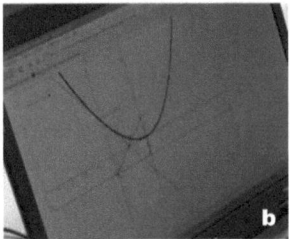

Fig. 8.6: Folding sheet (a) and GeoGebra situation (b) as present in the beginning of excerpt 3.2a, line 399
(a) points on the lower edge tagged with (from left to right): G, H, E, F with corresponding constructed points G', H', E', F' marked with a red cross

---

[132] To optimize comprehension, the changing situation on the screen is illustrated by pictures as well in this excerpt.
[133] At a later point in the working process, the students tag the intersection point between [PB] and its perpendicular through point A in the first printout with "M". This can be confused with the point M as given in the folding diagram, corresponding to the focal point of the parabola. In this transcript however, it is not yet referred to this intersection point by "M". To stay close to the tagging as present in the students' inscription attached in the appendix, it is referred to this point by "M_i". This index is used as a reminder that M_i means the 'intersection point' and to avoid confusion with the focal point that has been given as "M" in the folding sheet diagram.

8.1 Epistemic Functions of Gestures: Detailed Descriptions

While comparing the printout and GeoGebra representations, Lisa uses pointing gestures to *focus* on the entities gathered as important in the folding sheet diagram:

399 L: between the-
[[cen-t-r-e-] *(points at point M on the folding sheet)* and the point *(points at point G on the folding sheet)*
[‚the arbitrary point' *(Rosa grabs the mouse)* (.) P'

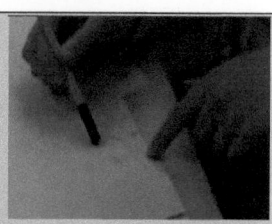

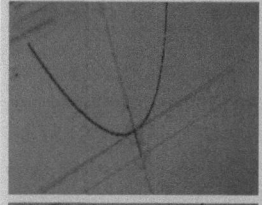

400 R: That means in (.) this case *(moves point P to the y-axis)* (.) ‚mh. (4sec)
401 L: (incomprehensible) *(Rosa moves point P to the left, then to the y-axis again)*
(4sec) ‚eh yeah do it like t-h-a-t once more-
402 R: *(holds point P on the y-axis)* like this'
403 L: N-O.

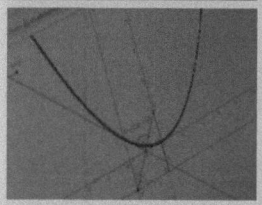

404 R: *(moves point P to the left again such that the situation is similar to the one displayed on the printout)*

like this'

405 L: y-e-s. let's somehow- *(Rosa lets the mouse go)*]¹³⁴ ‚uhm-

‚we have the center [[B-] *(points on point M on the folding sheet)* (.)
of the segment to the [P-]] *(points at point G on the folding sheet with her other hand)*

(0:35:27)
406 R: that means that is *(goes along the folded edge for point H, parallel to the lower edge of the sheet, with her finger and leaves it there)*

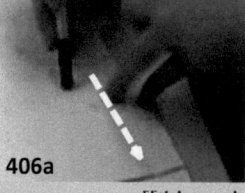
406a

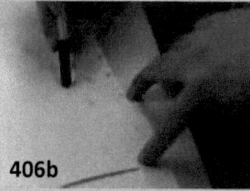
406b

[[this straight line.] (…)

---

[134] The pointing on point *G* started in line 399 is released.

407 L: what' (*still holds her fingers in the same position on the folding sheet, briefly looks at the screen, then at the folding sheet again*) (..)][135]

(0:35:33)
408 R: well-
[[,this this-] (*multiply goes along the x-axis on the screen with her finger*) with the numbers on it. (..)]

409 L: y-e-s. (.) ,that is][136]
(*Lisa takes her hands from the two points, alternately looks at the folding sheet and at the screen*) (7sec)

Lisa re-**gathers** the focal point and the point on the directrix and sources this gathering out by successively indicating the corresponding points M and G on the folding sheet. Starting her approach with "between", she makes clear that she refers to the distance between these two points. While indicating point G, she refers to it in speech as "arbitrary point' (.) P' " (399). By doing this, Lisa emphasizes this arbitrariness of the point but also the correspondence to the point as represented in the GeoGebra environment, where it is termed "P". Making this connection, Lisa implicitly justifies her indication of a point corresponding to point P on the lower edge in the folding diagram.[137] Sourcing out the gathered points M ("cen-t-r-e" in speech; focal point in general) and G ("arbitrary point' P" in speech; arbitrary point on the directrix in general) *specifies* their *locations* within the folding sheet diagram (*level 1*) and the indication is held across turn. Rosa varies the GeoGebra environment, first adjusting the case in which point P is straight beneath the focal point, and then moving point P a bit to the left and back to the y-axis again (400/401). Still holding the fingers on points G and M on the folding sheet, Lisa asks Rosa to move point P back to the left again so that similar situations are represented in the printout, on the folding sheet, and in the GeoGebra environment (402). Having held the indication during the variation of the GeoGebra environment, Lisa starts a new approach in line 405. She renews the indications of the points G and M on the folding sheet while re-gathering the points corresponding to points G and M in the GeoGebra-diagram in their verbal utterance: Different to before, she refers to the "center B" (405) and not merely to "the- cen-t-r-e-". With this, she verbally explicates the correspondence between the point M by *specifying* its *location* within the folding diagram and the point B within the GeoGebra environment, both representing the focal point within different representations. Again, she keeps the points M and G, the "centre" and the "arbitrary

---

[135] Rosa releases her pointing started in line 406.
[136] End of gesture unit started in line 399.
[137] The point P can be chosen arbitrarily on the lower edge and the actual chosen point realizes the situation resembling the situation represented in the printout the most. Furthermore, this similar situation has already been established in the folding situation by fixing the segment $\overline{GM}$ and its perpendicular through the point G', the point that corresponds to A in the folding diagram.

point", indicated (held) across turn while Rosa identifies another correspondence between the two representations: the horizontal folding line in the folding diagram corresponds to the x-axis in the GeoGebra environment (406-408). Her first approach to offering her thought is incomplete: "that means that is [[this] straight line.]" (406) is accompanied by a tracing of the folding line (406a). This tracing identifies "this straight line" within the folding diagram and thus *specifies* its location, but it lacks clarification for what is referred to by the "that". Accordingly, Lisa expresses her confusion by claiming for an explanation of what is meant (407: "what' "). Rosa in turn makes herself clear by tracing the x-axis as represented on the screen, compensating the imprecise verbal "well- [[,this this-] with the numbers on it. (..)]" (408) by gesturally *specifying what* and *where*. The understanding of this connection made between the horizontal folding line and the x-axis is confirmed by Lisa, who releases her indication (409).

Rosa's successive tracings of the two horizontal lines (406, 408) provide a short-time focus on them as well. Different to the focus set by Lisa in sourcing out and holding onto components of the diagram (399-405, 405-409), this *focusing* is accomplished by extraction. Rosa starts her approach in 406 with "that means" and expresses that she refers to a conclusion she has made, following from the indications made by Lisa. She holds these indications while Rosa adjusts the GeoGebra environment. This focusing is considered to avoid confusion since the points considered as important, "the centre" and "the arbitrary point" are not lost out of sight. Sikveland and Ogden write in this regard that "gesture holds provide a visible means for marking something as 'not yet dealt with', and their retraction as a way of displaying (literally) that the issue has been solved" (Sikveland & Ogden, 2012, p. 194). This retraction is identified in line 409 when Lisa agrees on the connection made before by Rosa. Retracting the focus on the two points in the folding diagram, she also retracts her approach started in line 399. However, *focusing* on the two points also sets a frame for a comparison of components between the two representations 'folding sheet' and 'GeoGebra environment'. After Lisa has already identified the correspondence for the two points (focal point M/B and point on directrix G/P), the focused frame prompts Rosa to connect the horizontal folding line and the x-axis.

The focusing gesture establishes a zoom into the diagram. This has two main benefits:

> *(1) It holds attention at components or specifies aspects considered as important while social interaction can change the topic for a short time.*
> *(2) It eases comparison and through this supports a connecting action.*

In the following example, we will see how the interviewer uses *focusing* by gesture as a didactic means:

Excerpt 3.2b, CF7-1036-1049 (CF7e18, 1:26:12 – 1:26:40)

While working on the task concerning the continued fraction, Lisa and Rosa have to substantiate their conjecture that "the decimal number approaches the two more and more" (conjecture fixed for subtask 2.2). Earlier in the working process, the interviewer made a mistake when seemingly writing down a transformation of a general element $\frac{a}{b}$. Rather than transforming the term, he actually presented how the successor of this element can be noted as difference to 2 (lines 686-701) (See Fig 8.7 and note sheet 4 as attached in the appendix on page 334.):

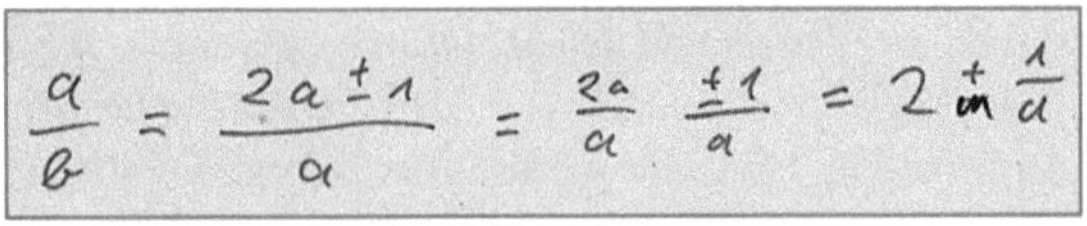

$$\frac{a}{b} = \frac{2a \pm 1}{a} = \frac{2a}{a} \frac{\pm 1}{a} = 2 \pm \frac{1}{a}$$

Fig. 8.7: Interviewer's fixed transformation of a general element $f(x) = \frac{a}{b}$ (note sheet 4)

It can be guessed that the interviewer actually aimed to give the hint that the element *following* a general element $\frac{a}{b}$ can be noted as $2 + \frac{1}{a}$, since the first equality had been fixed by the students earlier not as equality but as follows:

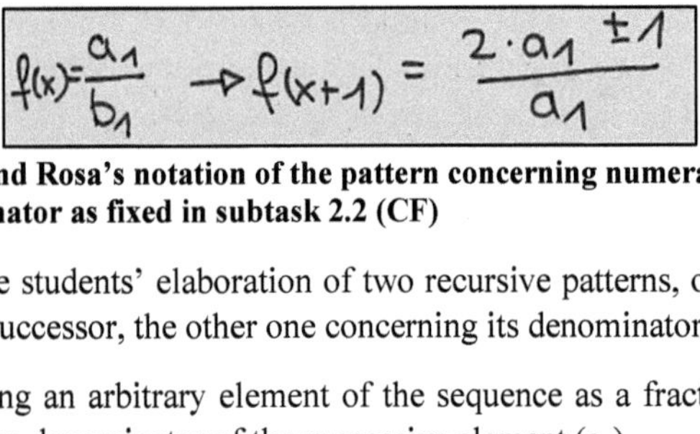

$$f(x) = \frac{a_1}{b_1} \rightarrow f(x+1) = \frac{2 \cdot a_1 \pm 1}{a_1}$$

Fig. 8.8: Lisa's and Rosa's notation of the pattern concerning numerator and denominator as fixed in subtask 2.2 (CF)

This represents the students' elaboration of two recursive patterns, one concerning the numerator of the successor, the other one concerning its denominator:

(1) Representing an arbitrary element of the sequence as a fraction, its numerator becomes the denominator of the successive element (a_1).
(2) The numerator of this successive element can be generated from the numerator of the preceding element by multiplying it by two and adding or subtracting 1. This addition and subtraction alternates for ongoing elements.

Their preceding elaboration of this pattern is also considered a reason why the students are not confused by the interviewer's inscription but accept his intervention. They understand his verbal explanation behind the background of the pattern they found (Fig. 8.8) and neglect his reference to equality. Right away, the interviewer's mistake does not have further consequences: In the following, the students use the term $2 + \frac{1}{a}$ as starting point. Applying the recursive pattern found in subtask 1.2 (see Fig. 8.17), they note $1 + \frac{2}{2+\frac{1}{a}} = 2 - \frac{1}{2a+1}$ (see 851-871, note sheet 5 on page 335). From then on they consider the following correspondence between two consecutive steps as:

8.1 Epistemic Functions of Gestures: Detailed Descriptions

$$2 + \frac{1}{a} \quad \rightarrow \quad 2 - \frac{1}{2a+1}$$

Fig. 8.9: Visualization of two consecutive elements of the sequence, where $2 + \frac{1}{a}$ is the successor of a general fraction $f(x) = \frac{a}{b}$ (note sheet 5)

However, the equation represented in Fig. 8.7 is fixed as lasting inscription produced by the interviewer, who has an authoritative status in this setting. This prelude needs to be considered when the inscriptions at hand are included in an analysis of this excerpt.

The notation displayed in Fig. 8.9 represents the elements each as difference to 2 and is used as a starting point to explore the development of this difference. The concrete differences are added in and placed beneath the table on work sheet 2, as can be seen in Fig. 8.10 (See also the original writing products of the students as attached in the appendix on page 331.):

$f(0)$	1	+1	1	$2-1$
$f(1)$	3	-1	3	$2+1$
$f(2)$	$\frac{5}{3}$	$+\frac{1}{3}$	1.666666667	$2-\frac{1}{3}$
$f(3)$	$\frac{11}{5}$	$-\frac{1}{5}$	2,2	$2+\frac{1}{5}$
$f(4)$	$\frac{21}{11}$	$+\frac{1}{11}$	1.909090909	$2-\frac{1}{11}$
$f(5)$	$\frac{43}{21}$	$-\frac{1}{21}$	2.047619048	$2+\frac{1}{21}$
$f(6)$	$\frac{85}{43}$	$+\frac{1}{43}$	1.976744186	$2-\frac{1}{43}$

Fig. 8.10: Table on work sheet 2 as filled out by the students in the beginning of CF7e18-1063

To connect the approach towards two, the link between the step visualized in Fig. 8.9, and the 'generic element' $f(x) = \frac{a}{b}$, fixed in a wrong way in the inscription displayed above (Fig. 8.7), needs to be made. The students struggle with the three different terms ($\frac{a}{b}$, $2 + \frac{1}{a}$, and $2 - \frac{1}{2a+1}$) and are not able to identify 'a' while tracing concrete examples (934, 981/983, 1001/1003, 1024). This is when the interviewer (I) uses gesture to *focus* on significant components within the inscriptions and with this promotes the students' epistemic process:

1036: I: (..) you start with that you (*first points at* $2 + \frac{1}{a}$ *on note sheet 5, then briefly at the table on work sheet 2*) [this ‚ehm on you a] [pick an [element-

1037 /R: (*low voice*) mh'

1038 I: ‚and you say that eh- ‚you claim then that would be the same as [two plus one divided by a.]] (*points at* $2 + \frac{1}{a}$ *on note sheet 5*)
1039 /R: (*low*) yes. (*Lisa nods*)

1040 I: [f-o-r example.
(*points at* $2 + \frac{1}{5}$ *next to the table on work sheet 2, at the level of f(3)*) [‚here] this is true.

1041 R: yes ‚that means [‚there the five is the a'
(*points at the term* $2 + \frac{1}{5}$ *as well*)
1042 I: (*nods*) exactly.
1043 R: mh.
1044 L: where is the five'
1045 R: ehm- in which]^{138} we just had

(1:26:35)

1046 /L: a-h-
[‚there we have the five.]^{139} (*points at the numerator of the fraction belonging to f(2) in the table on work sheet 2*) right'

1047 R: yes , when the a is here
1048 /L: than this is
1049 /R: ‚originally always [up there.]]^{140} (*points at* $\frac{a}{b}$ *in the equation on note sheet 4 (see Fig. 8.7)*

---

^{138} Releases her pointing started in 1041.
^{139} Interviewer releases his pointing started in 1041.
^{140} Lisa releases her pointing held since 1046.

8.1 Epistemic Functions of Gestures: Detailed Descriptions

The initial brief pointing at note sheet 5 and work sheet 2 (1036) reveals the Growth Point of the interviewer's approach to prompt the students. It reveals a short preview of how he aims to provide hints to the students of a similarity between the two terms $2 + \frac{1}{a}$ and $2 + \frac{1}{5}$ (1038/1040): He resumes the **connection** between an arbitrary element and the possibility to represent it as $2 + \frac{1}{a}$ (1036-1038). Subsequently, he offers the example of $2 + \frac{1}{5}$ next to the table on level of $f(3)$, specifying where "this is true" (1040) (see Fig. 8.11 below).

$f(2)$	$\frac{5}{3}$ $+\frac{5}{3}$	1.666666667	$2 - \frac{1}{3}$
$f(3)$	$\frac{11}{5}$ $-\frac{4}{5}$	2,2	$2 + \frac{1}{5}$
$f(4)$	$\frac{21}{11}$ $+\frac{1}{11}$	1.909090909	$2 - \frac{1}{11}$

Fig. 8.11: Detail of the table on work sheet 2

The successive reference first to $2 + \frac{1}{a}$ (1038) and then to $2 + \frac{1}{5}$ (1040) *specifies* the *locations* of the terms suggested to be compared and prompts a **connection** between them so that Rosa identifies the correspondence as "yes ,that means [,there the five is the a' " (1041). Simultaneously, she is also pointing to the provided example. The interviewer keeps on *focusing* on the term representing the concrete $2 + \frac{1}{5}$ while the students search for where to "get the five' " (1044). This *holding of the focus* simplifies a comparative scanning of the environment and with that a goal-oriented exploring. This leads to the **connecting** action in 1046, when the "five" is identified by Lisa as numerator of the fraction representing $f(2)$. She points to the numerator, *specifying* its *location* and holds her indication. This *focuses* on the representation of an element as a fraction and may in turn initiate Rosa's general statement reconsidering the notation $\frac{a}{b}$ for a general fraction (1049) (see Fig. 8.7).

After the end of this excerpt, the students elaborate their idea further by recalling the numerator of an element being the denominator of the successor within the concrete elements displayed on work sheet 1. This allows the students to trace where the a, the numerator of the 'generic fraction', appears in the consecutive elements.

*In this excerpt, the focusing of the interviewer on concrete **locations** eases goal-oriented exploring and comparative scanning of the diagram in order to gather a 'fitting' component. By this, **gathering** and **connecting** actions accomplished by the student are **prepared**.*

8.1.2.2 Exemplifying

Exemplifying means to give a concrete example by the use of gesture. This can enhance the epistemic process by visually supporting the concretizing of a structure in two ways: First, when *speech refers to an example* that is needed to be specified (see excerpt 3.2c) due to imprecise terms. Second, when *gesture refers to a concrete example* while speech expresses a general idea (see excerpt 3.2d):

Excerpt 3.2c, PA5-477 (PA5e7.2, 0:36:34 – 0:36:51)

Mike and Tim aim to justify their conjecture about the equality of the lengths of the segments $[AB]$ and $[AP]$. In the first scene of this episode, preceding the excerpt presented here, Mike communicates his observation of points P and B being reflected at the tangent line. His fragile and imprecise speech indicates that his idea is not yet elaborated (lines 457ff.). To explain it more in detail, encouraged by Tim's interest (460: "what do you mean"), he switches to the folding sheet but soon to the GeoGebra environment. In the following, he is able to put his idea into words, complemented by illustrative examples:

> **475** M: yes well ‚I just wanted to say that the tangent is the reflection' ‚and uhm-
> **476** /T: yes ‚that wa-
> (0:36:40)
> **477** /M: and (*moves his hand first directed to the printout, then towards the screen*) [[everything that] is on the [tangent]]
>
> ‚from those [[both points] are equally far away.] (*points towards the screen and moves the hand to the left and to the right*) (*Tim moves point P to the right*) (…)

477a

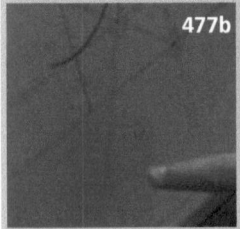
477b

> then they are probably also from [that point
>
> **478** /T: okay
> **479** /M: (*points at the intersection of the tangent on the screen and the y-axis, then at another point*)
>
> equally far away] or from [that one]]

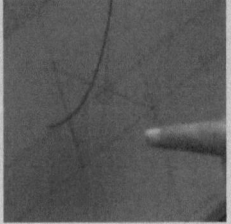

8.1 Epistemic Functions of Gestures: Detailed Descriptions   191

In 475/477, Mike is able to express what he "just wanted to say" (475), concluding a general structure as following from the idea he first mentioned in 457. This idea is repeated in 475, "the tangent is the <u>reflection</u>'", and implicitly provides a reason for a general statement about the relation between any point on the tangent line and two reflected points ("everything that is on the tangent ,from those both points are equally far away"). The "both points" mentioned in speech are *specified* by gesture as reflected on the tangent line (477a, *location* and *relation* of points being reflected on the tangent line, *level 2*). The following indication of possible points on the tangent line (477b, 479) *specifies* their *locations* and provides examples for the expressed idea. In former approaches (459, 473), Mike verbally referred to an example of a point "three centimeters to the right along the tangent line" and claimed that this point should still have the same distances from "the two points". He was not able to formulate this as a general structure before when he approached the seen structure from an illustrating example. This time, the examples are not the approach leading to the expression of a seen structure, but concretize it afterwards.

The benefit of the *exemplification* through gesture can thus be seen in:

(1) The short-time visualization of several possible cases for which the conjecture holds true, specified as embedded into the diagram.

(2) The simplification of speech. In the approaches made before he described an exemplified point by "three centimeters to the right along the tangent". This may be too concrete to make a general structure following from it. Using the terms "that point" and "that one", specified by gesture, suggests an arbitrariness of the points used for illustrating the idea.

The *exemplification* that is reconstructed in the next excerpt is different in that speech does not reveal a reference to a concrete example added by the use of gesture:

Excerpt 3.2d, CF7-5 (CF7e1, 0:01:52)

Lisa and Rosa just started to work on the arithmetic-analytic task (CF) when Rosa offers a structure seen within the denominator of the fraction, *exemplified* by referring to the fixed case of $f(3)$.

She refers to the representation of the first four continued fractions, as noted on work sheet 1:

$$f(0) = 1$$

$$f(1) = 1 + \frac{2}{1} = 1 + 2 = 3$$

$$f(2) = 1 + \frac{2}{1 + \frac{2}{1}} = 1 + \frac{2}{1+2} = 1 + \frac{2}{3} = \frac{5}{3}$$

$$f(3) = 1 + \frac{2}{1 + \frac{2}{1 + \frac{2}{1}}} =$$

Fig. 8.12: First four elements of the sequence $(f(x))_{x \in \mathbb{N}}$ defined by the given continued fraction, fixed on work sheet 1

5 R: for [[the one] [that stands down there] *(first points with the pen to the denominator of the fixed continued fraction of f(3) on work sheet 1, then circles the denominator of f(2))*[141] 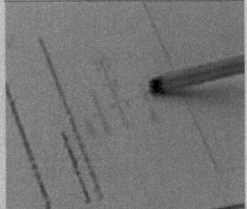 [always two divided by one comes along.]] *(points at two spots on the right hand side of the lowest denominator "1")* or am I wrong.

Rosa has seen a similarity between the continued fractions and remarks in how far two consecutive fractions differ: it is always added $\frac{2}{1}$ to the lowest denominator. This corresponds to a recursive *complementing*-structure. However, Rosa starts with the word "for" which suggests to *substitute* something "for" something else. This may indicate that the substitution-structure is already implicitly present, considering that some part of the fraction, the lowest denominator, can also be substituted by a term to get to the next continued fraction. While offering the complementing structure in speech, Rosa *specifies* the *what* and the *where* of "the one that stands down there" by referring to the concrete example of two consecutive elements $f(2)$ and $f(3)$ on *level 1* (See Fig. 8.13 below.).

---

[141] See Fig. 8.13 for concrete locations pointed at and for the extent of framing.

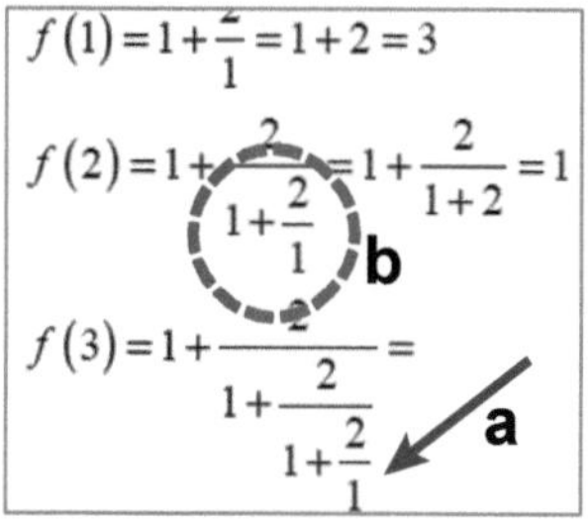

8.1 Epistemic Functions of Gestures: Detailed Descriptions

$$f(1) = 1 + \frac{2}{1} = 1 + 2 = 3$$

$$f(2) = 1 + \underbrace{\frac{2}{1+\frac{2}{1}}}_{\textbf{b}} = 1 + \frac{2}{1+2} = 1$$

$$f(3) = 1 + \frac{2}{1 + \frac{2}{1 + \frac{2}{1}}} = \quad \textbf{a}$$

Fig. 8.13: Indications in CF7e1-5: (a) "[[the one]", (b) "[that stands down there]"

The "down there" is localized within the continued fraction of $f(2)$, referring to "the one that stands down there" generally as 'the 1 standing in the lowest denominator'.

Illustrating a general structure by means of a concrete example enhances accessibility by providing visual *access to it. This makes possible to compensate for lacking words by specifying aspects or entities within the concrete example.*

8.1.2.3 Contrasting

The following example led to the identification of the *contrasting-function* to visualize the non-validity of a considered idea. This function allows the confrontation of one situation with another situation by visualizing their difference respecting one or more aspects. A hypothetical situation can be shaped ephemerally in speech to confront with a captured situation. Through this, argumentation can be facilitated: Speech is simplified, as the competing properties of the two compared situations or objects do not have to be expressed verbally. The following episode illustrates how such a *contrasting* gesture can support the linguistic act of refuting a conjecture by depicting the counterargument.

Excerpt 3.2e: PA5-284-286 (PA5e3.1, 0:22:22 – 0:22:28)

In the episode before, the students have been introduced to the GeoGebra environment and trace the curve so that the following diagram appears on the screen:

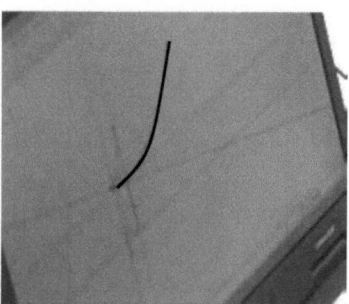

Fig. 8.14: GeoGebra diagram in the beginning of PA5e3 (trace thickened to optimize visibility)

They identify the curve to be an exponential function and to be shaped in the same way as the curve on the folding sheet. Initiated by a researcher present in the field, the students gather pairs of values to check their assumption concerning the graph being described by an exponential function. This makes them realize that a property of an exponential function, the y-values decrease from right to left, does not match the considered shape of the curve.

284 /M: it <u>can't</u> be that it is an exponential function' *(looks at Tim)*
285 T: right.
286 M: because ‚uhm
287 /T: then it [would be] *(points towards the screen[142])* <u>smaller</u> there

288 /M: [elsewise the number would be smaller and <u>smaller</u> *(points at the screen from right up to left down)* (.)]

(looks at Tim) the left side normally

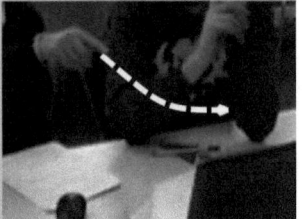

Mike's argument against the conjecture of the curve being an exponential function consists of the three utterances 284 (conclusion), 286, and 288 (substantiating conclusion). He claims that "it <u>can't</u> be that it is an exponential function" (284). What he means with "it" becomes clear in the context of discussing the type of function represented by the curve. Tim confirms this statement and Mike, furthermore, launches an explanation that substantiates why the exponential function is excluded to be represented by the curve. He starts with "because" (286) but hesitates. This makes Tim start an approach, imprecisely stating that "it would be smaller there" (287). This mentioning 'something becoming smaller' in combination with pointing towards the screen suggests that he has in mind the same reason that in turn is expressed by Mike (288). He uses imprecise speech ("the number") and leaves out aspects that are specified in gesture: The iconic reference to the curve of an exponential function is combined with the indexical reference to the screen. Through this, two main functions are fulfilled: Shaping the referential frame in the interaction space in front of the monitor (***level 2***), the gesture ***specifies*** the ***where***, the ***what***, and the ***how*** of the semantic content of speech. This allows "the number" to be interpreted as y-value and the direction of decrease to be from right to left, adding an

---

[142] The epistemic process in this scene will be reconstructed in an upcoming chapter. The entire scene is given here to present the gesture in 288 as integrated into the larger situation.

aspect that is needed to interpret the argument to its full account. Furthermore it superimposes the actual case with the one to exclude.[143] This *visualizes the contrast* in shape and by that illustrates the counterargument.

*The gesture benefits the epistemic process by visualizing that the actual case and the hypothetical case – the one to exclude – do not fit together. Contrasting, thus, requires a comparison of different cases and supports this comparison by hinting at differences that are specified by the use of gesture. This way, the use of the gesture **supports** the verbal **connecting** action.*

It has also been observed that the interviewer can use *contrasting* gesture to prompt a connecting action, just as has been seen for a focusing gesture before:

Excerpt 3.2f, PA5- 1673-1674 (PA5e16.5, 2:10:51 – 2:11:06)
Tim and Mike have almost solved the parabola task. The only issue left concerns the precise formulation of the definition of the parabola as geometrical locus (subtask 5). So far, they have worked out the definition as a set of points that all have the same distance "from a fixed point B and from a point P located on a straight line g"[144] (1655). The interviewer is not fully satisfied with that definition. For an arbitrary parabola, there is no such point P but a straight line which holds that any point has equal distance to this line and to the focal point. Using a gesture, he directs attention to the feature that point P is positioned directly beneath its corresponding point A on the parabola:

1673 I: so. [„where do you know now that *(points at point A on the screen)* [[„if A is here for instance'] „that you take

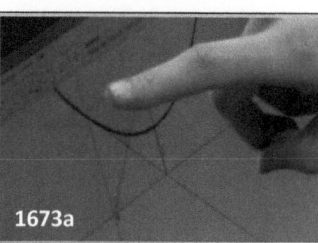

---

[143] From subtask 2 it is revealed that the students are aware of the fact that the GeoGebra environment represents the same curve as the one worked out on the folding sheet. The curve is not completely traced in the GeoGebra diagram, but its shape is known to both students to be symmetric. Against this background, the gesture illustrates the verbal argumentation.
[144] The directrix is called 'g' in the GeoGebra environment.

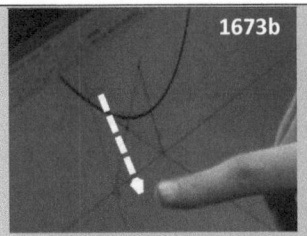

(points at point P on the screen) [exactly this point P]] and

(points at a point on the left hand side of the directrix) [not maybe]] ,one over there on the straight line.

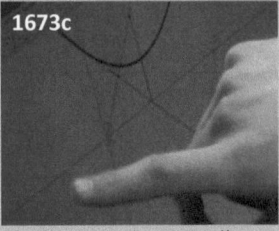

1674 M: because [it has the same] *(briefly points towards the screen)* ,x-coordinate. (..) *(briefly looks at the Interviewer)*

The interviewer prompts the students to determine point P unambiguously as a point on the directrix having the smallest distance from point A. Giving them a hint, he (maybe unconciously) suggests a connection in gesture, moving straight down from point A to point P on the screen. First, point A is located by gesture to *specify* its *location* within the concrete situation (*level 1*) (1673a: "[[,if A is here for instance']"), and then same is done with point P. The change of location is performed in a downward movement (1673b) and already suggests the *relation* between the two points as point P being located directly beneath point A. The example given then refers to an arbitrary other point on the directrix further to the left side (1673c): On *level 2*, the interviewer *specifies* the *location* of a point that lies on the directrix, not directly beneath point A but "over there on the straight line." (1673). The interviewer suggested the essential relation between the two points A and P by visually connecting them within the GeoGebra-diagram just before, but the reference to the potential point further to the left lacks such a visual connection. He implicitly demands for a distinguishing property and 'prompts' it visually by *specifying* both *locations*, a concrete one (1673b, *level 1*) and a hypothetical one (1673c, *level 2*): The two gestures *contrast* both points on the directrix with respect to their spatial *relation* to point A in a *diachronic* way; the actual one being located directly beneath point A, the one not having the defining property apparently lacking a connection to point A. Mike's response in line 1674, offering the common x-coordinate of points A and P as a connecting feature is most likely triggered by the interviewer's visual connection in the first example and its missing in the second one.

Different to before, the objects that are compared and *contrasted* are not represented at the same time but consecutively. They do not differ in shape that can be visualized as not fitting together (as in excerpt 3.2e) but in another property that concerns a specified relation: The point on the directrix corresponding to the point A on the parabola being located straight beneath point A.

The epistemic benefit of this contrasting through gesture consists in ***preparing a connection*** by making it ephemerally visible. In sum, we have seen that contrasting gestures can benefit the epistemic process in the following ways:

*Gestures can benefit the epistemic process by visualizing differences between two cases in a synchronic or in a diachronic way. Contrasting requires a comparison of different cases and supports or **initiates** this comparison by hinting at differences that are specified by the use of gesture. This way, the use of the gesture can support the verbal connecting action but can also **prepare** it.*

8.1.2.4 Making More Precise

A gesture can *make an utterance more precise,* so that the accomplishment of an epistemic action is enhanced. From speech it can be reconstructed that an epistemic action is accomplished, but its specificity cannot be valued without considering the gesture. It is the accomplishment of the gesture that completes the semantic content of speech by *specification*. This provides a real surplus of meaning without which the epistemic action would be incomplete.

To make this description comprehensible, the epistemic benefit provided by gesture through *making the verbal expression more precise* will be reconstructed in two illustrative examples:

Excerpt 3.2g, PA7-991/993 (PA7e11.2, 1:26:19 – 1:26:40)

Lisa and Rosa are challenged to find a description of the parabola similar to the one that is given for the circle (subtask 5 of the parabola task). They mention the general property of positive slope for increasing x-values (986-990: ",the points move with- increasing x- (…)" "more and more-" "increasing y."), as it is the case for the right side of the parabola as visible on the screen.

---
986 L: Uhm- (*both students look at the screen*) (13sec) ,the points move with- increasing- x- (…) ,no. (*rests on her elbows*) (..)
987 R: With increasing x
988 /L: (*synchronic*) sing x- (*Rosa looks at Lisa*) more and more-
989 R: increasing y.
990 L: (*laughs*) Exactly. (*Rosa grins, briefly looks at the Interviewer*) (4sec)

Becoming aware of referring to a particular case, Rosa revises her statement:

991 R: (*looks at work sheet 5*) yes or to be precise sometimes it is also- (*looks at the screen*) decreasing. (.) ,so it is also with stronger decreasing x (*moves her index finger from bottom right to top left twice in an arched curve*)

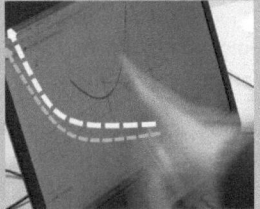

[[in this form] [more and more increasing (*holds the finger in the top left position*) y.] (*moves her finger back to the bottom right position*) (.)]
992 L: yes.

(1:26:28)
993 R: [and [there are also-

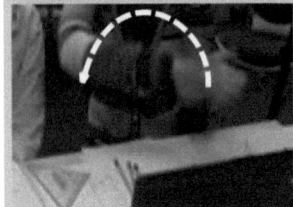

(*holds both hands together on the left side, then indicates a downwardly opened curve to the right with her hand*)
[those.]]
(*Lisa nods*) (...) ,my- (*both students look at the screen*) (35sec)

In 986-990, the structure is expressed with reference to a point moving along the curve (986: ",the points move with- increasing- x-"), starting from the axis of reflection and moving from left to right. Rosa now claims "to be precise" (991) and to consider the co-variation of the curve on the left hand side of the axis of reflection as well. Continuing the idea of the point moving along the curve, this movement is considered to be directed from right to left now, starting at the axis of reflection again. In speech, Rosa refers to the "stronger decreasing x" (991) and thus to the x-value becoming smaller. *Making more precise* that this consideration concerns the left branch of the curve (*specifying where*), she adds "in this form" (991) and *specifies* this form (*how* it is increasing) by moving the hand arc-shaped from right to left in an upwards-movement on *level 2* in front of the screen. As before, the-value increases, as becomes visible in the upwards-movement. Compensating the simple and syntactically incorrect speech ("it is also with stronger decreasing x in this form more and more increasing y."), gesture *makes more precise* the reference for Lisa, who expresses understanding and confirms Rosa's thought (992). The use of gesture thus supports understanding the utterance as **connecting** action

8.1 Epistemic Functions of Gestures: Detailed Descriptions

in which the pattern concerning the correspondence between x-value and y-value is relativized. In turn (993), Rosa provides another example of a possible kind of parabola. Her verbal utterance reveals that she offers an example (993: "and there are also-those."), while speech *makes more precise* that this is another example proving wrong the generality of the property that 'the points move with- increasing- x- increasing y-'. Lisa's confirmation expressed by nodding and the following 35 seconds without the students interacting, indicate that the specification by using gesture was sufficient to object to the idea.

The gesture *makes* the verbal utterance *more precise* in two different ways to benefit the epistemic process:

(1) In 991, it *makes more precise* which aspects are considered: gesture specifies the where of the direction and the how of shape and with that *makes more precise* the **connection** between similarity (shape) and difference (direction of movement).
(2) In 993, gesture *makes* the imprecise verbal formulation *more precise*. Verbally only referring to the term "those", gesture allows the example to be interpreted as representing a parabola that does not fulfil the considered property. It *specifies* this '*what*' that has been left imprecise by the verbal utterance. This specification 'what' follows from the *relation* to the axis and the shape (*how*) that have been *specified*. It is due to the concrete shaping in gesture that the example can be identified as a counter examples, sufficient to object the considered general idea.

Nevertheless, as presented in the next excerpt, the benefit of gestures that make the verbal expression more precise does not necessarily need to be found in the comprehension of an imprecise utterance.

Excerpt 3.2h, CF7-25 (CF7e2, 0:03:13)

Lisa and Rosa start to calculate the first elements of the continued fractions in subtask 1. They consider using the calculator to determine the corresponding decimal number but the interviewer prompts to calculate the terms till $f(7)$ manually. Looking at the continued fractions $f(0)$ till $f(3)$ that are given on work sheet 1 (see Fig. 8.12 on page 192), Lisa sees a structure that may simplify the calculation:

25 L:but [one can simply ‚theoretically [start below *(points with the left index finger to the lowest denominator of f(3) and simultaneously at the representation of f(1) on work sheet 1)* and then always take these numbers ‚right' (…)]] do you know what I mean' (.)

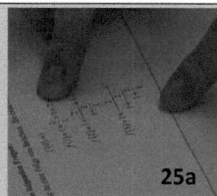

25a

(takes both hands off the sheet)

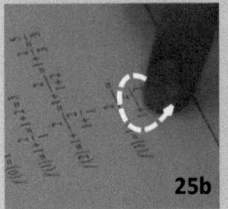
25b

because [this thing there]
(circles the lower part of the term of f(3), $1 + \frac{2}{1+\frac{2}{1}}$)

theoretically is exactly [[that]
(points at the term $1 + \frac{2}{3}$ *in f(2) on the work sheet (see Fig. 8.15 for exact indications))*

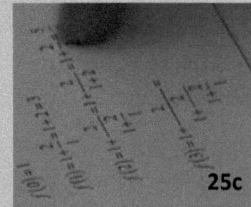
25c

and then simply

25d

[one plus two divided by one plus two thirds.]
(indicates the corresponding components of the term each in $1 + \frac{2}{1+\frac{2}{1}}$ *(left index finger) and in* $1 + \frac{2}{3}$ *(right index finger) (see also Fig. 8.16))*

26 R: (.) what' *(slightly shakes her head and rises her eyebrows)*

Lisa expresses that she sees a structure that determines a rule to calculate the next element. In speech, this idea is formulated imprecisely and reveals that her idea is a sketch rather than precisely thought through: "[one can simply ‚theoretically start [below and then always take these numbers ‚right' (…)]]" (25a). The word "start" suggests the reference to a procedure; the word "always" shows the generality in her idea that most likely concerns a recursive pattern to calculate the next element of the sequence. While speech alone does not provide further information on the components which she sets in relation, Lisa uses gestures to *make* her utterance *more precise*: The first pointing *specifies* "below" as referring to 'below the fraction bar' (*where*). The second pointing *specifies where* "these numbers" are located within the inscription. Both indications together *specify* both elements to be *related* (25a). However, the concrete identification leaves open what "these numbers" are in general but allows for several interpretations: The gesture may refer to the numbers that stand for $f(1)$, but also for the final result standing in the end of the calculation. The first interpretation seems unlikely since Lisa refers to "numbers", not to a single number to be 'taken'. The second interpretation also seems unlikely since it does not become clear which final result shall be 'taken' since the reference to the final result "3" for $f(1)$ requires a further specification of the extent of the "below". Both references lack the *specification what* and are left imprecise in that regard.

8.1 Epistemic Functions of Gestures: Detailed Descriptions

Lisa asks Rosa "do you know what I mean' " but does not receive any answer. This prompts her to amplify her idea with concrete reference to the elements $f(3)$ (25b) and $f(2)$ (25c) (see Fig. 8.15).

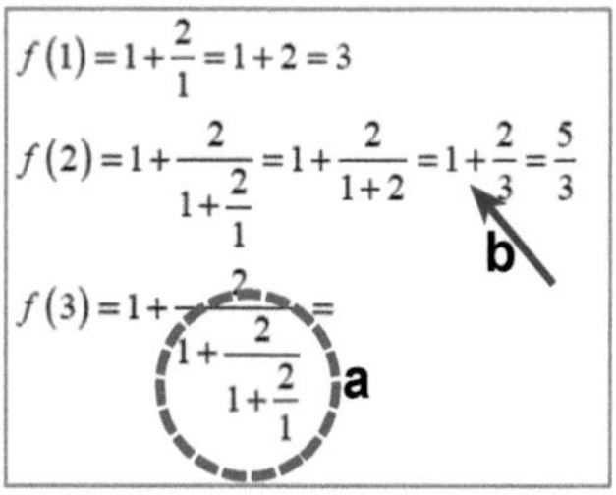

$$f(1) = 1 + \frac{2}{1} = 1 + 2 = 3$$

$$f(2) = 1 + \frac{2}{1+\frac{2}{1}} = 1 + \frac{2}{1+2} = 1 + \frac{2}{3} = \frac{5}{3}$$

$$f(3) = 1 + \frac{2}{1+\frac{2}{1+\frac{2}{1}}} =$$

Fig. 8.15: Indications in CF7e2-25:
(a) "[this thing there]" (25b), (b)"[that]" (25c)

Starting with "because" she begins to substantiate her idea: She refers to one "thing" being equal to another one and her gestures *specify what* is referred to more concretely (see Fig. 8.15): First, she *specifies* the whole denominator of $f(3)$, $1 + \frac{2}{1+\frac{2}{1}}$, by ephemerally framing it (*where* and *what*, 25b/Fig. 8.15a), second, she points to $1 + \frac{2}{3}$ (*what*, 25c/Fig. 8.15b). While the indication of $1 + \frac{2}{3}$ is held (red arrow in Fig. 8.16a), the relation between the terms is *made more precise* by tracing the components of the framed fraction $1 + \frac{2}{1+\frac{2}{1}}$ (light blue, dotted arrows in Fig. 8.16a) simultaneous to: "and then simply [one plus two divided by one plus", finishing with synchronous tracing of the "two thirds]" in both terms (Fig. 8.16b).

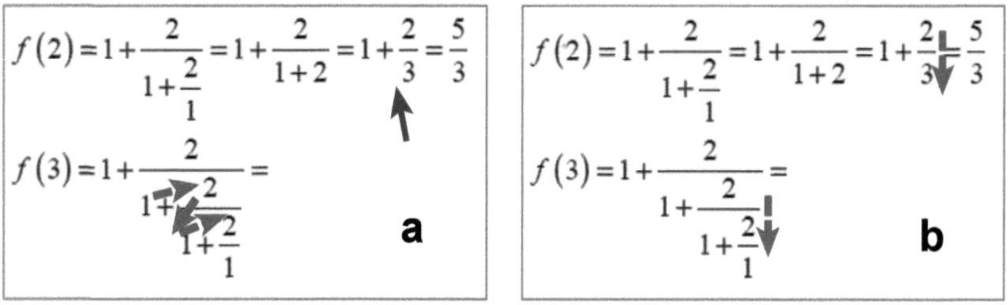

$$f(2) = 1 + \frac{2}{1+\frac{2}{1}} = 1 + \frac{2}{1+2} = 1 + \frac{2}{3} = \frac{5}{3}$$

$$f(3) = 1 + \frac{2}{1+\frac{2}{1+\frac{2}{1}}} =$$

a

$$f(2) = 1 + \frac{2}{1+\frac{2}{1}} = 1 + \frac{2}{1+2} = 1 + \frac{2}{3} = \frac{5}{3}$$

$$f(3) = 1 + \frac{2}{1+\frac{2}{1+\frac{2}{1}}} =$$

b

Fig. 8.16: Pointing (red arrow) and tracing (light blue, dotted arrow)
(a) "[one plus two divided by one plus two" (b) "two thirds]"

Lisa's gestures illustrate which components of the terms are related by *specifying* this *relation* on *level 1*. The specification *makes* the pattern mentioned in the beginning of the utterance *more precise*. However, the two times Lisa indicates a component related to the denominator of $f(3)$, she specified different locations: First, while saying "these numbers", she points to the 'final solution' for $f(1)$, second, referring to something being equal to the lower denominator of $f(3)$, she indicates the simplified fraction $1 + \frac{2}{1}$ in $f(2)$. This inconsistency in turn leads to Rosa claiming for specification ("what' ", 26) and reflects that Lisa did not make her idea clear in social interaction.

The epistemic benefit of the gesture that *make* the reference of imprecise speech *more precise* does not concern the understanding of the idea (benefit for the listener), but the act of expressing it (benefit for the speaker). Goldin-Meadow (2003) already stated that gesturing itself supports speaking by compensating lacking words or not elaborated formulations and with that, making the verbal expression more fluent. It saves the speakers cognitive effort while bringing out an idea that is not precisely thought through (Goldin-Meadow, 2003, pp. 145-166). Hence, in total, the main benefit of the gestures supporting the epistemic actions in these excerpts can be pinpointed as *making the meaning of an utterance more precise, such that the verbal expression is unburdened*. This way, even imprecise ideas can be shaped in a first, tentative version and **connecting** actions can be **prepared** or **supported**.

8.1.2.5 Gluing

Gestures can *glue together meaning* by the style they perform together with speech. As described in chapter 3.2.1, several gesture phrases may form a gesture unit. That is, an entire gesture, from its preparation phase up to its recovery, may encompass several strokes. Just as speech may connect several aspects in one sentence, gesture can do the same in a gesture unit. One gesture phrase is considered to be meaningfully assigned to one part of the verbal utterance (Kendon, 2004, p. 112). To support connecting actions, each gesture phrase of a gesture unit refers to one aspect that is connected, specifying different aspects of the same thing as drawing a larger picture of it. How this has been reconstructed from the data will be elaborated in the following illustrative examples:

Excerpt 3.2i: PA5-66 (PA5e1, 0:07:58)

This excerpt taken from Mike and Tim's elaboration of the parabola task has already been dealt with in chapter 7.1.1 (excerpt 1.1b, see also example 2.1a). Mike and Tim have constructed four points according to the folding construction. They connected them to shape a curve such that its representation on the right hand side of point M is fixed (see Fig. 7.2 on page 116). At this point, Tim connects the left hand side of the diagram with the shape of the curve as displayed to make a conjecture about the shape on the right hand side:

66 T: I would sa ,yeah right.

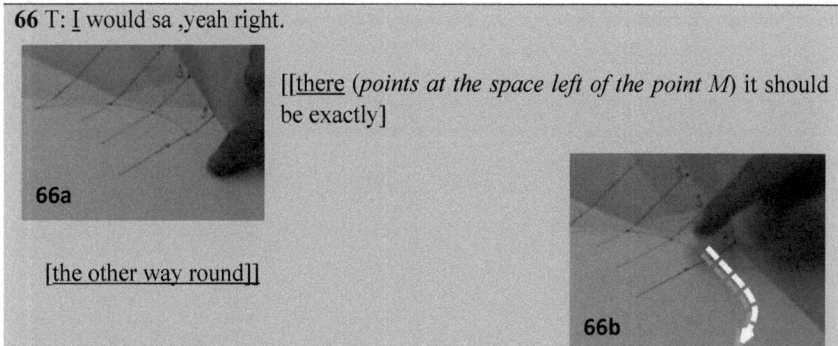

[[there (*points at the space left of the point M*) it should be exactly]

[the other way round]]

The two aspects of *location* ("there") and *style* ("the other way round") are *specified* to represent both sides of the curve as related. Clarifying the imprecise terms, gesture **supports** the **connection** made in speech that represents the right hand side in comparison to the fixed left hand side of the curve. Furthermore, gesture *specifies* the two aspects of *where* and *how*, each on its own, in two separate movements: The indication in the first gesture phrase (66a) specifies where the comparison refers to (the left hand side); the depiction in the second gesture phrase (66b) specifies how it is manifested in the shape of the curve. Both aspects are clarified individually. This simplifies the tracing of a newly introduced idea, both in speech and in gesture. The integration in a gesture unit supports the connecting action by visually 'gluing' together both aspects as being connected.

More generally, the following can be stated: *A gesture can support a connecting action by consecutively specifying different aspects in separated gesture phrases, 'glued together' within a gesture unit.*

In the example at hand, this concerns the offering of a newly considered idea as a connecting action. In connecting actions that are directed backwards respecting structure seeing, the representational gluing by gestures has also been observed to benefit reasoning actions by tracing the single entities, for example, in the next excerpt:

Excerpt 3.2k: PA7-1118-1120 (PA7e16, 1:38:09 – 1:38:18)

Lisa and Rosa are close to defining the parabola as geometrical locus (parabola task, subtask 5). They have figured out the definition of the 'parabola as a set of points that always have the same distance from points P and B' (1005 – 1008). Rosa adds for consideration that this distance is not the same for two different points, revealing her different perspective on the parabola[145] as function rather than as set of points. To defeat her

---

[145] The disparity in perspectives on the function has been a common epistemic gap between the students that needed to be overcome by all three pairs of students.

conjecture concerning the defining property of a point on the parabola, Lisa refers to a concrete example displayed in the printout situation.

> **1118** L: ,in any case (*both students laugh, Lisa points at point A on the first printout with a pen*) (.) ,uhm- (*the interviewer slides the folding sheet a bit to the back*) (..) (*Lisa first points at point B with the pen, then at point A and at the same time at point P with her left index*) [[,this- segment.]
>
>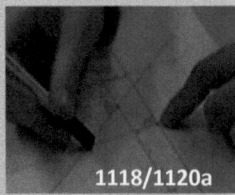
>
> **1119** R: yes' (*nods*)
>
> **1120** L: [A P.] (*nods*) is always as long as- (*points at point A with her left index and at point B with the pen*) [A B.]]
>
>

Lisa's initial pointing at point B is quickly corrected to indicate the segment $[AP]$ within the diagram (1118). The gesture is held in the final position of the pointing in 1118 and Lisa's hesitation gives Rosa a chance to confirm that she follows the explanation (1119). Lisa's following utterance summarizes the core of her proposition: "[A P.] is always as long as- [A B.]]" (1120), fully formulated in speech. The accompanying pointing gestures give a visual access to this conjecture, indicating the two segments considered being equal in length simultaneously to naming them in speech. The whole gesture unit consists of three strokes: the first one to ***specify*** the ***location*** of "this segment" (1118) in the concrete example (***level 1***), and the second and the third (1120a, 1120b) to source out the verbal connection. The ***locations*** of each of the two endpoints of the segments, A and P and A and B, are ***specified*** consecutively, co-timed to their naming in speech. This visually ***glues*** them as each connected to one segment and the two single segments are ***glued together*** as connected representationally. With this, the performance of gesture visually **supports** the **connecting** action accomplished in speech.

Differently to the example given in excerpt 3.2.i, the connection made in 3.2.k is not a newly established one, but justifies a structure that has already been mentioned (transcript lines 1005 – 1008). This backwards-connecting is one way to extend a seen structure, often linked to validating a conjecture. In this regard, the visual access provided by gluing gestures seems to underpin Lisa's verbal substantiation and with that supports the connecting action.

As has already been the case for the forming-functions, metaphorical gestures play a particular role also for the performing-functions:

8.1.2.6 Structuring Verbal Discourse

Gestures can visually represent the structure of discourse, using metaphorical references to argumentation steps. This has already been presented in section 7.1.2.2, considering the representational character of such gestures. At this point, an epistemic perspective will be considered for analyzing the same excerpt.

Excerpt 3.2m, IN7-297 (IN7e3, 0:29:33)

Lisa and Rosa solve the 'king's consultants'-task (IN7). They already hypothesized how to behave after the nth ring of the bell, considering how many marked hats they see and how the other consultants behaved at the $(n-1)^{th}$ ring of the bell. After checking their strategy for some concrete cases, they decided it to be valid. In the following excerpt, they investigate whether there may exist another strategy by assuming the case that all consultants wear a marked hat, considering the generic example of five consultants sitting in the circle:

---

297 R: *(takes the black pen)* the at the fourth ring they <u>still</u> see- *(writes: "4th ring: no")* (4sec) none ‚oh- *(scratches out "no")* ‚there I have always written nobody *(writes "nobody")* (...) *(writes "⇒" in a new line)* (0:29:44) ‚that means there have to ‚be-

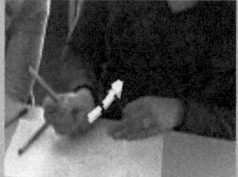 

(briefly raises her hand and turns it forward in one fast movement)

[more than four]

because they are only five everybody has one. *(writes: "all")*

Rosa connects the number of hats she sees with the reaction of the other consultants. Based on the fact that nobody signals at the fourth ring of the bell ("at the fourth ring they <u>still</u> see none"), she infers the number of marked hats in the game (",that means there have to ,be more than four").[146] While this is considered as **connecting actions** that can be reconstructed from the verbal utterance, Rosa's accompanying gesture metaphorically represents this as tracing a path from the back (premise) forth in an upwards arc by turning the wrist forward. This can be interpreted twofold: On the one side and within the concrete context, this may represent the aspect of 'more', using the grounded metaphor 'more is up' on a virtual number line (Lakoff & Johnson, 1980, pp. 15-16). Underlying this, also the metaphorical mapping of PROOF IS A JOURNEY can be identified (Edwards, 2010, pp. 233-234), based on the image schema of source-path-goal

---

[146] See excerpt 1.2c in section 7.1.2.2 for evidence of this interpretation.

(see chapter 4.3.2). With this, the 'goal' corresponds to the conclusion within this mapping. This metaphorical meaning does not develop within the social interaction, so the gesture is not considered to explicate meaning by specifying aspects of a mathematical object. Moreover, it is deeply embodied in the everyday-concept of concluding from a given premise, not necessarily related to a specific concrete topic. It supports the reasoning action by illustrating it in a more conceptual way not referring to a mathematical object but to the structure of reasoning on a meta-level.

Metaphorical gestures can give visual access to the structure of a reasoning action by making use of the source-path-goal schema.

Furthermore, this may foster the epistemic process implicitly by supporting the organzation of the utterance: According to the *Information Packaging Hypothesis (IPH)* (Kita 2000, p. 163), gestures help speakers to "package" spatial information into units appropriate for verbalization. (Alibali, Kita, & Young 2000, p. 593). By the use of gestures, information is parsed into entities more convenient to put into words, "consequently, the collaboration between the two modes provides speakers with wider possibilities to organize thought in ways suitable for linguistic expression" (Kita 2000, p. 180). Assuming the IPH, Rosa's use of the 'conclusion-gesture' may, thus, take part in *organizing thought* and allows abstracting from the generic example.

8.1.2.7 Summarizing Overview on the Performing-Functions

The five presented *performing-functions* of gestures have been considered important to describing how students' use of gestures may benefit the performance of epistemic actions. They have been detected throughout the data independent from the concrete students and the concrete tasks. Nevertheless, their concrete performance is crucially situated and dependent on factors like the *mathematical topic*, the *representations* provided, or the *shared knowledge* on a mathematical object. General features and benefits can be condensed as follows:

Focusing: Focusing takes place when a gesture holds attention to a specific entity or an aspect localized within a representation. It emphasizes this aspect or entity against the rest of the representation and helps to concentrate on it. Synchronic focusing eases comparison by limiting the view to the aspect to be compared; diachronic focusing works as a reminder of an aspect or an object to consider, holding on to a representation of it. Both are considered to prepare and to support connecting actions.

Exemplifying: When gestures refer to an example to illustrate a general structure, they exemplify it. This can appear to happen in two different ways: First, the exemplification can support the verbal reference to an example by specifying its components by the use of gesture. Second, the gesture can refer to an example to concretize a general structure referred to in speech. It may simplify speech by leaving the concrete reference to components of a structure to gesture so that speech can be imprecise (and with that also less concrete). Exemplifying can help to concretize a structure by supporting the verbal exemplification or by realizing it itself.

Contrasting: Contrasting gestures visualize distinctive aspects of mathematical objects by confronting their representations. At least one of these representations is ephemerally shaped by means of a gesture that specifies the aspects that differ. The contrasting-function eases comparison: Synchronic contrasting performed on level 2 against the background of the actual case may support counter argumentation as connecting action, diachronic contrasting may prompt a connecting action by suggesting differing features of the actual and a hypothetical case.

Making more precise: In situations in which gestures make a mathematical idea more precise, their specifying function provides a concrete benefit for the epistemic process instead of merely specifying imprecise speech. Although this might be the case as well, it is directed towards the accomplishment of an epistemic function. On the one hand, it helps the speaker express a new idea: It makes the meaning of an utterance more precise, so the verbal expression is unburdened and less effort is needed for its correct formulation. On the other hand, the listener is challenged to combine verbal and non-verbal representation in order to understand the speaker's utterance. This may enhance the shared use of both modes of expression. Nevertheless, this point is also suggested to cause epistemic gaps if the shared understanding is not made explicit.

Gluing: Single aspects or entities to which is referred in gesture phrases can be concatenated by being 'glued together' in a gesture unit. This allows splitting up references to different aspects while still representing them as visually connected. First, this can support a connecting action by representing each aspect on its own simultaneously to its mentioning in speech. This way, less information needs to be processed at a time to understand which entities are connected. Second, gluing-gestures can also realize a connecting action by visually suggesting the connection of entities referred to in one gesture unit.

Structuring verbal discourse: Gestures can visualize the structure of a reasoning-action by metaphorically representing the connection between condition and conclusion as a 'path from source to goal'. As gluing, it links single components in one gesture, this time in a metaphorical way.

8.2 Extensive Reconstruction of a Larger Scene: Merging the Possible Variety of Gestures' Contribution to the Epistemic Process

To describe the epistemic functions, each one has been presented as analytically separated in the analyses so far. Actually, they rarely appear on their own, but are combined to benefit the epistemic process. This becomes apparent in the following detailed analysis of a larger scene:

CF7E5.1, "why does that work'" (CF7e5, 0:13:03 – 0:14:17)
Lisa and Rosa work out the task dealing with the continued fraction. They have determined the elements $f(3)$ to $f(6)$ and found a recursive pattern to get from one element to the next one, as requested for subtask 1.2:

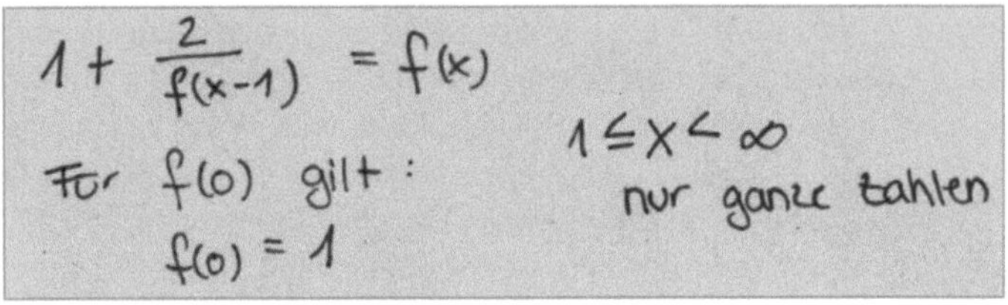

$$1 + \frac{2}{f(x-1)} = f(x)$$
$$\text{For } f(0) \text{ gilt}:$$
$$f(0) = 1$$
$$1 \leq x < \infty$$
$$\text{nur ganze Zahlen}$$

Fig. 8.17: Lisa's and Rosa's solution for subtask 1.2 of the CF-task. The values considered for x are noted on the right hand side as "$1 \leq x < \infty$" and below, more specifically, "only integer numbers".

The development of this substitution structure is based on Lisa's observation that the denominator of one fraction term always corresponds to the fraction that represents the preceding element (see excerpt 3.2h). Furthermore, the inscription added on the right hand side gives a more detailed description of the values that can be chosen for x: "x is- (..) larger one or equal (.) and smaller (.) smaller than infinity'" (178), "and only even ehm and only integer numbers." (179). They now start to deal with subtask 1.3 on work sheet 2 in which the students are requested to explain how and why the pattern works:

> **180** I: Oke. *(Lisa browses to work sheet 2)* (…)
> **181** R: *(sighs and laughs)* yes-
> **182** L: (..) oh lord. why does that work'
> **183** R: (19sec) *(both put the pen aside and put the first two note sheets in order in front of them)*

8.2 Extensive Reconstruction: Gestures' Contribution to the Epistemic Process

(0:13:30)
well [[because-]
(moves the left hand forward)

[we]]
(moves both hands together)

(takes work sheet 1 again)
184 L:(..) it's just like that

(0:13:36)
185 R: [because [this thing]

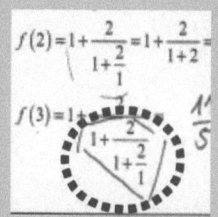

(points at and circles the representation of f(3) on work sheet 1)
,always again

[,just becomes one

(the right hand moves in steps downwards towards above the sheet)

186 /L: that appears-

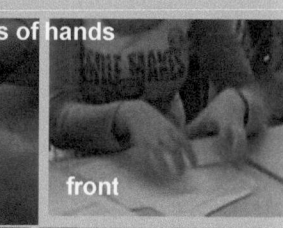

positions of hands
side   front

187 /R: ,one more] ,that is

[this one]] *(briefly points to the front)* (.)

[,that just becomes one
188 /L: this one term becomes-

(0:13:44)
189 /R: ,one two up there in addition] *(briefly points at the fraction of f(1) on work sheet 1)*
[[and this two up there] [we always divide by that before-]] *(movement of the hands in leaps back and forth)*

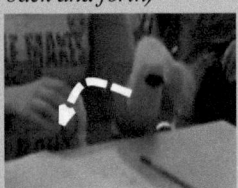

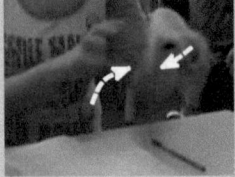
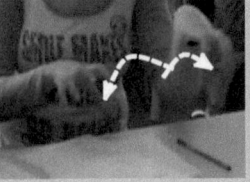

[[,and then that becomes] *(both hands are turned forward simultaneously)*

[,also always- one more-]]
(0:13:51)
(both hands turn around each other, both students laugh) (.)

8.2 Extensive Reconstruction: Gestures' Contribution to the Epistemic Process

190 L: *(laughing)* and that write down now comprehensibly.
191 R: quite. ,that is why I stopped.
192 L: *(whispering)* well. *(Rosa places note sheet 1 further away in front of her on the table and takes a second note sheet)* (9sec) yes.

(0:14:08)
193 R: well that is
[[every tree]

is in [itself a tree]] *(moves both hands in front of the body)* and

$$f(1) = 1 + \frac{2}{1} = 1 +$$

[[this thing] *(points at the representation of f(1) on work sheet 1)*

[is again a thousand times-] *(moves her right hand again in leaps in a circle directed anti-clockwise)*
194 /L: exactly.

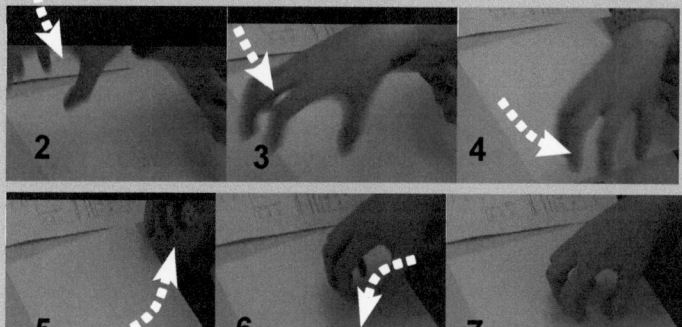

195 /R: ,in itself again.]
(9sec) *(takes a pen and writes "1.3" on the new note sheet 2 and circles it)*

To provide a better comprehension of the semiotic and the epistemic content of this scene, it will be decomposed into smaller excerpts. These are chosen to be semantically consistent, that is, each excerpt contains an action that is in itself completed as a speech act. Nevertheless, the single excerpts are always analyzed within the larger context of the entire scene.

8.2.1 Semiotic Analysis: Epistemic Functions of Gestures

Excerpt 4.1a: lines 180-187: "because..." - First Offer of a Justification

Lisa utters that she considers this task to be difficult, starting a verbal repetition of the task description with "oh lord" (182). After a pause of 19 seconds, Rosa starts to formulate an answer to this question (183-187):

> Well [[because-] *(moves the left hand forward)* [we]] *(moves both hands together) (takes work sheet 1 again)* (183)
> [because [this thing] *(points at and circles the representation of f(3) on work sheet 1)* ,always again
> [,just becomes one *(the right hand moves in steps downwards towards above the sheet)* one more] (185-187)

In line 183, she starts a first approach and places her hands in the gesture space, palm directed downwards and fingers loosely spread (Fig. 8.18).

Fig. 8.18:   Hands shaped in line 183

That this gesture reveals the Growth Point of her idea of explaining why the structure works does first become apparent in line 185. First, Rosa steps back from this general reference in the gesture space, making a fresh start to **make more precise** what she refers to in her idea: Verbally, she mentions that 'this thing always becomes one more' and **makes** the single aspects of this idea **more precise** by **sourcing** them **out** consecutively. "This thing" (185) is specified as denominator of the continued fraction $f(3)$ by **extracting** it as single a component within the entire term, ephemerally framed by the use of gesture (Fig. 8.19).

8.2 Extensive Reconstruction: Gestures' Contribution to the Epistemic Process

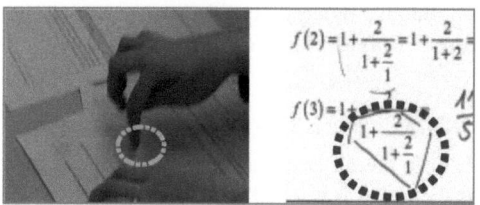

Fig. 8.19: Indication simultaneous to "[because [this thing]" (185)

What is meant by that it 'always becomes one more' is **depicted** in a metaphorical gesture co-timed to "[becomes ,only one ,one more]". The hand is formed as in line 183, palm directed downwards and fingers loosely spread (Fig. 8.20).

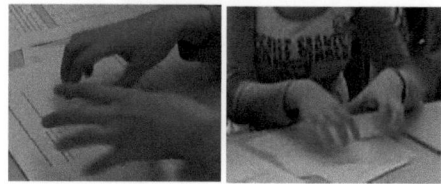

Fig. 8.20: Shape and position of hands in the second part of line 185, similar to the shape in line 183

In this gesture, the right hand metaphorically 'grasps' the component that has just been **extracted** as a component from the term representing the continued fraction $f(3)$ (see Fig. 8.19). At the same time, the left hand is placed in the interaction space above the table, probably referring to the rest of the term a bit left above the right hand. While the left hand is held, the right hand moves down in steps (Fig. 8.21):

Fig. 8.21: Movement of the right hand co-timed to "[becomes only one ,one more]" (185/187)

This refers to the inscription 'growing' downwards with every step. At this point the right hand may also recall the concrete entity **extracted** before, the denominator of $f(3)$, $1 + \frac{2}{1+\frac{2}{1}}$. Holding the left hand and moving the right hand downwards thus makes visible that this fraction can be found in every following fraction as a bottom part of the growing denominator. Another interpretation concerns the **depiction** of successively adding the extracted fraction in the bottom. Furthermore, the representation in gesture **makes more**

precise the idea of an iterative process, considering successive elements. However, in the following extract she elaborates her idea further, as becomes apparent in speech as well as in gesture.

Excerpt 4.1b: lines 187-189: "that is..." - Elaboration of the Idea

As before in line 185, Rosa reveals her idea more spontaneously in gesture as in speech. She briefly points to the upper horizontal interactions space in which the gestures in excerpt 4.1a have been accomplished, co-expressing a spatial reference to "[this one]]". While pointing with her right hand, the left hand is still placed and shaped similarly to how it has been placed and shaped before; as if metaphorically 'grasping something' (Fig. 8.22):

Fig. 8.22:   Virtual indication of "[this one]]" (187)

The reference of this virtual indication is ambiguous: In the following, Rosa first refers to the fraction of $f(1)$ on work sheet 1, and then specifies her reference to the "two up there" by indicating concretely the two as numerator (Fig. 8.23).

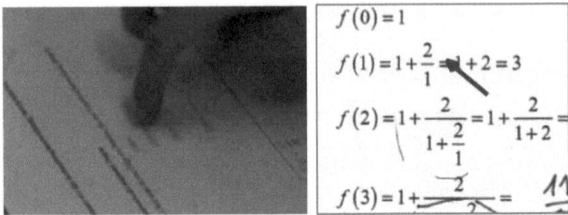

Fig. 8.23:   Indication co-timed to "one two up there]" (189)

It is thus not clear whether the idea that she expressed in the first gesture (187), by pointing in the upper interaction space, concerned the reference to the numerator of an arbitrary fraction, or the concrete element $f(1)$ to start the substitution with, since "$1 \leq x < \infty$", as has been noted for subtask 1.2 (Fig. 8.17). However, another interpretation allows for the possibility that it is referred to both at once, probably not even consciously. This interpretation is derived from the second part of Rosa's utterance in line 189:

[[and this two up there] [we always divide by that before-]] *(movement of the hands in leaps back and forth)*

[[,and then that becomes] *(both hands are turned forward simultaneously)* [,also <u>always</u>- one more-*]] (both hands turn around each other, both students laugh)* (.) (189)

Co-timed to "[[and this two up there] [we always divide by that before]]", Rosa **makes more precise** the explanation of the substitution pattern started in line 185. That the idea she expresses now is linked to the one expressed in line 185 (and to the Growth Point detected in line 183) is concluded from the catchment identified by the similarity of the gesture. Both hands are placed within the gesture space, suggesting 'grasping' something. This time, however, the gesture is shaped in a more explicit manner: While the palm is still directed downwards, the fingers are as well, but not loosely spread as before (see Fig. 8.24a). Her more accurate gesture may reveal that Rosa shaped the idea that she aims to express more accurately now as well and/or that she uses gestural representation more consciously. Furthermore, the concrete movements of the gesture differ to the one performed synchronously to the utterance in lines 185/187 (Fig. 8.21), suggesting the visual representation of a slightly changed idea.

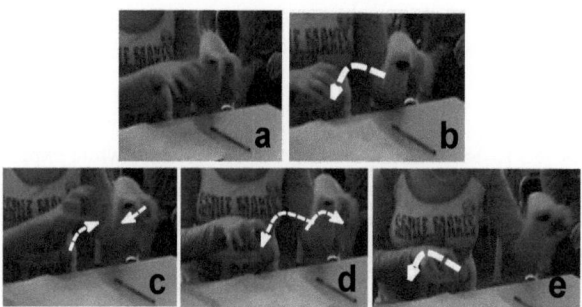

Fig. 8.24a-e: **Gesture co-timed to "[[and this two up there] [we always divide by that before-]]" (189)**

Again, the gesture metaphorically **depicts** the substitution of the denominator: While the left hand **sources out** the held reference to the numerator, the right hand 'grasps' the denominator from the fraction virtually represented in the gesture space. By this, the 'grasping' metaphorically **extracts** the numerator and denominator as two components of the 'body diagram'. In Fig. 8.24c-e, the reference to the substitution of the denominator by the fraction before is refined, **depicting** how the right hand 'grasps' where the left hand has formerly been positioned as numerator. That **makes more precise** that the entire preceding fraction is 'grasped and put' in the numerator of an element placed lower in the list of fractions virtually placed in the gesture space. In turn, Rosa adds that "[[,and then that becomes] [,also <u>always</u>- one more]]", performing two gestures that seem similar at first sight but reveal core differences of the idea:

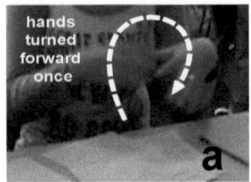

Fig. 8.25a: "[[,and then that becomes]"
Fig. 8.25b: "[,also a<u>lways</u>- one more]]" (189)

First, the hands are synchronously turned forwards around the wrist once, co-timed to "[[,and then that becomes]". This forwards movement may metaphorically refer to a forthcoming point in time, visually referring to the "then" as a successive step of the sequence. The discrete component of 'one step' is complemented by the reference to a continuous process, metaphorically depicted by the hands turning around each other.[147] This gesture **illustrates the generality** in the pattern that becomes expressed in speech ("[,also always- one more]]"). Both metaphorical references specify a style of the development of the fractions: first iteratively, referring to a single step, then continuously, as infinite process. Both aspects are **glued** together in a gesture unit, realizing the general structure of the process.

As the students utter then (190/191), it is a hard task writing down the thought explained by Rosa in lines 183 to 189. Rosa's extensive use of gesture while expressing her idea already suggests that it is not easy to be captured in words alone. Moreover, she implicitly mentions that she is not yet done with her explanation but stops because the explication does not seem to be comprehensible to her (191). In a pause of nine seconds preceding the last excerpt of this scene, Rosa arranges her thoughts to express the explanation of 'how and why the pattern works' in a different, metaphorical way:

Excerpt 4.1c, lines 193-195: "well that is… – Metaphorical Elaboration of the Idea
Rosa establishes the 'tree-metaphor' to express that every continued fraction again contains the similar continued fraction:

well that is [[every tree] is in [itself a tree]] *(moves both hands in
front of the body)*
and [[this thing] *(points at the representation of f(1) on work sheet 1)*
[is again a thousand times-] *(moves her right hand again in leaps in
a circle directed anticlockwise)* ,in itself again.]* (193/195)

The statement of 'a tree being in itself a tree again' is suggested to rather refer to 'every limb of a tree being a tree for itself again'. The use of metaphors in this concerns 'something being the same in itself again'. It maps the source domain of the metaphor, the tree

---

[147] See also excerpt 3.1g on page 180.

with its limbs being trees, on the target domain, the mathematical representation of continued fraction with its denominators being continued fractions again. However, it is Rosa's use of gesture that makes this mapping explicit by **sourcing out** the references to be represented equally within the gesture space.

Fig. 8.26: Metaphorical representation of the tree-metaphor ("[[every tree] is in [itself a tree]]", 193)

First, she refers to the 'tree being in itself a tree' by placing the left hand in the gesture space, palm directed to the left side, fingers loosely spread, and the right hand loosely open placed close to the palm of the left hand less spread (Fig. 8.26). Co-timed to speech, the reference to the tree (right hand) being a tree in another tree (left hand) got a physical form through **sourcing** it **out** to the eyes. With a similar gesture (Fig. 8.27), she then **depicts** the infinite iterative process of 'putting' the continued fraction of $f(1)$ in the denominator of the next fraction:

Fig. 8.27: Position of the hands in the beginning of the gesture co-timed to "[is again a thousand times-] ,in itself again.]" (193/195)

The left hand is held as in the gesture accompanying the verbal tree-metaphor (Fig. 8.26), while the right hand reminds the 'grasping' of the denominator that has been used before in 185 and 189 (Fig. 8.20, Fig. 8.24). **Sourcing out** both references, the one to the tree-metaphor and the one to the 'grasping and putting'-metaphor thus **makes more precise** the correspondence between them and, furthermore, the correspondence to the idea of recursive substitution explained before in this scene. The combination of both, the held left hand (tree) and the moving right hand (fraction) enable Rosa to integrate also the idea of infinite iteration:

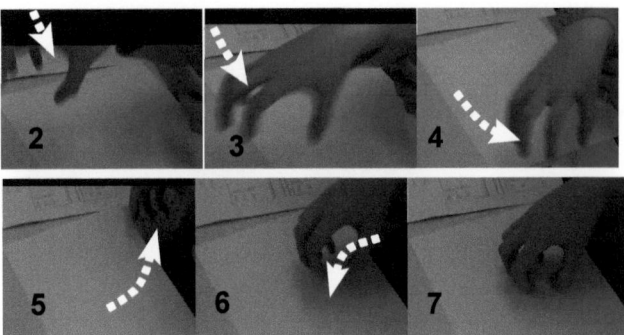

Fig. 8.28: Movement of the right hand co-timed to "[is again a thousand times-] ,in itself again.]" (193/195)

The fraction **extracted** from the inscription and 'grasped' by the right hand is 'put' in one step of the sequence after the other, hence iteratively in every next element. The movement in a circle (Fig. 8.28) **depicts** a geometric construct without an end, metaphorically refers to the process being infinite and extends the verbally uttered "thousand times" to 'infinitely'. Lisa expresses confirmation (194: "exactly.") and Rosa reveals that she explained her idea sufficiently by implicitly suggesting to formulate the idea such that it can be inscribed, starting with fixing "1.3" on a new note sheet.

Summary of Epistemic Functions of Gestures in This Scene:

Rosa **sources out** the "this thing" mentioned in speech (185) by indicating the fraction of $f(1)$ and with this makes more precise the reference she refers to in her idea. She **extracts** it by circling the term and with that makes possible its figurative 'grasping' in turn (185). This enables the **depiction** of 'adding something below' and **makes more precise** the reference to an iterative process in which each element of the sequence follows the other. To elaborate her idea, Rosa makes use of a reference in gesture she had already **sourced out** in line 185 and with this recalls the extracted fraction. The numerator and the denominator are **sourced out** into the gesture space and a metaphorical **depiction** of 'grasping the denominator of the preceding fraction and putting it below, as next element' makes it simpler for Rosa to express her idea **more precisely** as 'substitution'. Furthermore, it is not only referred to a single step, but to substitution as a recursive pattern deriving stepwise the elements in a continuous process. The aspects of 'iteration' and 'continuous process', **illustrated generally** in gesture, are **glued** together by two gesture phrases constituting one gesture unit (189). Putting the idea in different words, Rosa establishes a metaphor to condense her idea to 'the fraction is in itself again infinitely again'. The reference to the metaphor and the reference to the fraction used before are **sourced out** into the gesture space and its similarity in shape establishes the

8.2 Extensive Reconstruction: Gestures' Contribution to the Epistemic Process 219

correspondence between them in a catchment. Combining in this gesture the reference to the extracted fraction (185), to the substitution (189) and to the metaphor (193), Rosa can metaphorically represent the substitution of the fraction in itself as a continuous stepwise process. Different than before, the aspects are not glued together referring to one idea after the other, but merged in a metaphorical depiction of infinity.

Remark: The Associated Sign 'Grasping Hand'

As becomes eye-catching, an important feature of the gesture use in this scene is the establishment of an associated sign, termed the 'grasping hand'. It develops from successive extraction, outsourcing, and depicting and is associated to a component extracted from the inscribed continued fraction: In line 185, the denominator is extracted and grasped by the right hand; in line 189 also the 'grasping' left hand representing either the two as numerator or the preceding fraction becomes explicit. The use of this gesture can be traced throughout the scene, developing the idea of substitution as a recursive pattern and simplifying its representation (see Table 8.2). During this process, Rosa repeatedly changes between the different levels of reference. Her references on level 1 provide a shared basis for the epistemic process within the social interaction. Repeatedly, they are used as a common anchor point for elaborating different variations of the 'grasping'-idea on level 3. Associated signs may combine forming- and performing-functions in order to **recall** the information bundle associated to the 'grasping'-gesture in its use.

Table 8.2: Development and use of the associated gesture 'grasping hand'

Line	Associated Sign	Current Information Bundle	Immediate Mathematical Object
183	Gesture on level 3: **Growth Point 1**	None	None: Growth Point
185a	Extraction on level 2	- component of the term that is considered to be 'grasped'	

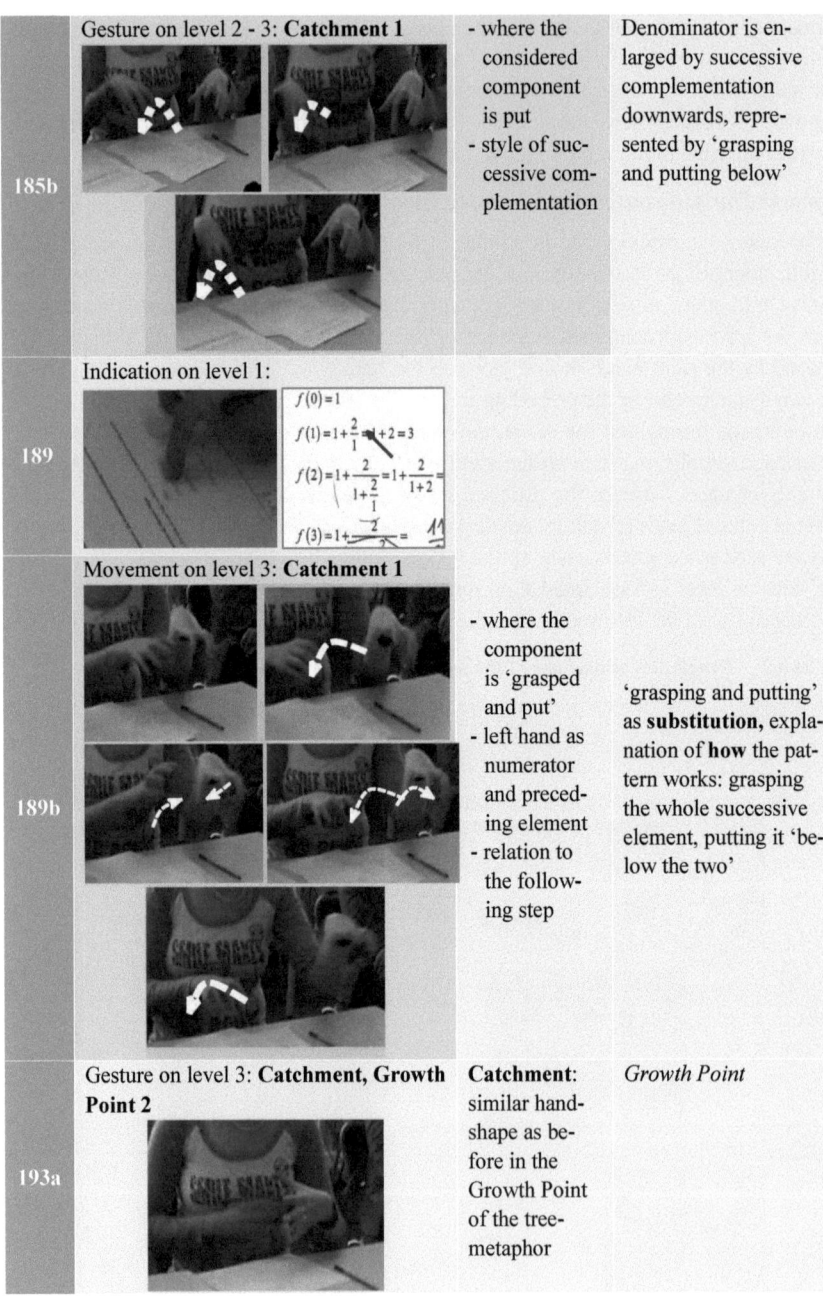

8.2 Extensive Reconstruction: Gestures' Contribution to the Epistemic Process

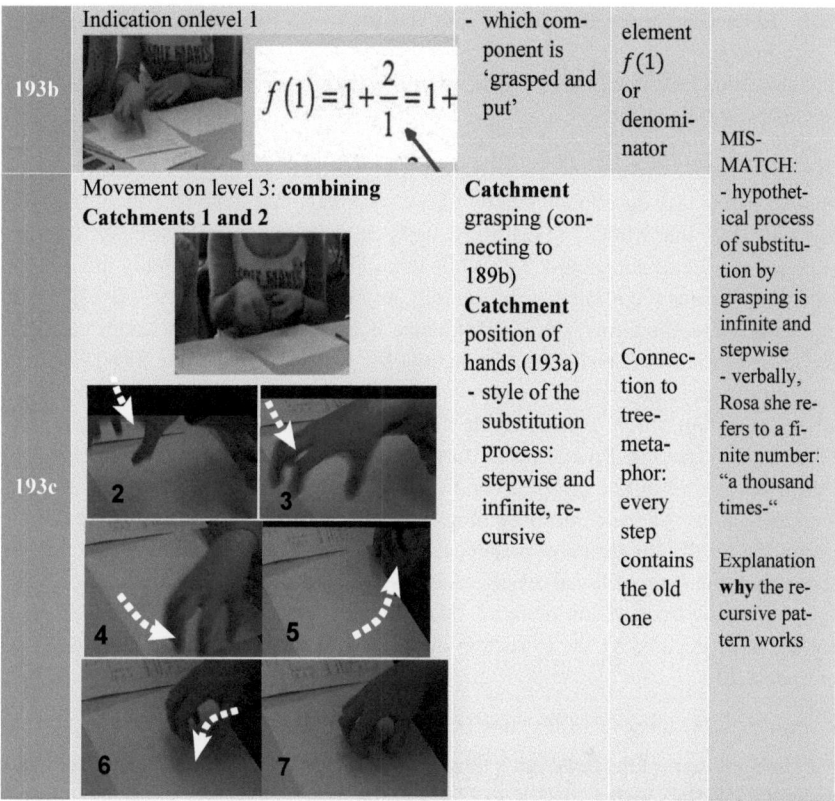

This tracing of signs associated to the development of the idea to explain and justify how and why the recursive pattern works reveals the power of the catchments tied together by gestures using the 'grasping hand'. More and more aspects are added or varied respecting the concrete implementation of the grasping hand, so the initial idea of 'grasping and putting' is refined more and more in the use of gesture. This allows a simultaneous reference to **how** and **why** the recursive pattern works in line 193, recalling the substitution in an infinitive iterative process in combination with the compositional structure of the continued fraction.

The semiotic analysis however did not comprise the reconstruction of the epistemic process so far. Epistemic functions are defined to be directly involved in the accomplishment of epistemic actions and their analysis already suggests that the performance of gestures crucially influences the epistemic process in this scene. This will be captured in the following section.

8.2.2 Reflection on the Benefit of Using Gestures with Respect to the Epistemic Process

The interaction has already been described in chapter 8.2.1 so that it will be restricted to the epistemic actions here.

On excerpt 4.1a: lines 180-187: "because..." - First offer of a reason

Initiated by the task description and by Lisa's seemingly desperate repetition of it, "oh lord. why does that work' " (182), Rosa starts an approach of justifying the structure proposed as solution for subtask 1.2. While this approach is started verbally in line 185, gesture anticipates it co-timed to the verbally imprecise "well [[because-] [we]]" (183, Fig. 8.18). Here, the Growth Point of the idea is expressed in gesture and reveals the seed of the idea that is elaborated throughout the whole scene. In line 185-187, Rosa develops the basic structure of the reasoning in this scene that is identified as an elaborating connection action. By specifying the reference to the fraction as extracted from the inscription (see Fig. 8.19), she clarifies her verbal utterance and makes more precise what is represented by the 'grasping hand'. From speech, this can be identified as leading to verbally uttering a structure concerning the composition of the fraction, used as a first approach to justify the recursive pattern. Gesture makes it more precise though, making visible that this compositional aspect concerns the increasing extent of the inscription representing the fraction. Furthermore, this generally depicted structure (Fig. 8.21) has already been prepared by the extraction of the term in the beginning of line 185 (Fig. 8.19 and 8.20).

On excerpt 4.1b: lines 187-189: "that is..." - Elaboration of the idea

In speech, the start of the backwards directed connection to the structure just referred to is started by Rosa's verbal "that is" (187). In gesture, she links this to the general depiction of 'always adding in the bottom of the fraction' (Fig. 8.21) by taking up the associated sign of the 'grasping hand' (Fig. 8.24a-e) as a situationally conventionalized sign that is connected to the idea represented in line 185. Furthermore, the depiction is varied to realize the reference to a modified structure: always taking the preceding element and putting it in the denominator to compose the following element. That this is actually seen as embedded in a continuous iterative process to reason the recursive pattern, becomes apparent in gluing 'one general step' (Fig. 8.25a) in a 'continuous process' (Fig. 8.25b).

On excerpt 4.1c, lines 193-195: "well that is... – Metaphorical elaboration of the idea

The second approach to reasoning as connecting action directed backwards respecting a structure starts verbally in line 193. It is synchronically realized as sourced out in gesture and both, the verbal and the gestural realization finish at the same moment in line 195.

The use of the associated sign 'grasping hand' prepares, realizes and supports epistemic actions all at once. In its first use (183) the seed of a larger, significant idea that is elaborated within the interplay of speech, inscription and gesture can be detected. This elaboration is identified in the catchments indicated by the recurring use of this associated sign. As we have seen here, it can lead to giving visual access to a structure (185) and in turn can be used to reflect on it or to modify it (189, 193).

Remark: Link to the epistemic process

This scene is an example for the large influence that gestures can have on the epistemic process. An analytical distinction of gesture and speech led to an extended condensed process diagram of the EDE, based on the notation of Bikner-Ahsbahs (2006). The epistemic actions that can be reconstructed from the verbal utterances are represented by black lines and those realized exclusively by the use of gesture by light blue lines. Signs are combined to visualize relations between epistemic actions within the process.[148] The separation of the verbal and non-verbal contribution to the epistemic process does not only make apparent the direct epistemic benefit of the use of gesture in this scene, but also visualizes how the verbal and the non-verbal elaboration of the epistemic actions are deeply intertwined.

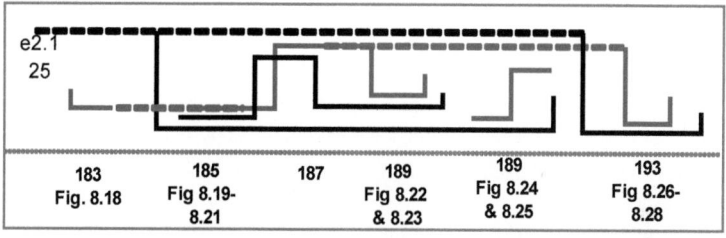

Fig. 8.29:   Condensed process diagram of Scene CF7e5.1

The structure which is aimed to be justified in this scene has been seen by the students in scene CF7e2.1, starting in line 25. This is indicated on the upper left hand side as continuation linked to that utterance. The light blue lines make visible that gestures reveal the epistemic process differently, anticipating and complementing the epistemic process reconstructed from speech. The number of the concrete line and of the picture representing the gesture give reference where further information to the performance and benefit of the gesture are provided. As can be seen from the light blue lines, the gestures reveal an epistemic process parallel to the one reconstructed from the verbal utterances.

---

[148] The dotted line indicates a reference to a previous epistemic action that may have been interrupted or that is simply reconsidered. The superposition of epistemic actions represents their temporal relation, their visual "embedding" their contextual relation. For example, the elaboration of the structure in lines 185 till 189 comes along with seeing and elaborating another structure, in speech as well as in gesture.

In line 185, the compositional structure of the continued fraction is suggested by introducing the grasping hand in combination with the downward movement. The grasping hand is now a recurring feature to refine the idea in lines 189 and 193, connecting it with other aspects (see also Table 8.2).

8.3 Summary

As we have seen in this chapter, gestures can benefit the epistemic process by fulfilling different epistemic functions. When gestures fulfill *forming-functions*, their epistemic benefit arises from making a visual representation of a mathematical object or idea accessible. This can be accomplished by *sourcing* its reference *out* to visual representation on one of the three referential spaces (see chapter 7.2), by *depicting* a 'body diagram' on levels 2 or 3,[149] or by *extracting* components or relations between them from inscription. While these three epistemic functions concern symbolic-iconic gestures whose meaning arises from inscription, metaphoric gestures can *illustrate generality* of a concretely uttered idea. All three forming-function accomplished by iconic-symbolic gestures highlight on different benefits concerning the epistemic process. Sourcing out can adjust common knowledge about representations of the verbal reference. Depicting and extracting gestures refer to a mathematical object or idea in an iconic way so that the ephemeral representation shaped by the gesture is condensed to the essential relations, emphasizing the core idea. Depicting gestures may not only represent a diagram by relating entities, but also allow *dynamic* developments to be captured. The iconic character of depicting gestures may recall formerly expressed ideas by evoking a similarity. This may induce catchments in which it is tied on with earlier actions within the epistemic process. It turned out that by those and other benefits, gestures do not only *support* epistemic functions that are carried out in the verbal utterance, but have also been found to *prepare* epistemic actions. They can provide conditions that prompt epistemic actions that become explicit in speech. Furthermore, they can also *realize* epistemic actions themselves, providing an indicator for what is left implicit within the social interaction.

Although the epistemic benefits that can be achieved by gestures crucially depend on the concrete situation, some especially effective ways of using gestures to enhance the epistemic process have been identified. The *performing-functions* describe how gestures have been used in a characteristic way leading to an effect, so they catalyze the epistemic process at that very moment. Without claim for completeness, six performing-functions have been reconstructed throughout the data: Gestures can be used to temporarily *focus* on a represented entity to literally hold the attention on to it. When a gesture supports the concretizing of a seen structure by giving a concrete example, the gesture *exemplifies*. *Contrasting*-gestures give visual access to differences between an actual and a hy-

---

[149] That is, an icon of relations of which at least a part is embodied by the use of gesture.

pothetical case, facilitating their comparison. Gestures can *make* the references to mathematical objects *more precise* when epistemic actions are accomplished. Aspects specified by different gesture phrases can be *glued* together in a gesture unit, suggesting a semiotic connection between them. A metaphorical gluing can *structure the verbal discourse* and with this, support argumentation.

Although the epistemic functions of gestures are presented as analytically separated, their advantages becomes especially salient when considering how they play together to enrich the epistemic process. They actually are rarely performed in separation, so the epistemic benefit of the gestures is considered to be grounded in the interplay of various forming- and performing-function.

An extended version of the condensed process diagram, which is a diagram using signs to condense the epistemic actions accomplished in a scene (see Fig. 8.29), makes the extent of gestures' influence on the epistemic process visible. These also provide a possibility to compare this influence for different epistemic processes as will be done in chapter 9.

9 Towards the Role of Gestures: A Comparison

On Differences and Similarities Concerning Gestures' Contribution to Social Processes of Constructing Mathematical Knowledge

The categories presented in chapters 7 and 8 have been developed, tested, and refined throughout the data. In chapter 7, this concerns the within-functions as embracing two dimensions of categories: The specifying-function captures how gestures contribute to shaping the immediate mathematical object by enriching the semantic content of speech, the referential levels add a spatial component that suggests a degree of conceptualization expressed in gesture (the concrete, the potential and the general, see Fig. 7.11). Furthermore, concrete aspects specified by the gesture have been worked out to define subcategories (chapter 7.1.1). Similarly, the main categories of forming- and performing-functions have been defined for the epistemic functions of gestures (8.1), both categories split up in subcategories that provide a more detailed description of how gestures benefit the accomplishment of epistemic actions. I also found that gestures prepare, support, but also themselves *realize* epistemic actions; the *condensed process diagrams* provide a possibility to compare how many more epistemic actions are accomplished in gesture before being elaborated in speech, or even exclusively in gesture.

This chapter represents the horizontal and vertical comparison, that is, between different pairs of students and between the elaborations of different tasks by the same pair (see section 6.2.3.3). Similarities and differences are worked out to shed light on gestures' contribution to social processes of constructing mathematical knowledge.

9.1 Horizontal Comparison: One Task, Different Pairs of Students (PA5, PA7, PA6)[150]

The three pairs of students spent different amounts of time on elaborating the parabola task. While it took Rosa and Lisa 107 minutes to work out the parabola task (PA7), Mike and Tim (PA5) finished after 132 minutes, and Kris and Simon (PA6) after 168 minutes. Because of these differences, it is of no sense to compare total amounts but only percentage values. For example, the lowest relative amount of epistemic-dense episodes has been detected in the PA6-data, deviating from the other two data sets (see Table 9.1):

---

[150] The order of data sets as chosen for developing the functions of gestures (see section 6.2.3.2) is retained.

Table 9.1: Overview on the epistemic-dense episodes identified within the three PA-data sets

	PA5	PA7	PA6
Number of EDEs	16	16	24
Total Duration of EDEs (in Minutes)	43:16	30:58	37:59
Relative Duration of EDEs with Respect to the Duration of the Entire Working Process	32.8%	30.0%	22.6%

As has been mentioned in chapter 7.1.2, 91.4% of the mathematical illustrators, the gestures considered for this study, specified aspects of the verbal utterance. The horizontal comparison shows that this average value is even surpassed for two of the three pairs (Table 9.2).

Table 9.2: Overview on the frequencies of gestures in the epistemic-dense episodes of the PA-data sets

	PA5	PA7	PA6
Total Amount of Mathematical Illustrators (Interviewer's Gestures in Brackets)	155 (8)	126 (19)	274 (5)
Specifying-Gestures (Total and Relative)	149 90.3%	119 94.4%	269 98.2%

While Kris and Simon (PA6) produced the largest amount of specifying-gestures, they had the lowest percentage of epistemic-dense episodes. This shows that it does not hold 'the more the merrier' for specification in gesture. Moreover, the qualities of the gestures become relevant when conjecturing on the epistemic benefit of gestures.

The Within-Functions Differ Between the Three Pairs of Students

Only 57.2% of these specifying-gestures have been identified as mono-specifying, that is, 42.8% of them specified more than one aspect. The total amount of aspects specified is thus even larger than the percentage amount listed above. As is displayed in Fig. 9.1, the amount of gestures specifying one, two, three, or four aspects differs remarkably between the three data sets:

9.1 Horizontal Comparison: One Task, Different Pairs of Students (PA5, PA7, PA6)

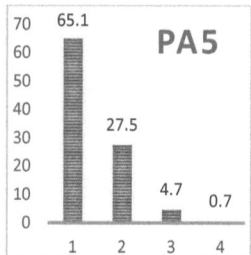
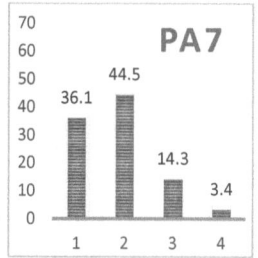
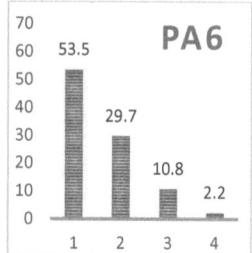

Fig. 9.1: *Horizontal Comparison:* **Percentage of gestures that specify one (1), two (2), three (3) or four (4) aspects[151]**

Only slightly more than 36% of Rosa and Lisa's gestures (PA7) specified only one aspect, while they used many gestures specifying two aspects (44.5%). This only gives a brief idea about the *effectiveness* of the students' gestures concerning their representational function, transporting additional meaning to the verbal utterance. However, even more interesting are the results on the concrete within-functions of the students' gestures, that is, the specifying-functions with respect to the three referential levels. The following diagrams display which aspects have been specified on levels 1, 2, and 3 by the three pairs of students in the epistemic-dense episodes of the parabola tasks:

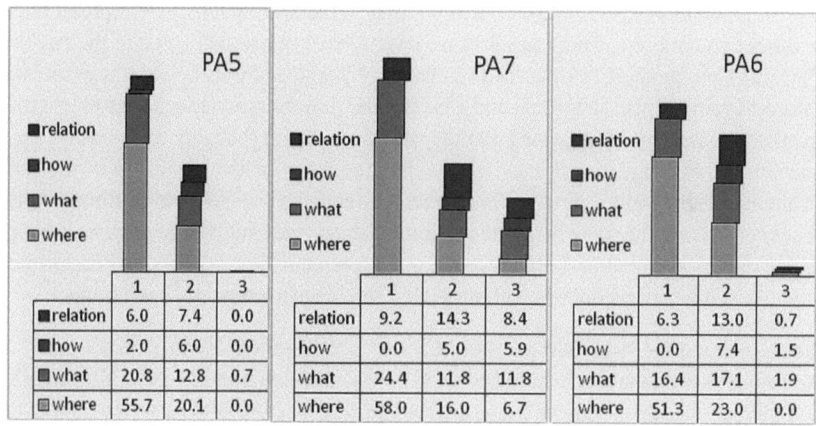

Fig. 9.2: *Horizontal comparison:* **Amount of specifying-gestures specifying a certain aspect on a certain referential level (percentages given in the table below)**

Again, the use of gestures in PA7 reveals a noticeable feature: Compared to the other two data sets, the amount of aspects specified on level 3 has been remarkably high

---

[151] Values do not add up to 100% because of gestures for which the meaning of the gesture phrase cannot be understood without considering the entire gesture unit as glued together in one movement. These "partial aspects" rate up to 3.7% in the PA6-data. However, they are not considered in this comparison.

(32.8%, compared to 0.7% for PA5 and 4.1% for PA6). Given that the amount of multi-specifying-gestures has also been identified to exceed the respective values for PA5 and PA6 (see Fig. 9.1), this high value of *aspects* specified on level 3 needs to be seen against the percentage of *gestures* that specify on level 3:

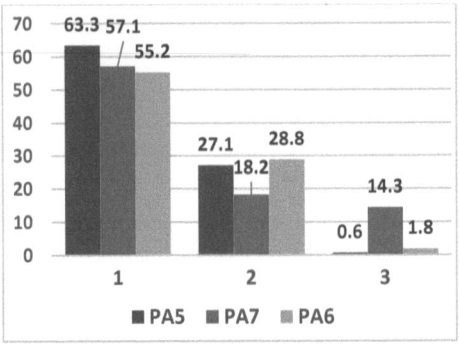

Fig. 9.3: *Horizontal comparison:* Amount of specifying-gestures on each of the three referential levels (in %)[152]

While all three pairs of students specify with a similar rate on level 1, Rosa and Lisa perform much more specifying-gestures on level 3, the level that is underrepresented in the other two data sets. These gestures on level 3 have been established in the packing of information bundles, as described in chapter 7.2, and can thus be seen as associated gestures. From the fact that Rosa and Lisa use the gesture space to a larger extent than the other two pairs, it is suggested that they more often, and probably more easily, share conceptual knowledge as expressed in associated gestures on level 3. However, this does not automatically imply a direct influence on epistemic actions. Moreover, the epistemic process of Rosa and Lisa is benefitted by establishing meaning of free gestures through the packing of information bundles on all three referential levels. This facilitates interacting on mathematical objects without giving reference to concrete inscriptions.

The Associated Gesture 'Shape of the Curve' is 'Suggested' by the Task

Interestingly, the associated gestures seem to not develop completely arbitrarily: All three pairs of students establish the associated gesture 'shape of the curve' to refer to the curve within their working process.[153] This actually is not a big of surprise, given that the representations provided prompt a strong graphical association. This result raises the question whether some associated signs/gestures are established more naturally for a

---

[152] Values do not add up to 100% since gestures on levels 1/2 and 2/3 are excluded.
[153] Beside, Lisa and Rosa also establish a gesture referring to the 'axis of reflection' (See chapter 7.2.1, example 2.1b.) and Kris and Simon establish another one that has been interpreted as '*distance-gesture*': For this, the thumb and index- or middle-finger indicate two endpoints of a segment, loosely forming a C-shape. Kris and Simon share this gesture on levels 1 and 2, but do not transport it into the gesture space.

specific task or learning environment than others, and questions how the development of associated gestures can be initiated more effectively. Keeping in mind that three data sets provide a very small sample, this hypothesis should first be tested by means of individual data sets, probably in further studies.

There is No Remarkable Difference in the Impact of Gestures on the Epistemic Actions

Interestingly, the horizontal comparison of gestures' *influence on the epistemic actions* accomplished by the students as displayed in the condensed process diagrams does not reveal any remarkable conspicuities.[154] However, this fact may be remarkable itself, considering that they seem to have used gestures differently, as can be concluded from the observations described in the latter sections. The condensed process diagrams of the epistemic-dense episodes make visible that in all three working processes, the use of gestures realized epistemic actions that led to structure seeing or parts of its extension. It is suggested that the sample of three pairs of students does not provide sufficient evidence to make statements on considerable differences in this regard.

9.2 Vertical Comparison: One Pair of Students, Different Tasks (PA7, CF7, IN7)

As has been described in chapters 5.4.2, 6.1.3, and 6.2.2, all three tasks differ in various aspects, such as mathematical topic, representations provided, and mathematical activities requested by the task. The rate of the in-total-duration of epistemic-dense episodes in relation to the whole working processes ranged up to 38.5% for the induction task:

Table 9.3 Overview on the epistemic-dense episodes identified within the three data sets concerning Rosa and Lisa's elaborations of the three tasks

	PA7	CF7	IN7
Number of EDEs	16	19	8
Total Duration of EDEs (in Minutes)	30:58	34:20	22:26
Relative Duration of EDEs with Respect to the Duration of the Entire Working Process	30.0%	33.0%	38.5%

Only 86.6% of Rosa and Lisa's gestures in the elaboration of the arithmetic algebraic task (CF7) specified one or more of the four aspects and only 78.1% did so in their elaboration of the task dealing with logical reasoning (IN7). Found below the average of

---

[154] For spatial reasons, the diagrams are not presented here but attached in the appendix, on pages 309-310.

91.4%, these rates however still indicate that there is a lot of information hidden in the gesture modality.

Table 9.4: Overview on the specifying-gestures in Rosa and Lisa's epistemic-dense episodes

	PA7	CF7	IN7
Total Amount of Mathematical Illustrators (Interviewer's Gestures in Brackets)	126 (19)	164 (12)	96 (0)
Specifying-Gestures (Total and Percentage)	119 94.4%	142 86.6%	75 78.1%

From the way the tasks are designed, it has been expected that the parabola-task prompts a great amount of gestures used to indicate parts and graphical entities within the representations, hence gestures on level 1. Surprisingly, the amount for these gestures is still high for the induction-task (IN7), in which a word problem has been given without any further graphical or numerical representation and data provided:

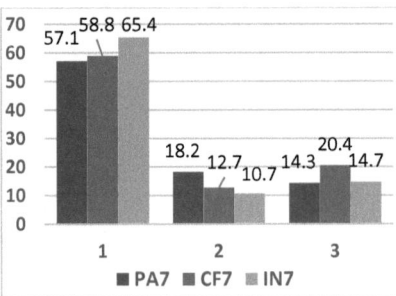

Fig. 9.4: *Vertical comparison:* **Amount of specifying-gestures on each of the three levels (in %)**[155]

In addition, the amount of mono-specifying-gestures is extraordinarily high for the elaboration of the induction task:

---

[155] Values do not sum up to 100% since gestures on level 1/2 and 3/4 are excluded.

9.2 Vertical Comparison: One Pair of Students, Different Tasks (PA7, CF7, IN7)

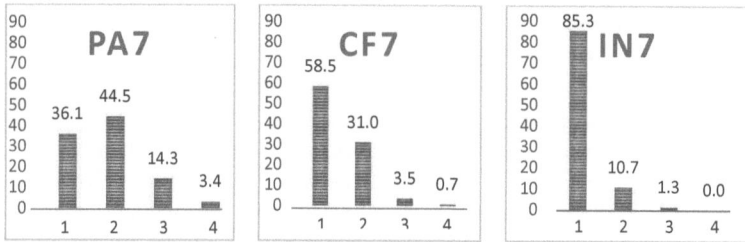

Fig. 9.5: *Vertical comparison:* **Percentage of gestures that specify one (1), two (2), three (3), or four (4) aspects**

Further information on the aspects actually specified is given in the following Fig. 9.6:

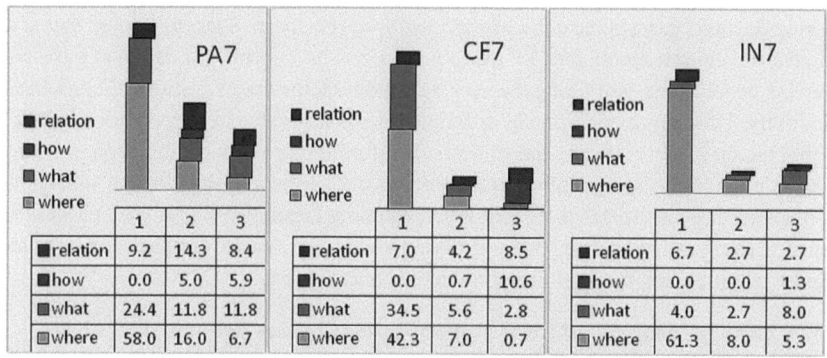

Fig. 9.6: *Vertical Comparison:* **Amount of specifying-gestures specifying a certain aspect on a certain referential level (percentage given in the table below)**

In the Induction Task, a Beneficial Gesture Use Requires to Create Inscriptions first.

As we can see now when comparing the coding-results for the IN-data to those of the other two data sets (see Fig. 9.6), most specifying-gestures in the IN7-EDEs give additional information on the 'where'. More precisely, this concerns concrete references within a fixed diagram. Actually, Rosa and Lisa produced many inscriptions themselves and indicated entities of it, specifying their location by using gestures. Using pointing gestures, they sourced out references to specific consultants considered within the inscriptively represented situation. By this, they shared these references in social interaction while elaborating concrete cases. However, this would not have become possible without first producing diagrams to refer to. This shows up the importance of not using only gesture alone when sharing diverse cases in social interaction, but to also source out inscriptions as a stable and lasting shared source of reference. Similar observations have been made for the parabola task when Rosa tries to source out a relation of relations (*rate of change of the local rate of change*) into the gesture space (see section 8.1.1.1, excerpt 3.1b). In the case of the induction-task, there are also several pieces of information to be represented at the same time when considering a concrete case: (i) the number of consultants sitting in the circle, (ii) the number of consultants wearing a marked hat, and (iii) the consultant whose perspective is considered. All three components need to be related to shape a concrete case and to decide whether the consultant knows about his own status at a given ring of the bell (and therefore raises his hand). Even more, the own knowledge on the number of hats seen needs to be related to the reaction of this other consultant. At each stage, the situation can be varied to consider different situations (own position, number of the ring of the bell, and behavior of the others) and to conclude about different implications (number of marked hats visible, one's own behavior). While the basic situation (number of consultants, position of consultants with marked hats, one's own position) has been recorded in inscription, especially in the beginning of exploring the task, they source out *varying conditions* (perspective taken, marked hats visible for the chosen consultant) by using gesture to share them as a common basis for drawing conclusions. Not fixing the varying condition, the use of gestures sill preserves flexibility. This way, exploration of different cases is facilitated by specification 'where' within the diagram created by the students. But furthermore, this combination of fixing the basic situation in inscription and sourcing out further information by indicating through gesture also seems to benefit argumentation: Gesture provides visual access to concrete cases and by that, supports justifying within these cases. Gestures reduce complexity by grounding the argumentation in the concrete case.

The 'Origin of Meaning' of Level 3-Specifications Differs Between the Tasks
Another remarkable fact that becomes visible in Fig. 9.4 concerns the high rate of specifying-gestures on level 3 for the elaboration of the continued fraction-task (CF7; 20.4%, compared to 14.3% and 14.7% for PA7 and IN7 respectively). Furthermore, Fig. 9.6 reveals that the quality of these specifications also differ: Almost half of the gestures performed on level 3 in the CF7-data specified a style (how). Different to the geometric-algebraic task, for which graphical entities have been transported up to level 3, the arithmetic-analytic task did not deal with any graphical representations that are expected to be shaped iconically within the gesture space to represent the mathematical object. Moreover, Rosa and Lisa established associated gestures to 'embody' ideas in a metaphorical way. An example has been presented in chapter 8.2, where the 'grasping hand' becomes established as a gesture used to 'extract a part of the inscription and put it somewhere'. It reveals the strategy of complementing the denominator (Fig. 8.21) or the one of substituting the denominator of a fraction by the previous element (Fig. 8.24a-e). The way (style) in which this gesture 'grasps and puts' (See, for example, Fig. 8.21, Fig 8.24a-e, and Fig 8.28.) is identified to specify the strategy explained by Rosa to be followed as conducted *iteratively,* and the process itself to be considered as *infinite* (Fig. 8.28). Similar to the grasping-hand, there have been found some more associated ges-

tures in the CF7-epistemic-dense episodes that are considered to be grounded in the inscriptions: For example, the 'increasing numbers'-gesture has been presented in section 7.2.2.2 as established as shared by Rosa and Lisa. Furthermore, the idea of the limit as value that is approached in a *potentially* infinite process may prompt the performance of dynamic gestures on level 3 when referring to the limit of the sequence given by the continued fraction. The following excerpt shows such a specification of style:

Excerpt 5.1a: CF7-908 (CF7e17, 1:18:14)

In subtask 2.2, Rosa and Lisa stated two conjectures about their observations on the elements of the sequence as filled in the table in subtask 2.1.[156] The first one is that "the decimal number approaches the two more and more" and the second one concerns the pattern that describes how to derive the decimal fraction of a general element $f(x+1)$ from the decimal fraction of the preceding one, $f(x)$. They write down the second conjecture as represented in Fig. 9.7:

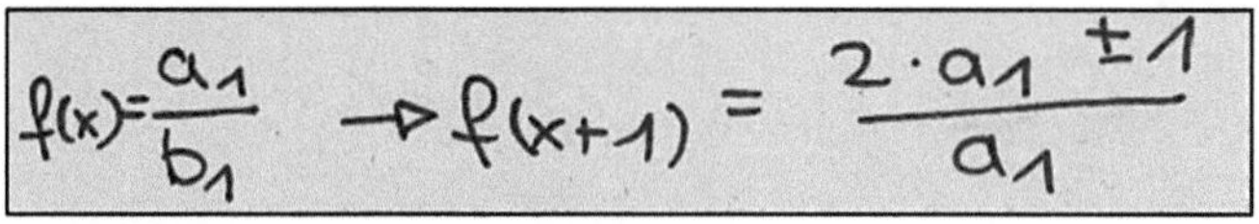

$$f(x) = \frac{a_1}{b_1} \rightarrow f(x+1) = \frac{2 \cdot a_1 \pm 1}{a_1}$$

Fig. 9.7: Lisa and Rosa's notation of the pattern concerning the numerator and denominator as fixed for subtask 2.2 (CF7)

To work out subtask 2.3, they are challenged to substantiate the conjectures they stated. Another recursive pattern has been found out to be noted as[157]

$$f(x) = \frac{a}{b}, \text{ then } f(x+1) = 2 + \frac{1}{a}.$$

Using another pattern they found before,[158] they can write $f(x+2) = 1 + \frac{2}{2+\frac{1}{a}}$ and transform this to $2 - \frac{1}{2a+1}$, successor of f(x+1) (note sheet 5). In the beginning of the episode that this excerpt is taken from, the interviewer prompts the students to focus on the terms $2 + \frac{1}{a}$ and $2 - \frac{1}{2a+1}$ (893). Lisa visualizes the connection between them by producing the following diagram on note sheet 5 (894):

---

[156] See the students' written products as attached in the appendix.
[157] They first considered, more generally, $f(x) = 2 \pm \frac{1}{a}$. To keep focused, they restricted their approach to one of the two possible cases.
[158] Solving subtask 2.1 they found the recursive pattern $f(x) = \frac{2}{f(x-1)}$ to derive one element from its predecessor (see Fig. 8.17).

$$2 + \frac{1}{a} \rightarrow 2 - \frac{1}{2a+1}$$

Fig. 9.8: Visualization of two consecutive elements of the sequence, where $2 + \frac{1}{a}$ is the successor of a general fraction $f(x) = \frac{a}{b}$

This is what Rosa refers to in the following, reasoning that the sequence approaches two from below more and more.

908. /R: [,and because of this we are [always] [no we are not]
we are always [[closer from below] *(the right hand first drops and is then moved upwards in a wave-like movement)* to the two. (..)]
so [from] *(moves the arm from left to right once)* [,because it is <u>minus</u> that.]] *(points at the two in $2 - \frac{1}{2a+1}$)*

The gesture we focus on is the one co-timed to "we are always [[closer from below] to the two (..)]". The interpretation of the gesture requires the consideration of the co-expressive speech, referring to approaching the limit value "two" from below. This idea of approaching a limit is already a metaphoric one, grounded in the everyday-concept of moving closer to something (see Fig. 9.9). Illustrating this idea of movement, gesture specifies the style (how) as steady and dynamic in contrast to the iterative, stepwise process suggested by the sequence.

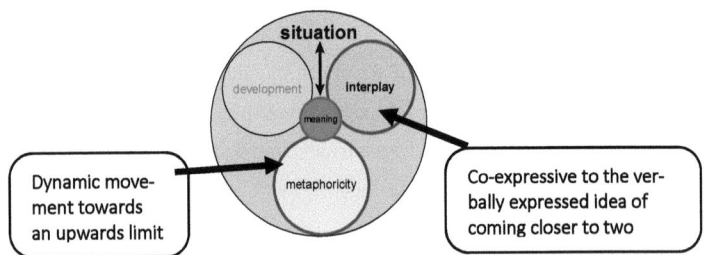

Fig. 9.9: Components influencing the interpretation space for disclosing the students' interpretation of the gesture CF7-908

Shortly after, this gesture not grounded in inscription but in the verbally uttered metaphorical idea underlying the concept of limit is modified to refine the idea to consider two subsequences:

9.2 Vertical Comparison: One Pair of Students, Different Tasks (PA7, CF7, IN7) 237

914. /R: ,that means we

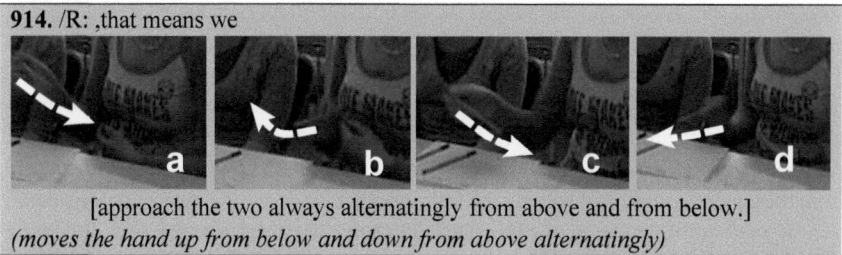

[approach the two always alternatingly from above and from below.]
(moves the hand up from below and down from above alternatingly)

This time, all three components are combined to provide meaning expressed by the gesture:

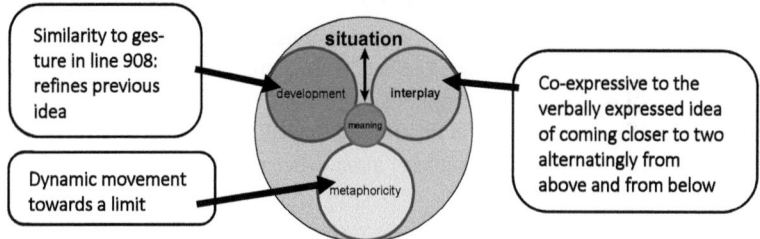

Fig. 9.10: Components influencing the interpretation space for disclosing the students' interpretation of the gesture CF7-914

Different to the gesture presented in sections 7.1.2.2 (excerpt 1.2c)/8.1.2.6 (excerpt 3.2m) and in section 8.1.1.4 (excerpt 3.1g), the link between gesture and speech is considered to be explicit for the "approaching two" gesture: Here, the metaphor concerns the fictive movement towards an object, is expressed in speech and illustrated by an iconic-physical gesture. In the other excerpts, a metaphoric meaning of the multimodal utterance is suggested only in gesture while the verbal utterance refers to a rather concrete idea. That way, *the gesture-speech interplay in the semiotic composition leaves the grounding of possible meaning in an image schema implicit, relying on experience concerning spatial relation and motion that is not situated in the mathematical task (see chapter 4.2.1). Because of this implicitness of the metaphor, the gesture is not considered to specify the verbal utterance.*

This observation bridges back to a table presented earlier in this section (Table 9.4) and provides a possible explanation for the low rate of specifying-gestures identified for the task on logical reasoning, IN7. In the epistemic-dense episodes of this task, there have been found numerous gestures that refer to metaphorical ideas without establishing an explicit link to the semantic content of speech in social interaction. The path-gesture (section 8.1.2.6, excerpt 3.2m) in this context can also be considered as representing the path from one step to the next one. Considering the basic step as a concrete case, the rule constituting this generally valid path is the second important component in the logic of carrying out an inductive proof.[159] The 'infinite process' gesture (section 8.1.1.4, excerpt 3.1g) expresses the general character of this step rather than presenting additional information.[160] Without being explicitly linked to a mathematical object or idea, these metaphorical gestures are not considered associated signs whose meaning develops within the social interaction. These gestures rather refer to an underlying idea that is detached from the concrete context of solving the task. Another kind of using gesture is considered to refer to the differentiation of cases needed to solve the induction task. The following excerpt gives an example of such a metaphorical gesture use, identified eight times in the EDEs of Rosa and Lisa's elaboration of the task on logical reasoning:

Excerpt 5.1b: IN7-52-54 (IN7e1.1, 0:10:41)

Lisa and Rosa work out the task "The king's consultants". They investigate the situation of seeing two marked and two unmarked hats, as is captured inscriptively in the diagram shown in Fig. 7.3 (on page 118). From this, they found that in this situation, nobody can decide about their own status at the first ring of the bell. To step back to what is already known, Rosa sums up two cases that have already been considered and in which the test can be passed for sure:

52. R: At least- *(looks at Lisa, then at the note sheet again)* we pass the test in two cases now. *(looks at the interviewer first, then at Lisa)* (...)
(indicates a circle in the air)
[[,I have the cross] or-
53. /L: *(at the same time)* yes-
54 /R: *(moves the hand to her left)*

[only someone else has that.]
(holds her index in the air to the front, looks at the interviewer) (...)]

cross. *(holds her finger to her mouth)*

Rosa explicates the two cases which can be decided on their own status with certainty, assuming one marked hat is existent in the circle. In line 52, she draws a circle on the

---

[159] This concerns the validity of the induction step that is, the possibility to conclude that an assumption holds true from the fact that it holds true for its predecessor. The first important component is the existence of a basic case to start the induction with. See the short analysis of the task in chapter 6.1.3.
[160] While the proving strategy of mathematical induction is also suggested for the arithmetic algebraic task, the tasks differ in their explicit handling of mathematical objects: In the CF-task, the continued fractions are the focus of attention; in the IN-task the mathematical topic stays implicit. The actual impact of this difference on the performance of gestures may be worth investigating in further research, qualitatively as well as quantitatively.

right side of the gesture space while verbally referring to the case of 'having the cross'.[161] The concrete movement of shaping a circle may remind the situation of 'sitting in a circle' so that the gesture-speech-utterance can be paraphrased as 'amongst all consultants, I am the one wearing the marked hat'. The pointing gesture accomplished in line 54 is an abstract one (McNeill, 1992, p. 173), probably directed towards an imagined consultant. Simultaneously, she verbally explicates the contrary case that "only someone else has that." (that only someone else wears the marked hat). Moreover, the indexical reference within the gesture space changes from right (line 52) to left (line 54), accompanying the successive consideration of two exclusive cases explicated in speech. This metaphorical use of gesture space has already been observed by Cienki and Müller (2008, p. 491) and also by McNeill (1992, p. 155), visually representing the differentiation of two different conditions (Cienki & Müller) or attributes (McNeill).

Similar to the two metaphorical gestures mentioned earlier (sections 8.1.1.4 and 8.1.2.6), this kind of gesture gives a reference to the logical design of the proof to be carried out.

The Direct Epistemic Impact of Gestures Differs Between the Tasks
That the kind of gestures made while elaborating the induction-task does not actively influence the epistemic process is also hypothesized from the comparison of the condensed process diagrams of the epistemic-dense episode. These are presented in Fig. 9.12, Fig. 9.13, and Fig. 9.14 on the following pages. They shall be compared as a whole and due to that no further information to the content of the episodes or references to line numbers are of interest here.[162] The black lines represent the epistemic actions as originally reconstructed from speech act analysis, and the light blue lines are added when an epistemic action has been reconstructed in addition from the analysis of gestures. The main signs used to describe the reconstruction of the epistemic processes have already been introduced in section 6.2.3.1. They are combined to represent relationships between epistemic actions. The dotted lines indicate when a previous epistemic action is continued: Such a continuation can be made within the same episode, but also across episodes, e.g. when it is referred back to a structure seen in an earlier episode in order to elaborate it. In this case, the line number of the reference is given. The spatial arrangement of the pictograms within one episode visualize how they are linked as related at the same time. For example, entities are often gathered and connected immediately. This is represented by visually "embedding" the gathering within the connecting action. In the same way, epistemic actions can be accomplished while another structure is elaborated.[163]

The condensed process diagram of the EDEs of Lisa and Rosa's elaboration of the parabola-task is presented in Fig. 9.12. As can be seen, there are mostly connecting actions

---
[161] With 'having the cross' they refer to 'wearing a marked hat'.
[162] Originally, the diagrams have been sorted according to subtasks. This is not needed at this point since the vertical comparison deals with tasks that are different anyway. The comparison of the diagrams can thus only lead to statements on the total amount of epistemic actions realized by gestures.
[163] See also the example given in chapter 8.2.2.

reconstructed exclusively from gestures (16 times). Furthermore, the connecting actions reconstructed from gesture cover different types: There are accumulating connecting actions standing for themselves, but also those leading directly into structure seeing actions (see Fig. 9.11), and third, connecting actions in which a seen structure is elaborated.

Fig. 9.11: Gesture-connecting action leading into structure seeing

Gestural gathering actions have been reconstructed solely in combination with other epistemic actions, as concretizing a seen structure or as being embedded in a (verbally accomplished) connecting action.

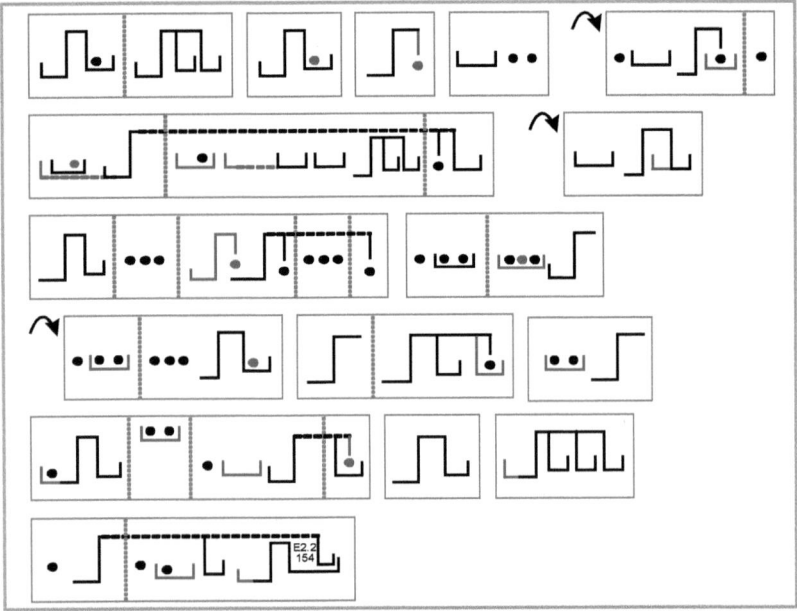

Fig. 9.12: Condensed process diagram of the epistemic-dense episodes (PA7)

In comparison, the reconstruction of the epistemic process concerning the arithmetic algebraic task appears to be much more complex in respect to the interplay of epistemic actions accomplished in speech and gesture (see Fig. 9.13). Again identifying many connecting actions, them being in the majority is not as eye-catching as in the PA7-EDEs. Furthermore, the gestural connecting action as leading into structure seeing, identified four times in the PA7-data, does not occur even once in the CF7-data. Instead, there are

9.2 Vertical Comparison: One Pair of Students, Different Tasks (PA7, CF7, IN7)   241

connecting actions that are accomplished (almost) simultaneously to the structure seeing action. The first phase gives an example for this (representing excerpt 3.2d in section 8.1.2.2). It is also remarkable that five structure seeing actions, that is five generalities considered, are expressed in gesture (see also chapter 8.2.2), but gathering as basic action rarely is (only once).

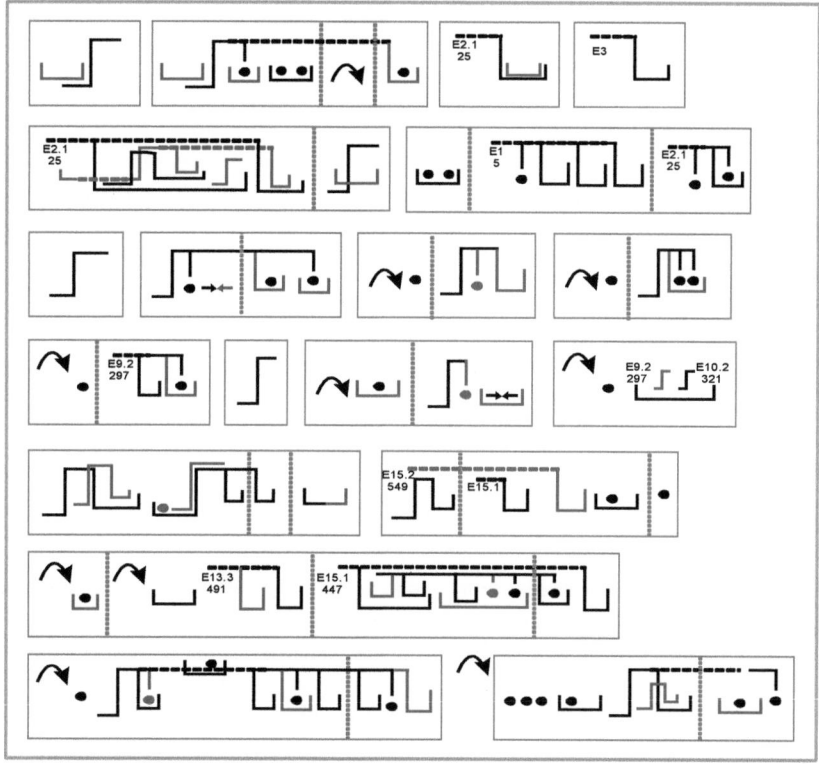

Fig. 9.13: Condensed process diagram of the epistemic-dense episodes (CF7)

As has already been mentioned in the beginning of this section, gestures' contribution to the epistemic process is scarce in the IN7-data. This becomes also visible in the condensed process diagram of the EDEs of IN7 (Fig. 9.14). The students explicate their ideas mainly through speech and inscription, using gesture moreover as an index that refers to ideas also expressed verbally. Metaphoric gestures like those described in the previous section may tell *the observer's eye about a student's idea from a "knower's" perspective*. However, in the social interaction between the learners, the link between the metaphorical gesture and the mathematical content does not become explicit.

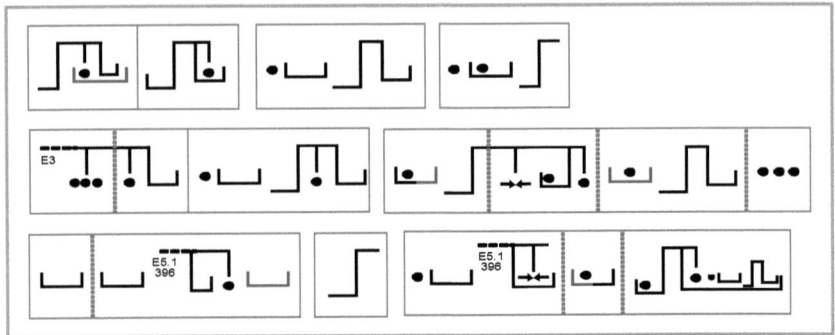

Fig. 9.14: Condensed process diagram of the epistemic-dense episodes (IN7)

9.3 Summarizing Conclusions Resulting from the Horizontal and the Vertical Comparison

The three main observations made in the horizontal comparison (between different pairs of students working on the same task) are:

- *The within-gestures differ between the three pairs of students.*
- *The associated gesture 'shape of the curve' is suggested by the task.*
- *There is no remarkable difference in the impact of gestures on the epistemic actions.*

We have seen that some gestures seem to be more effective than others when constructing mathematical knowledge in social interaction. While one associated gesture has been developed by all three pairs of students, one pair extensively made use of it to share general ideas and observations on a conceptual level, completely detached from inscription. The elaboration and use of the associated gesture on different referential levels seemed to have made their gesture use more effective. The associated gesture may have been suggested by the task, embodying a central idea inherent in it, but its core potential seems to unfold when the gesture becomes established on referential level 3.

Although the within-functions of gestures revealed differences in the students' use of gestures, the realization of epistemic actions exclusively by means of gestures appeared to be similar for all three pairs. This similarity is suggested to be grounded in the design of the task.

Comparing how the same pair of students uses gestures when elaborating different tasks (vertical comparison) led to suggesting that:

9.3 Summarizing Conclusions Resulting from the Horizontal and the Vertical Comparison 243

- *The amount of conditions that is needed to be set in relation when elaborating the IN-task requires to fix a basic setup in inscription. Then, concrete indications of components of this inscription helps to consider differing conditions.*
- *The 'origin of meaning' of level 3-gestures differs between the tasks.*
- *The direct epistemic impact of gestures differs between the tasks.*

Again, the first two observations concern the representational function of gestures, while the third one deals with the potential of gestures to shape the epistemic process by realizing epistemic actions.

The learning environment dealing with the limit of a sequence determined by a continued fraction (CF) and the one dealing with mathematical induction (IN) are less graphical in nature than the parabola-task (PA). Although this was expected to influence the use of gestures and, more generally, of non-verbal signs, it was not clear before what this influence may look like. The results described in this chapter not only suggest that graphical tasks may prompt a development of associated gestures across all three levels of reference, but also that for less graphical tasks, the associated gestures may be moreover grounded more implicitly in image schemata (see chapter 4.2.1). This implicitness may be one reason for the diversity in the epistemic impact of gestures: In the elaboration of the parabola-task, the gestures seem to complement the epistemic process that becomes reconstructable from speech. For the CF-task, it seems that gestures provide an epistemic process that is parallel to the one revealed in speech. The 'gestural epistemic actions' are intertwined with the 'verbal epistemic actions' while the gesture-speech distribution within the epistemic process appears to be 'ordered' in the PA7-data. The gestures Rosa and Lisa used when elaborating the induction task supported epistemic actions but did not realize them. This may be related to the metaphorical character of the level 3-gestures. Not being situationally conventionalized within the social interaction, they have less power to represent ideas that are not mentioned in speech.

Bringing this together, it is hypothesized that the *use of gestures, as well as their contribution to the social process of mathematical knowledge construction depends on the concrete task, composed by*

- *the mathematical subject it deals with,*
- *the mathematical activities requested to solve the task,*
- *the representations provided by the task.*

While this may be considered a trivial finding, it however is remarkable that the task seems to have a larger influence on the epistemic impact of gestures than the choice of students. In order to make probable that they solve the task successfully, they have been chosen from 10th grade and being high-performing in mathematics. Leaving this aside, there are no conspicuous similarities between the three pairs.

Concluding from the results gained in this study, gestures' contribution on the epistemic process may be dependent on the concrete task rather than on the concrete individuals.

Learning from high-performing students how they use gestures in an efficient way may thus provide a step towards using gestures as a didactic means. However, the horizontal and the vertical comparison included only three samples each. In order to test and extend these hypotheses, individual data needs to be analyzed in further studies, including high-performing and non-high-performing students.

10 Reflection, Synopsis, Perspectives

A Review of Gestures' Contribution to the Construction of Knowledge in Epistemic-Dense Episodes

Within this study has been illuminated how gestures may foster the epistemic process in various ways:

Gestures can accomplish different functions *within* the semiotic composition that describe how a referential frame for the interpretation of the immediate object is shaped. This is done by *clarifying* and *specifying* the *where*, the *what* and/or the *how* of the verbal utterance. Based on these aspects, gestures can also *indicate, visualize* or *concretize relations*. By these specifications, gesture provides an additional informational source that may influence the mutual interpretation in social interaction. Information expressed by the use of these specifying-gestures appears to be shared on three spatial levels of reference. On these *referential-levels*, current inscription is involved in influencing the interpretation of the immediate object in different ways: On *level 1*, the gesture grounds the reference concretely in an inscription by indicating it through pointing. Gestures performed on *level 2* refer to possibilities, representing something (iconically) against the background of a fixed diagram. This background influences the interpretation of the gesture in this concrete performance. Using gestures on level 2, potential situations or dynamic aspects of an object can be represented. Detached from a fixed inscription, gestures can refer to mathematical ideas freely on *level 3*, the gesture space. Being reduced to the core, general aspects, this engagement is considered to be established within social interaction to provide a reference taken-as-shared:

1. In the working process, signs can be enriched with meaning to establish *information bundles* that are linked to the meaning laden signs. In the packing of these bundles, speech and gesture co-expressively *associate* additional information with a non-verbal sign. For example, the meaning of a gesture can be elaborated in catchments from concrete inscription to level 3 as a 'body diagram' so that the representation of relations becomes detached from the concrete. Through this shift, it is concentrated on some aspects while leaving aside others. The prior specification on the concrete, however, allows still keeping them in mind. Catchments provide one example of such recalled contexts, in which the associated sign is an *associated gesture*.

2. Gestures can *prepare* and *support* epistemic actions, and they can even *realize* them. In these cases, gestures fulfill an *epistemic function*. Two main categories of epistemic functions can be distinguished; *forming*-functions and *performing*-functions. A *forming-function* refers to the representation of the immediate mathematical object. By *sourcing out, extracting*, and *depicting*, epistemic actions can be augmented by providing visual access to the mathematical object. Furthermore, metaphorical gestures can give visual access to general ideas in connection with a concrete case. Diverse *performing-functions* have a more specific beneficial effect on an epistemic action. The effect of the visual access that is accomplished by the gesture is directed towards acting epistemically. For example, an iconic gesture can be performed against the background of a fixed DGS-situation (on level 2), so that the ephemeral representation is compared to the fixed representation. By this, differences between them can be specified and become more salient through the *contrasting*-function of the gesture. This can enhance verbal argumentation as *connecting action* in a reasoning-situation by providing visual access to the counter argument. Other performing-functions are *focusing* on aspects, *exemplifying* the general, *making* ideas *more precise, gluing* aspects together to establish a connection, and *structuring the verbal discourse*. The main difference between the forming- and the performing-function concerns their integration in the epistemic action: Forming-function are related to the mathematical object on which the epistemic action is performed, while performing-functions are rather linked to the epistemic action itself.

The construction of knowledge is assumed to be observed when structures are seen or extended in social interaction. Ways in which gestures directly contribute to this can now be summarized as follows:

3. The use of associated signs is suggested to support the epistemic process by recalling information bundles as a connection of mathematical meaning. This simplifies the illustration of connecting actions and by this provides visual access to *seeing* structures. Furthermore, when the meaning of a recurring gesture becomes situationally conventionalized within the process, the gesture can become a basic signs that literally stands for a mathematical idea. By this, their use can also refer to this idea and *realize structure-seeing actions* or *their extension*. This is related to the next conclusion:

4. A seen structure can be revealed in the use of gesture before it is expressed in a verbal utterance. In this can be identified a *potential for structure-seeing*.

5. In situations in which seen structures are reasoned, disproved, or concretized, gestures can indicate non-verbal arguments by *exemplifying* or *contrasting*.

In relation to the two models on which the analyses are based, the following methodological implications can be stated:

I. The Semiotic Bundle Model:

   The within-functions of gestures can be understood as relations within the semiotic bundle. Through their accomplishment, gestures crucially take part in enriching other signs with mathematical meaning in a synchronic, but also in a diachronic way.

II. The GCSt-Model:

   II.1. Epistemic actions can also be realized through gestures. This can help to get a better comprehension of 'spontaneously'[164] accomplished actions of connecting or structure-seeing.

   II.2. Stating a final hypothesis, an explanatory description is given about how structure-seeing can be accomplished as process traversing three steps: In the first step, a "seed"[165] for structure-seeing is revealed as a "vague idea" (Davydov, 1972/1990) in the possibly unconscious use of a non-verbal sign in the social interaction. This is what gesture studies call a *Growth Point* of a new idea. *Potential structure-seeing* can be elaborated to become a *significant idea* by the use of *catchments*. In the third step, the structure can be formulated verbally as general rule or regularity. Speech may still be imprecise or fragile. The importance lies in explicitly considering the structure as significant. At that point, structure-seeing is also expressed in the verbal utterance. In the terms of gesture studies, the *Growth Point* is *"unpacked"* then.[166]

In sum, the results show that gestures contribute to the social construction of mathematical knowledge in much larger extent than has been suggested so far. They are not only accessories supporting the verbal utterance, but can actively affect the process. The vertical comparison of gestures' contribution to the epistemic processes concerning different tasks suggests that possibilities offered by the use of gestures depend on the

---

[164] 'Spontaneous' in a sense means that these epistemic actions are not prepared by other epistemic actions that are reconstructable exclusively from the verbal utterances.
[165] In the context of mathematical epistemic processes "seeds for constructing actions" are first mentioned by Kidron et al. (Kidron, Bikner-Ahsbahs, Cramer, Dreyfus, & Gilboa, 2010, p. 173).
[166] A similar result from an individual point of view has recently been presented by Dreyfus et al. (2014): They describe a case of 'unpacking a Growth Point' in their investigation of the epistemic role of gestures as described in chapter 2. Based on a comparison of two analyses within different theoretical approaches, they suggest that "the growth point can constitute the beginning of a building process, and the catchment can be a signal of the consolidation process" (ibid., p. 150). Hence, unpacking the Growth Points through catchments seems to be relevant for individual and collective epistemic processes.

mathematical subject and activities on the one hand, and on the material provided on the other hand. The horizontal comparison however led to the assumption that there may be gestures that develop more naturally from the task than others, given that one associated gesture developed in all three epistemic processes concerning the parabola-task. Whether similar overlaps can also be observed for other tasks and for a larger number of students working on the parabola-task needs to be tested with further data. If this turns out to be true, it is recommended to take this into consideration for *designing a task*, respectively elaborating a task in the mathematics classroom.

Nevertheless, it also needs to be considered a critical perspective on the contribution of gestures: First to mention, the idea to which gesture may refer does not need to be conscious. The aspects specified in gesture do not necessarily reveal what is explicitly expressed, but show up possibilities of the gestural representation. In the representation of the results, the word 'can' is thus consciously chosen to refrain from any condition-consequence-hypotheses related to the use of gestures. This also makes it evident that an epistemic action solely realized in gesture cannot be seen as necessarily influencing the construction of mathematical knowledge in social interaction. To build on a 'gesturally gathered' entity or on a 'gestural connection', the meaning of the gesture needs to be made explicit to be taken as shared. This is seen as a necessary condition to elaborate a first, tentative idea and to work out a significant idea as emanating from a potential seed for structure-seeing.

This study offers tools that can sensitize for the role of gestures in social mathematical learning processes. This concerns both the analysis of gestures for the purpose of research as well as the application of gestures as a didactic means in the mathematics classroom. To know about the potential (and limits) of gesture as means of expression may influence the conscious use of gestures' representational and epistemic functions. This can help to enhance epistemic processes developing within the mathematics classroom. For example, gestures can be considered to provide an additional resource in inclusive mathematics classrooms, integrated as a visual approach when verbal expressions are not sufficient. Nevertheless, it is suggested that in the inclusive classroom, the students' needs that have to be considered by the teacher are even more diverse as in non-inclusive classroom; this also has to be considered when using gestures as a didactic means. For this reason, more fundamental research is needed on how students use gestures to get access to a mathematical idea and to construct knowledge, and on how non-high-achieving students use gestures. This is suggested to allow for a more effective and informative horizontal comparison of gestures used by different students. Research of this kind can in turn provide the ground for follow-up projects that investigate the application of gestures as a didactic means, for example conducted as Design-Based-Research-Studies (Cobb, Confrey, diSessa, Lehrer, & Schauble, 2003; Prediger, et al., 2012) in which also mathematical subjects different to the subjects considered in this study can be taken into account. For this purpose, also gestures without any specifying representational function may be considered further. As has been seen, they may refer to the logical structure of an utterance. This confirms and refines results just recently pointed out by Arzarello and Sabena (Arzarello & Sabena, 2014). Based on a case study they suggest that "gestures may also play specific roles in providing a logical structure to argumentation processes" (ibid., p. 99). Beside others, they consider the "striking synergy between specific linguistic functions and the features of co-timed gestures [...], for example logic connectives co-timed with abstract pointings" (ibid., p. 100) and gestures making use of the "opposition in space expressing the logical opposition of the two cases" (ibid., p. 100). Gestures of this kind came especially to the fore in the elaboration of the induction task: They represented the path from one step to the next and the distinction of different cases that needed to be reviewed. These are the two main considerations needed to be made in order to solve the task. This finding raises the question whether there may be typical types of gesture use with respect to different types of reasoning structures. With regard to the mathematics classroom, Arzarello and Sabena suggest that gestures may foster students' argumentation skills by structuring mathematical arguments (ibid., pp. 99-101). Further research on this topic may thus lead to the elaboration of the didactical role of gestures in argumentation.

The results gained in this study illuminate the importance of taking into account gestures when observing social mathematical learning processes. Ideas may become revealed in gestures not only *at an earlier point in time*, but sometimes even *exclusively*. Considering merely the verbal discourse may thus lead to failing to notice the *potential apart from what is verbally explicated*: the 'mathematics in our hands'.

11 References

Alibali, M. W. (1999). How children change their minds: Strategy change can be gradual or abrupt. *Developmental Psychology*, 127-145.

Alibali, M. W., & diRusso, A. (1999). The function of gesture in learning to count: More than keeping track. *Cognitive Development, 14*, 37-56.

Alibali, M. W., & Goldin-Meadow, S. (1993). Gesture-speech mismatch and mechanisms of learning: What the hands reveal about a child's state of mind. *Cognitive Psychology*, 468-523.

Alibali, M. W., Kita, S., & Young, A. J. (2000). Gesture and the process of speech production: We think, therefore we gesture. *Language and Cognitive Processes, 15*(6), 593-613.

Arzarello, F. (2006). Semiosis as a multimodal process. *Revista Latinoamericana de Investigacion en Matemática Educativa, Numero Especial*, 267-299.

Arzarello, F., & Edwards, L. (2005). Gesture and the construction of mathematical meaning (research forum 2). In H. L. Chick, & J. L. Vincent (Eds.), *Proceedings of the 29th Conference of the International Group for the Psychology of Mathematics Education. 1*, (pp. 122-145). Melbourne, Australia: PME.

Arzarello, F., & Paola, D. (2007). Semiotic game: The role of the teacher. In J. Woo, H. Lew, K. Park, & D. Seo (Eds.), *Proceedings of the 31st conference of the International Group for the Psychology of Mathematics Education, 2* (pp. 17-24). Seoul, Korea: PME.

Arzarello, F., & Sabena, C. (2011). Semiotic and theoretic control in argumentation and proof activities. *Educational Studies in Mathematics, 77*, 189-206.

Arzarello, F., & Sabena, C. (2014). Analytic-structural functions of gestures in mathematical argumentation processes. In L. D. Edwards, F. Ferrara, & D. Moore-Russo, *Emerging perspectives on gesture and embodiment in mathematics* (pp. 75-124). Charlotte, NC: IAP-Information Age Publishing.

Arzarello, F., Bikner-Ahsbahs, A., & Sabena, C. (2009). Complementary Networking: Enriching Understanding. In V. Durand-Guerrier, S. Soury-Lavergne, & S. Lecluse (Eds.), *Proceedings of the 6th Congress of the European Society for Research in Mathematics Education* (pp. 1545-1554). Lyon, France. Retrieved 06 04, 2014, from: http://ife.ens-lyon.fr/publications/edition-electronique/cerme6/plenary-01-bikner.pdf

Arzarello, F., Ferrara, F., Robutti, O., Paola, D., & Sabena, C. (2005). Shaping a multidimensional analysis of signs. In H. L. Chick, & J. L. Vincent (Eds.), *Proceedings of the 29th conference of the International Group for the Psychology of Mathematics Education. 1*, (pp. 127-131). Melbourne University, Australia: PME.

Arzarello, F., Robutti, O., Paola, D., & Sabena, C. (2009). Gestures as semiotic resources in the mathematics classroom. *Educational Studies in Mathematics, 70*, 97-109.

Austin, J. L. (1962). *How to do things with words.* Oxford: Oxford University Press.

Bauersfeld, H. (1988). Interaction, construction, and knowledge: Alternative perspectives for mathematics education. In T. Cooney, & D. Grouws (Eds.), *Effective mathematics teaching* (pp. 27-46). Reston, VA: National Council of Teachers of Mathematics and Lawrence Erlbaum Associates.

Bauersfeld, H. (1992). Classroom cultures from a social constructivist's perspective. *Educational Studies in Mathematics, 23*, 467-481.

Bavelas, J. B. (1994). Gestures as Part of Speech: Methodological Implications. *Research on Language and Social Interaction, 27*(3), 201-221.

Bavelas, J. B., Chovil, N., Lawrie, D. A., & Wade, A. (1992). Interactive Gestures. *Discourse Processes* (15), 469-489.

Beck, C., & Jungwirth, H. (1999). Deutungshypothesen in der interpretativen Forschung [Construal hypotheses in interpretative research]. *JMD, 20*, 231-259.

Beck, C., & Maier, H. (1994). Zu Methoden der Textinterpretation in der empirischen mathematikdidaktischen Forschung [On methods of text interpretation in empirical research in mathematics education]. In H. Maier, & J. Voigt (Eds.), *Verstehen und Verständigung. IDM 19, Untersuchungen zum Mathematikunterricht* (pp. 43-76). Köln: Aulis.

Behrens, D. (2013). *Mathematische Erkenntnis durch Gesten erlangen und teilen. Eine Fallstudie zur Funktion von Gesten im Erkenntnisprozess einer arithmetisch-analytischen Aufgabe* [Gaining and sharing mathematical insights through gestures. A case analysis on the function of gestures in an epistemic process of an arithmetic-analytic task]. Unpublished master's thesis. Bremen: Universität Bremen.

Behrens, D., Krause, C. M., & Bikner-Ahsbahs, A. (2014). "Ich zeig' uns was, was du nicht siehst" - Zur epistemischen Rolle von Gesten ["I show us something that you don't see" - On the epistemic role of gestures]. In J. Roth, & J. Ames (Eds.), *Beiträge zum Mathematikunterricht. Vorträge auf der 48. Jahrestagung der Gesellschaft für Didaktik der Mathematik (GDM) 2014 in Münster* [Proceedings of the 48th annual conference of the German society for mathematics education in Münster, Germany] (pp. 149-152). Münster, Germany: WTM Verlag. Retrieved 11 25, 2014, from: https://eldorado.tu-dortmund.de/bitstream/2003/33057/1/BzMU14-4ES-Behrens-352.pdf

Bergmann, J. R. (2004). Ethnomethodology. In U. Flick, E. v. Kardorff, & I. Steinke (Eds.). *A companion to qualitative research* (pp. 72-80). London: Sage Publications.

Bikner-Ahsbahs, A. (2004). Interest-dense situations and their mathematical value. *(Topic Study Group 24 (Students' motivations and attitudes towards mathematics and its study) of the International Congress for Mathematics Education).* Retrieved 06 05, 2014, from: http://www.math.uni-bremen.de/didaktik/ma/bikner/articles/article1.pdf

Bikner-Ahsbahs, A. (2005). *Mathematikinteresse zwischen Subjekt und Situation. Theorie interessendichter Situationen -Baustein für eine mathematikdidaktische Interessentheorie [Interest in mathematics between subject and situation. Theory of interest-dense situations - building blocks for a theory on interest from the perspective of mathematics education].* Hildesheim, Berlin: Franzbecker.

Bikner-Ahsbahs, A. (2006). Semiotic sequence analysis - Constructing epistemic types. In J. Novotná, H. Moraovsá, M. Krátká, & N. Stehliková (Eds.), *Mathematics in the centre. Proceedings of the 30th conference of the International Group for the Psychology of Mathematics Education. 2* (pp. 161-168). Prague (Czech Republic): PME.

Bikner-Ahsbahs, A., Dreyfus, T., & Kidron, I. (2010). "General Epistemic Need" - ein Motor für Erkenntnisentwicklung? ["General Epistemic Need" - an engine of *Beiträge zum Mathematikunterricht. Vorträge auf der 44. Jahrestagung der Gesellschaft für Didaktik der Mathematik (GDM) 2010 in München.* [Proceedings of the 44th annual conference of the German society for mathematics education in Munich, Germany] Retrieved 11 25, 2014, from: http://www.mathematik.tu-dortmund.de/ieem/cms/media/BzMU/BzMU2010/BzMU10_BIKNER-AHSBAHS_Angelika_Erkenntnisentwicklung.pdf

Bikner-Ahsbahs, A., & Prediger, S. (2014) (Eds.), *Networking of theories as a research practice in mathematics education.* New York: Springer.

Bikner-Ahsbahs, A., Kidron, I., & Dreyfus, T. (2011). Epistemisch handeln können - aber wie? [Knowing to act epistemically – but how?]. In R. Haug, & L. Holzäpfel (Eds.), *Beiträge zum Mathematikunterricht. Vorträge auf der 45. Jahrestagung der Gesellschaft für Didaktik der Mathematik (GDM) 2011 in Freiburg.* [Proceedings of the 45th annual conference of the German society for mathematics education in Freiburg, Germany] (Hauptvortrag [Invited lecture]) Retrieved 11 25, 2014, from: http://www.mathematik.tu-dortmund.de/ieem/bzmu2011/_BzMU11_1_Einfuehrungen-Hauptvortraege/BzMU11_BIKNER_Angelika_Epistem.pdf

Bikner-Ahsbahs, A., Sabena, C., & Arzarello, F. (2014). "Lost in translation" - Semiotisch-theoretische Kontrolle beim argumentative Problemlösen. ["Lost in translation" - semiotic-theoretic control in argumentative problem solving] *Beiträge zum Mathematikunterricht. Beiträge zur Jahrestagung der Gesellschaft für die Didaktik der Mathematik (GDM) 2014 in Münster.* [Proceedings of the 48th annual conference of the German society for mathematics education in Münster, Germany] Retrieved 11 25, 2014, from: https://eldorado.tu-dortmund.de/bitstream/2003/33077/1/BzMU14-4ES-Bikner-Ahsbahs-366.pdf

Bikner-Ahsbahs, A., Sabena, C., Arzarello, F., & Krause, C. M. (2014). Semiotic and theoretic control within and across conceptual frames. In C. Nicol, P. Liljedahl, S. Oesterle, & D. Allan (Eds.), *Proceedings of the joint meeting of PME 38 and PME-NA 36. 2* (pp. 153-160). Vancouver, Canada: PME. Retrieved 01 11, 2015, from: http://www.pmena.org/proceedings/PMENA%2036%20PME%2038%202014%20Proceedings%20Vol%202.pdf

Bjuland, R., Cestari, M. L., & Borgersen, H. E. (2008). The Interplay Between Gesture and Discourse as Mediating Devices in Collaborative Mathematical Reasoning: A Multimodal Approach. *Mathematical Thinking and Learning, 10,* 271-292.

Broaders, S., Cook, S., Mitchell, Z., & Goldin-Meadow, S. (2007). Making children gesture brings out implicit knowledge and leads to learning. *Journal of Experimental Psychology, 136,* 539-550.

Butcher, C., & Goldin-Meadow, S. (2000). Gesture and the transition from one- to two-word speech: when hand and mouth come together. In D. McNeill (Ed.), *Language and gesture* (pp. 235-257). Cambridge: Cambridge University Press.

Chen, C., & Herbst, P. (2013). The interplay among gestures, discourse and diagrams in students' geometrical reasoning. *Educational Studies in Mathematics, 83,* 285-307.

Cienki, A., & Müller, C. (2008). Metaphor, gesture, and thought. In R. W. Gibbs (Ed.), *The Cambridge handbook of metaphor and thought* (pp. 483-501). New York: Cambridge University Press.

Clark, H. H., & Brennan, S. E. (1991). Grounding in Communication. In L. B. Resnick, & S. D. Teasley (Eds.), *Perspectives on socially shared cognition* (pp. 127-149). Washington, DC: American Psychological Association.

Cobb, P., Confrey, J., diSessa, A., Lehrer, R., & Schauble, L. (2003). Design Experiments in Educational Research. *Educational Researcher, 32*(1), 9-13.

Cramer, J. (2011). Everyday argumentation and knowledge construction in mathematical talk. In M. Pytlak, T. Rowland, & E. Swoboda (Eds.), *Proceedings of the 7th Congress of the European Society for Research in Mathematics Education.* Rzeszów, Poland: University of Rzeszów. Retrieved 11 25, 2014, from: http://www.cerme7.univ.rzeszow.pl/WG/1/CERME7_WG1_Cramer.pdf

Cramer, J. (2010). Induktion durch vollständiges Zeigen. Schlussweisen in Argumentationsprozessen [Induction by complete showing. Concluding methods in argumentation processes]. In: *Beiträge zum Mathematikunterricht. Vorträge auf der 44. Jahrestagung der Gesellschaft für Didaktik der Mathematik vom 8.-12.3.2010 in München* [Proceedings of the 44th Annual German conference for Mathematics Education in Munich, Germany] (pp. 229-232) Münster: WTM.

Davydov, V. V. (1972/1990). *Soviet studies in mathematics education: Vol. 2. Types of generalization in instruction: Logical and psychological problems in the structuring of school curricula.* Reston, VA: NCTM.

de Ruiter, J. P. (1995). Why do people gesture at the telephone? In Biemans M, Woutersen M (Eds.), *Proceedings of the Center for Language Studies Opening Academic Year '95-96* (pp. 49-56). Nijmegen: Center for Language Studies.

de Ruiter, J. P. (2000). The production of gesture and speech. In D. McNeill (Ed.), *Language and gesture* (pp. 248-311). Cambridge: Cambridge University Press.

de Saussure, F. (1995). *Cours de linguistique générale.* Paris: Payot.

Denzin, N. K. (2009). The elephant in the living room: or extending the conversation about the politics of evidence. *Qualitative Research, 9*(139). doi:10.1177/1468794108098034

Di Pellegrino, G., Fadiga, L., Fogassi, L., Gallese, V., & Rizzolatti, G. (1992). Understanding motor events: A neurophysical study. *Experimental Brain Research, 91*, 176-180.

Dodge, E., & Lakoff, G. (2005). Image schemas: From linguistic analysis to neural grounding. In B. Hampe (Ed.), *From perception to meaning: Image schemas in cognitive lingustics* (pp. 57-91). Berlin: Mouton de Gruyter.

Dreyfus, T. (2012). Constructing abstract mathematical knowledge. In S. J. Cho (Ed.), *Proceedings of the 12th International Congress on Mathematical Education.* Seoul, South Korea: ICME. Retrieved 23 11, 2014, from: http://www.icme12.org/upload/submission/1953_f.pdf

Dreyfus, T., Hershkowitz, R., & Schwarz, B. B. (2001). The Construction of Abstract Knowledge in Interaction. In M. v. Heuvel-Pannhuizen (Ed.), *Proceedings of the 25th conference of the International Group for the Psychology of Mathematics Education.* 2, pp. 377-384. Utrecht: PME.

Dreyfus, T., Sabena, C., Kidron, I., & Arzarello, F. (2014). The Epistemic Role of Gestures. In A. Bikner-Ahsbahs, & S. Prediger (Eds.), *Networking of theories as a research practice in mathematics education* (pp. 127-152). New York: Springer.

Duval, R. (2002). The cognitive analysis of problems of comprehension in the learning of mathematics. *Mediterranean Journal for Research in Mathematics Education, 1*(2), 1-16.

Duval, R. (2006). A cognitive analysis of problems of comprehension in a learning of mathematics. *Educational Studies in Mathematics, 62,* 103-131.

Edwards, L. D. (2009). Gestures and conceptual integration in mathematical talk. *Educational Studies in Mathematics, 70,* 127-141.

Edwards, L. (2010). Doctoral students, embodied discourse and proof. In M. F. Pinto, & T. F. Kawasaki (Eds.), *Proceedings of the 34th conference of the International Group for the Psychology of Mathematics Education* (pp. 329-336). Belo Horizonte, Brazil: PME.

Edwards, L., Ferrara, F., & Robutti, O. (2013). Complementary theoretical perspectives on mulimodality. In B. Ubuz, C. Haser & M. A. Mariotti (Eds.), *Proceedings of the 8th Conference of the ERME.* Antalya, Turkey. Retrieved 10 07, 2014, from: http://cerme8.metu.edu.tr/wgpapers/WG16/WG16_Edwards.pdf

Ekman, P., & Friesen, W. V. (1969). The repertoire of nonverbal behavior: Categories, origins, usage, and coding. *Semiotica, 1,* 49-98.

Ernest, P. (2006). A semiotic perspective of mathematical activity: the case of number. *Educational Studies in Mathematics, 61,* 67-101.

Forceville, C. (2009a). A Course in Pictorial and Multimodal Metaphor. Lecture 1. Retrieved 06 06, 2014, from: http://projects.chass.utoronto.ca/semiotics/cyber/cforceville1.pdf

Forceville, C. (2009b). Non-verbal and multimodal metaphor in a cognitive framework: agendas for research. In C. J. Foceville, & E. Urios-Aparisi (Eds.), *Multimodal metaphor* (pp. 19-44). Berlin: de Gruyter.

Gallese, V., & Lakoff, G. (2005). The brain's concepts: The role of the sensory-motor system in conceptual knowledge. *Cognitive Neuropsychology, 22*, 455-479.

Gallese, V., Fadiga, L., Fogassi, L., & Rizzolatti, G. (1996). Action Recognition in the premotor cortex. *Brain, 119*, 539-609.

Gilboa, N., Dreyfus, T., & Kidron, I. (2011). A construction of a mathematical definition - the case of parabola. In B. Ubuz (Ed.), *Proceedings of the 35th conference of the International Group for the Psychology of Mathematics Education. 2*, (pp. 425-432). Ankara, Turkey: PME.

Goldin-Meadow, S. (2003). *Hearing Gesture: how our hands help us think.* Cambridge, MA: Harvard University Press.

Goldin-Meadow, S. (2010). When gesture does and does not promote learning. *Language and Cognition, 2*(1), 1-19.

Goldin-Meadow, S., Cook, S. W., & Mitchell, Z. A. (2009). Gesturing Gives Children New Ideas About Math. *Psychological Science, 20*(3), 267-272.

Goldin-Meadow, S., Nusbaum, H., Garber, P., & Breckinridge Church, R. (1993). Transitions in learning: Evidence for simultaneously activated strategies. *Journal of Experimental Psychology: Human Perception and Performance, 19*, 92-107.

Gullberg, M. (2003). Gestures, referents, and anaphoric linkage in learner varieties. In C. Dimroth, & M. Starren (Eds.), *Information structure and the dynamics of language acquisition* (pp. 311-328). Amsterdam: Benjamins.

Gullberg, M. (2010). Methodological reflections on gesture analysis in second language acquisition and bilingualism research. *Second Language Research, 26*(1), 75-102. doi:10.1177/0267658309337639

Hammersley, M. (2007). The issue of quality in qualitative research. *International Journal of Research & Methods in Education, 30*(3), 287-305. Retrieved 07 17, 2014, from: http://www.marjee.org/pdfs/Hammersley_issue.pdf

Hershkowitz, R., Schwarz, B. B., & Dreyfus, T. (2001). Abstraction in Context: Epistemic Actions. *Journal for Research in Mathematics Education, 32,* 195-222.

Hoffmann, M. H. (2001). *Peirces Zeichenbegriff: seine Funktionen, seine phänomenologischen Grundlegungen und seine Differenzierungen [The Peircean concept of signs: its function, its phenomenological basics, and its differentiation].* Retrieved 02 19, 2014, from: http://www.uni-bielefeld.de/idm/semiotik/Hoffmann-Peirces_Zeichen.pdf

Hoffmann, M. H. (2003). Einleitung: Warum Semiotik? [Introduction: Why semiotic?] In H. G. Hoffmann (Ed.), *Mathematik verstehen - Semiotische Perspektiven [Understanding mathematics – Semiotic perspectives]* (pp. 1-18). Hildesheim: Franzbecker.

Hoffmann, M. H. (2005). *Erkenntnisentwicklung. Ein semiotisch-pragmatischer Ansatz [Knowledge development - A semiotic and pragmatic approach].* Frankfurt am Main: Klostermann.

Hoffmann, M. H., & Roth, W.-M. (2005). What you should know to survive in knowledge societies. On a semiotic understanding of 'knowledge'. *Semiotica, 156,* 101-142.

Hoffmann, M. H., & Roth, W.-M. (2007). The complementation of a representational and an epistemological function of signs in scientific activity. *Semiotica, 1/4*(164), 1-24.

Hookway, C. J. (1985). *Peirce.* London: Routledge.

Iverson, J. M., & Goldin-Meadow, S. (1998). Why people gesture when they speak. *Nature, 6708,* 228.

Iverson, J. M., & Goldin-Meadow, S. (2001). The resilience of gesture in talk: gesture in blind speakers and listeners. *Developmental Science, 4*(4), 416-422.

Janßen, T. (2010). *Epistemische Aufbauhandlungen und die Konstruktion mathematischen Wissens. Theorieweiterentwicklung durch Vergleich zweier Modelle* [Epistemic assembling actions and the construction of mathematical knowledge. Further-developing theory through the comparison of two models]. Unpublished master's thesis. Bremen.

Johnson, M. (1987). *The body in the mind: The bodily basis of meaning, imagination, and reason.* Chicago: University of Chicago Press.

Jungwirth, H. (2003). Interpretative Forschung in der Mathematikdidaktik - ein Überblick für Irrgäste, Teilzieher und Standvögel [Interpretative research in mathematics education - An overview for vagrants, partial migrants, and resident birds]. *ZDM, 35*(5), 189-200.

Kelle, U., & Kluge, S. (2010). *Vom Einzelfall zum Typus [From the individual case to the type]*. Wiesbaden: VS Verlag.

Kendon, A. (1980). Gesticulation and speech: Two aspects of the process of utterance. In M. R. Key (Ed.), *The relation between verbal and nonverbal communication* (pp. 207-227). The Hague: Mouton.

Kendon, A. (1988). How gestures can become like words. In F. Poyatos (Ed.), *Crosscultural perspectives in nonverbal communication* (pp. 131-141). Toronto: Hogrefe.

Kendon, A. (2000). Language and Gesture: Unity or duality? In D. McNeill (Ed.), *Language and gesture* (pp. 47-63). Cambridge: University Press.

Kendon, A. (2004). *Gesture: visible action as utterance*. Cambridge: Cambridge University Press.

Kidron, I., Bikner-Ahsbahs, A., & Dreyfus, T. (2011). How a general epistemic need leads to a need for a new construct: A case of networking two theoretical approaches. In M. Pytlak, & T. &. Rowland (Eds.), *Proceedings of the 7th Congress of the European Society for Research in Mathematics Education* (pp. 2475-2485). Rzeszów: University of Rzeszów.

Kidron, I., Bikner-Ahsbahs, A., Cramer, J., Dreyfus, T., & Gilboa, N. (2010). Construction of knowledge: need and interest. In M. M. Pinto, & T. F. Kawasaki (Eds.), *Proceedings of the 34th conference of the International Group for the Psychology of Mathematics Education. 3* (pp. 169-176). Belo Horizonte, Brazil: PME.

Kita, S. (2000). How representational gestures help speaking. In D. McNeill (Ed.), *Language and gesture* (pp. 162-185). Cambridge, UK: Cambridge University Press.

Knoblauch, H. (2010). *Wissenssoziologie [Sociology of knowledge]*. Konstanz: UVK.

Krause, C. M. (2012). Arten des Zeichengebrauchs und ihre Rolle im mathematischen Erkenntnisprozess [Modes of sign use and their role in mathematical epistemic processes]. In M. Ludwig, & M. Kleine (Eds.), *Beiträge zum Mathematikunterricht. Vorträge auf der 46. Tagung für Didaktik der Mathematik vom 5.3. bis 9.3.2012 in Weingarten.* [Proceedings of the 46th Annual conference of the German society for mathematics education in Weingarten, Germany]. Retrieved 4. 14., 2014, from: http://www.mathematik.uni-dortmund.de/ieem/bzmu2012/files/BzMU12_0208_Krause.pdf

Krause, C. M. (2015, in press). Gestures as part of discourse in reasoning situations: Introducing two epistemic functions of gestures. In: *Proceedings of the 9th Congress of the European Society for Research in Mathematics Education (CERME 9 in Prague 2015).*

Krause, C. M., & Bikner-Ahsbahs, A. (2012). Modes of sign use in epistemic processes. In Tai-Yih Tso (Ed.), *Proceedings of the 36th conference of the International Group for the Psychology in Mathematics Education. 3* (pp. 19-26). Taipei, Taiwan: PME.

Kress, G. (2001). Sociolinguistics and social semiotics. In P. Copely (Ed.), *The Routledge companion to semiotics and linguistics* (pp. 66-82). London: Routledge.

Krummheuer, G., & Brandt, B. (2001). *Paraphrase und Traduktion. Partizipationstheoretische Elemente einer Interaktionstheorie des Mathematiklernens in der Grundschule [Paraphrase and traduction. Participational theoretical elements of an interaction theory of learning mathematics in elementary school].* Weinheim und Basel: Beltz Verlag.

Krummheuer, G., & Naujok, N. (1999). *Grundlagen und Beispiele Interpretativer Unterrichtsforschung [Fundamental principles and examples of interpretative classroom research].* Opladen: Leske + Budrich.

Lakoff, G., & Johnson, M. (1980). *Metaphors we live by.* Chicago: University of Chicago Press.

Lakoff, G., & Johnson, M. (1999). *Philosophy in the flesh: The embodied mind and its challenge to western thought.* New York, NY: Basic Books.

Lakoff, G., & Núñez, R. (1997). The Metaphorical Structure of Mathematics: Sketching Out Cognitive Foundations for a Mind-Based Mathematics. In L. English (Ed.), *Mathematical reasoning: Analogies, metaphors, and images* (pp. 21-89). Hillsdale, NJ: Erlbaum.

Lakoff, G., & Núñez, R. (2000). *Where mathematics comes from: How the embodied mind brings mathematics into being.* New York: Basic Books.

Lamnek, S. (2005). *Qualitative Sozialforschung [Qualitative social research].* Weinheim, Basel: Beltz.

Marghetis, T., Edwards, L. D., & Núñez, R. (2014). More than mere handwaving: Gesture and embodiment in expert mathematical proof. In L. D. Edwards, F. Francesca, & M.-R. Deborah (Eds.), *Emerging perspectives on gesture and embodiment in mathematics* (pp. 227-246). Charlotte, NC: Information Age Publishing.

McNeill, D. (1992). *Hand and mind: what gestures reveal about thought.* Chicago: University of Chicago Press.

McNeill, D. (2005). *Gesture and thought.* Chicago: University of Chicago Press.

McNeill, D., & Duncan, S. (2000). Growth Points in Thinking-for-Speaking. In D. McNeill (Ed.), *Language and gesture* (pp. 141-161). Cambridge: Cambridge University Press.

Merton, R. K. (1968). *Social theory and social structure.* New York: Free Press.

Mittelberg, I., & Waugh, L. R. (2009). Metonymy first, metaphor second: A cognitive-semiotic approach to multimodal figures of speech in co-speech gesture. In C. Forceville, & E. Urios-Aparisi (Eds.), *Multimodal metaphor* (pp. 329-356). Berlin/New York: de Gruyter.

Montredon, J., Amrani, A., Benoit-Barnet, M.-P., Chan You, E., Llorca, R., & Peuteuil, N. (2008). Catchment, growth point, and spatial metaphor: Analyzing Derrida's oral discourse on deconstruction. In A. Cienki, & C. Müller (Eds.), *Metaphor and gesture* (pp. 171-194). Amsterdam/Philadelphia: John Benjamins.

Müller, C. (1998). *Redebegleitende Gesten: Kulturgeschichte, Theorie, Sprachvergleich [Gestures with speech. Cultural history, theory, comparison].* Berlin: Berlin Verlag A. Spitz.

Nemirovsky, R. (2003). Three conjectures concerning the relationship between body activity and understanding mathematics. In N. A. Pateman, B. J. Dougherty, & J. T. Zilliox (Eds.), *Proceedings of the 27th conference of the International Group for the Psychology of Mathematics Education. 4* (pp. 113-120). Hawaii, USA: PME.

Nöth, W. (2000). *Handbuch der Semiotik [Handbook of semiotics].* Weimar: Metzler.

Núñez, R. (2000). Mathematical idea analysis: What embodied cognitive science can say about the human nature of mathematics. In T. Nakahara, & M. Koyama (Eds.), *Proceedings of the 24th conference of the International Group for the Psychology of Mathematics Education, 1* (pp. 3-22). Hiroshima, Japan: PME.

Núñez, R. (2008). A fresh look at the foundations of mathematics. Gesture and the psychological reality of conceptual metaphor. In A. Cienski, & C. Müller (Eds.), *Metaphor and gesture* (pp. 93-114). Amsterdam: John Benjamins.

Núñez, R. E., Edwards, L. D., & Matos, J. F. (1999). Embodied cognition as grounding for situatedness and context in mathematics education. *Educational Studies in Mathematics, 39*, 45-65.

Parrill, F., & Sweetser, E. (2004). What we mean by meaning: Conceptual integration in gesture analysis and transcription. *Gesture 4*(2), 197-219.

Peirce, C. S. (1931-1958). *Collected papers (CP)* (Volumes I-XIII). (Eds. P. W. C. Hartshorne, & P. Weiss), Cambridge, Massachusetts: Harvard University Press.

Peirce, C. S. (1977). *Semiotics and significs (SS)*. (Ed. Charles Hardwick). Bloomington IN.: Indiana University Press.

Pike, K. (1967). *Language in relation to a unified theory of the structure of human behavior*. The Hague: Mouton.

Prediger, S., Link, M., Hinz, R., Hussmann, S., Ralle, B., & Thiele, J. (2012). Lehr-Lernprozesse initiieren und erforschen [Initiating teaching-learning-processes]. *MNU, 65*(8), 452-457.

Priwitzer, J. (2010). *Epistemische Basishandlungen in Modellen zur Konstruktion mathematischen Wissens* [Basic epistemic actions in models of constructing mathematical knowledge]. Unpublished master's thesis. Bremen.

Przyborski, A., & Wohlrab-Sahr, M. (2010). *Qualitative Sozialforschung: Ein Arbeitsbuch [Qualitative social research: An exercise book]* . München: Oldenbourg Verlag.

Radford, L. (2002). The Seen, the Spoken and the Written: a semiotic approach to the problem of objectification of mathematical knowledge. *For the Learning of Mathematics, 22(2)*, 14-23.

Radford, L. (2003). Gestures, speech and the sprouting of signs. *Mathematical Thinking and Learning, 5*(1), 37-70.

Radford, L. (2006). Tres tradizioni semiotiche: Saussure, Peirce y Vygotskij [Three semiotic traditions: Saussure, Peirce, and Vygotskij]. *Rassegna*, 34-39.

Radford, L. (2010). The eye as a theoretician. Seeing structures in generalizing activities. *For the Learning of Mathematics, 30*(2), 2-7.

Randell, J. (1977). Some leading ideas in Peirce's semiotics. *Semiotica, 19*, 157-178.

Rasmussen, C., Stephan, M., & Allen, K. (2004). Classroom mathematical practices and gesturing. *Journal of Mathematical Behavior, 23*, 301-323.

Reichertz, J. (2004). Abduction, induction and deduction in qualitative research. In U. Flick, E. v. Kardorf, & I. Steinke (Eds.), *A companion to qualitative research* (pp. 159-164). London: Sage Publications.

Reynolds, F. J., & Reeve, R. A. (2002). Gesture in collaborative mathematical problem-solving. *Journal of Mathematical Behavior, 20*, 447-440.

Rizzolatti, G. (2005). The mirror neuron system and its function in humans. *Anatomy and Embryology, 210*(5-6), 419-421.

Rizzolatti, G., & Craighero, L. (2004). The Mirror-Neuron System. *Annual Review of Neuroscience, 27*, 169-192.

Röpke, I. (2011). Watching the growth point grow. In: *Proceedings of the 2nd Conference on Gesture and Speech in Interaction.* Retrieved 6 13, 2014, from: http://coral2.spectrum.uni-bielefeld.de/gespin2011/final/Roepke.pdf

Roth, W.-M., & McGinn, M. K. (1998). Inscriptions: Towards a theory of representing as social practice. *Review of Education Research, 68*(1), 35-59.

Sabena, C. (2007). *Body and signs: a multimodal semiotic approach to teaching–learning processes in early Calculus.* PhD-Dissertation. Turin: Turin University.

Schütz, A. (1971). *Gesammelte Aufsätze. Bd. 1. Das Problem der sozialen Wirklichkeit [Collected papers, series 1. The problem of social reality].* Den Haag: Nijhoff.

Schwarz, B. B., Dreyfus, T., & Hershkowitz, R. (2009). The nested epistemic actions model for abstraction in context. In B. B. Schwarz, T. Dreyfus, & R. Hershkowitz (Eds.), *Transformation of knowledge through classroom interaction* (pp. 11-41). London, UK: Routledge.

Seeger, F. (2006). Ein semiotischer Blick auf die Psychologie des Mathematiklernens [A semiotic perspective on the learning mathematics]. *Journal für Mathematikdidaktik* (3/4), 265-284.

Sikveland, R. O., & Ogden, R. (2012). Holding gestures across turn: Moments to generate shared understanding. *Gesture, 12*(2), 166-199. Retrieved 11 28, 2014, from: http://eprints.whiterose.ac.uk/75687/1/Sikveland_Ogden_Gesture_2013.pdf

Soeffner, H.-G. (2004). Social Scientific Hermeneutics. In U. Flick, E. v. Kardorff, & I. Steinke (Eds.), *A companion to qualitative research* (pp. 95-100). London: Sage Publications.

Steelker-Weithofer, P. (2003). Semiotische Aspekte der Mathematik [Social aspects of mathematics]. In R. Posner, K. Robening, & T. A. Sebeoek (Eds.), *Semiotik: Ein Handbuch zu den zeichentheoretischen Grundlagen von Natur und Kultur, Vol. 3* (pp. 2569-2587). DeGruyter Mouton.

Steinke, I. (2004). Quality Criteria in Qualitative Research. In U. Flick, E. von Kardorff, & I. Steinke (Eds.), *A companion to qualitative research* (pp. 184-190). London: Sage Publications.

Sweller, J. (1988). Cognitive load during problem solving: Effects on learning. *Cognitive Science, 12*, 257-285.

Sweller, J. (1994). Cognitive Load Theory, Learning Difficulty, and Instructional Design. *Learning and Instruction, 4*, 295-312.

Talmy, L. (2000). *Toward a cognitive semantics: Concept structuring systems, Volume 1.* Cambridge, MA: The MIT Press.

van Leuwen, T. (2004). *Introducing social semiotics: An introductory textbook.* London: Routledge.

Varela, F. J., Thompson, E. T., & Rosch, E. (1991). *The embodied mind: Cognitive science and human experience.* Cambridge, MA: MIT Press.

Voigt, J. (1995). Thematic patterns of interaction and sociomathematics norms. In P. Cobb, & H. Bauersfeld (Eds.), *The emergence of mathematical meaning: Interaction in classroom culture* (pp. 163-201). Hillsdale, NJ: Erlbaum.

Vygotsky, L. S. (1978). *Mind in society.* Cambridge, MA: Harvard.

Yoon, C., Thomas, M. O., & Dreyfus, T. (2011). Grounded Blends and Mathematical Gesture Spaces: Developing Mathematical Understandings via Gestures. *Educational Studies in Mathematics, 78*(3), 371-393.

12 Glossary

Overview on Some Notions and Theoretical Concepts As They Are Used in this Book

Notions that are newly coined within this study are emphasized by being written in bolt letters.

12.1 What is Meant by…

Associated signs, Rather than symbolically 'standing' for some mathematical
Associated gesture idea or object, signs can become *associated* with them within social interaction. That is, the mathematical meaning of the sign is situationally grounded in its use during the social construction of the mathematical object it represents for the students. This meaning does not need to be conventionalized and does not even need to be "mathematically true". It is moreover *situationally conventionalized* and represents the current stadium of an idea associated with this sign. Following this idea, an associated gesture is a gesture that became an associated sign as being situationally conventionalized. **(chapter 7.2.1)**

Basic sign — Basic signs are gestures that get a certain meaning in the social interaction within the mathematics classroom. They can be used later on in a similar context without further explanation. Associated gestures can become basic signs when their use refers symbolically to a mathematical object rather than in an iconic way to the inscription the associated gesture developed from. **(chapter 4.3.3, chapter 7.2.3)**

Catchments — "A catchment is recognized when two or more gesture features recur in at least two (not necessarily consecutive) gestures" (McNeill, 2005, p. 116). Those features refer to the appearance of the gesture and can be revealed, for example, in its form, movement, or dynamism. A catchment is not the gesture itself, it comprises those situations that are identified as being related by the appearance of similar gestures and their discursive link that can be deduced from the gesture. The idea behind this is that similar gestures refer to similar, or related references. By identifying the catchment-gestures produced by a person can thus be reconstructed *which* meanings are related for him (contexts in which the gesture appears). Furthermore it can be interpreted *what* this relation consist of by taking into account the recurring aspect. **(chapters 3.2.2, 4.3.3)**

Co-expressivity	Co-expressivity of speech and gesture concerns the assumption that gesture and speech refer to the same idea. **(chapter 3.2.2)**
Embodied Cognition	The theory of embodied cognition, also referred to as the theory of the embodied mind, adopts the assumption that all of human's thinking and knowing is coined by the bodily experience in the physical world. Within this study, the embodied mind paradigm is used as one methodological pillar to explain how a mutual construction of mathematical knowledge is possible at all. **(chapter 4.2)**
Epistemic function	The *epistemic functions* of gestures concern the ways in which gestures can contribute to the accomplishment of the three epistemic actions 'gathering', 'connecting', and 'structure-seeing'. **(chapter 8)**
Epistemic process	The social learning process of constructing mathematical knowledge, reconstructed in social interaction in the accomplishment of the three epistemic actions **gathering**, **connecting**, and **structure-seeing**. **(chapter 4.1)**
Gestures	Gestures are considered "idiosyncratic spontaneous movement[s] of the hands and arms accompanying speech" (McNeill, 1992, p. 37), "being done for the purposes of expression rather than in the service of some practical aim" (Kendon, 2004, p. 15). A gesture encompasses three main phases: preparatory, stroke, and recovery. The only obligatory

	phase is the stroke, the main part and movement of the gesture. **(chapter 3.2)**
- Gesture dimension (McNeill, 1992)	Gestures can have deictic, iconic, metaphorical, or beat features. **(chapter 3.2.3)** *Deictics*: Also referred to as pointing gestures. *Iconic*: Refers to an object or activity by a representation in similarity. *Metaphoric*: Like iconic gestures but referring to an abstract idea. *Beats*: Movements like 'beating the air'. Often associated with emphasizing parts of the verbal utterance Whether a gesture is considered to be iconic or metaphoric depends on its relation to the co-occurring speech (→ see *co-expressivity*)
- Gesture classification (Edwards, 2009)	Edwards distinguishes two kinds of iconic gestures. Her classification provides the basis for the definition of metaphorical gestures as proposed for this study in the context of mathematical talk, taking into account that mathematical objects are considered being abstract. (→ see also *gesture, gesture dimension*) *Iconic-symbolic*: "Rather than referring to a concrete object in and of itself, the [iconic-symbolic] gesture refers to a symbolic, written inscription, which in turn represents a specific mathematical entity or procedure" (ibid., p. 138). **(chapter 3.2.3)** *Iconic-physical*: Iconic-physical gestures on the other side refer to actually concrete material or actions such that its use or they themselves are being represented by the gesture. (Edwards, 2009) *Metaphoric: In mathematical talk, I considered those gestures metaphoric that represent a mathematical idea without being iconic-physical or iconic-symbolic.*
Gesture phrase, Gesture unit	Smallest units of analysis in this study: One *gesture unit* may consist of several strokes (→ see *gesture*) between which the hand is held in preparation for the next stroke. As *gesture*

	phrase is defined one part of the gesture unit meaningfully assigned to one part of the verbal utterance, constituted by gesture stroke and pre- and post-stroke hold (Kendon, 2004, p. 112). **(chapter 3.2.1)**
Gesture space	The space in the air in front of the body, roughly framed by the shoulders and the hips, respectively the table when sitting (McNeill, 1992, p. 86). **(chapter 3.2.1, chapter 7.1.3)**
Growth Point (GP)	A minimal (non-verbal) unit in which an idea becomes visible. This Growth Point often can only be identified in retrospective since per definition, the idea is not made explicit in its Growth Point. When the idea can be grasped to be explicated in speech, the Growth Point is "unpacked". **(chapter 3.2.2)**
Illustrators	Movements of the hands and arms having a meaning contextually embedded in verbal discourse. In this study, illustrators are considered. **(chapter 3.2.1)**
- Topic & interactive gestures	"Topic gestures depict semantic information directly related to the topic of discourse, and interactive gestures (a smaller group) refer instead to some aspect of the process of conversing with another person." (Bavelas, Chovil, Lawrie, & Wade, 1992, pp. 472-473). In this study, only topic gestures related to the topic of mathematical discourse are considered. **(chapter 3.2.1)**
Inscription	As inscriptions are understood those "signs that are materially [fixed] in some medium, such as paper or computer monitors" (Roth & McGinn, 1998, p. 37). **(chapter 4.4)**
Language	Grounded on the assumption that gesture and speech form one integrated system in language formation, I understand language as comprising them both. In this regard, 'speech' refers to spoken language such as 'verbal utterance' refers to the corresponding part of the utterance, composed by gesture and speech. The term 'language' is also used referring to a specific language in a linguistic sense, carried out by rules of production and understanding. **(chapter 3.2.2)**

Gesture-speech match/mismatch	A gesture-speech-mismatch is identified when gesture and speech convey different information. This difference may concern contradicting ideas but also complementary information. **(chapter 3.2.2)**
Metaphor	"The essence of metaphor is understanding and experiencing one kind of thing in terms of another" (Lakoff & Johnson, 1980, p. 5). Following this, *conceptual metaphors* refer to one concept in terms of another: For this are distinguished *source domain* and *target domain*, where the concept in the target domain is understood in terms of the source domain. According to Núñez, mathematical concepts "don't exist in any real perceivable world. They are metaphorical in nature" (Núñez, 2008, p. 103). Fundamental mathematical concepts, such as addition or subtraction, have their source domain in the real world; these are called *grounding metaphors*. Understanding sophisticated mathematical ideas requires linking mathematical ideas in different mathematical domains, including also different semiotic registers to represent them. The latter are called *linking metaphors*. **(chapter 4.2.1)**
Multimodality	Multimodality refers to the "idea that communication and representation always draw on a multiplicity of semiotic modes of which language may be one" (Kress, 2001, pp. 67-68). According to Arzarello, analyzing learning processes need to take into account the intertwining of multimodal semiotic resources used. **(chapter 4.2.2, chapter 4.3)**
Semiotic Bundle (SB)	A semiotic bundle (SB) consists of "(i) A collection of semiotic sets and (ii) "A set of relationships between the sets of the bundle" (Arzarello, 2006, p. 281). These relationships can concern the signs that are used simultaneously (analyzed in a synchronous analysis) and also relationships of signs used at different times, considering a diachronic analysis of the semiotic bundle as it evolves in time. **(chapter 4.3)**
Semiotic game	The teacher may take up a sign introduced by a student to work out its meaning in collaboration. By this, the teacher may help him to express his idea more profoundly by refining a link

between the student's sign and his verbal utterance. The teacher takes part in shaping the sign's meaning by using it and simultaneously paraphrasing the students' verbal expression. In the particular use of gestures, Arzarello et al. (2009) show up the didactic potential of semiotic games using gestures: If the teacher recognizes a gesture-speech-mismatch in a students' utterance of a thought in classroom interaction, he may detect a fruitful approach to an idea in this although this idea does not need to be conscious to the student. If the verbal formulation is not yet elaborated, the teacher can paraphrase it while simultaneously using the gesture introduced by the student. Through this, the idea represented implicitly in gesture is linked to it and is enriched with mathematical meaning. **(chapter 4.3.3)**

Semiotic resource	Following Arzarello et al. (2009) as semiotic resources are considered those signs that are used by the students or the teacher to carry out the working process. According to van Leuwen, "semiotic resources have a meaning potential, based on their past uses, and a set of affordances based on their possible uses, and these will be actualized in concrete social contexts where their use is subject to some form of semiotic regime" (van Leuwen, 2004, p. 285). *(footnote in* **chapter 4.3***)*
Semiotics	Semiotics is the science of signs. In this work, a Peircean tradition is adopted to take into account that a sign does not per se refer to a mathematical object. It has no intrinsic meaning but can be interpreted differently, depending on various influential factors. **(chapter 3.1)**
Sign	Following Peirce, a "sign, or representamen, is something which stands to somebody for something in some respect or capacity" (CP, 2.228). What is termed 'sign' always needs to be seen as being related to an object that is represented, inducing an interpretant that is itself a sign representing the object with respect to a certain perspective. The object expressed in a sign is always an *immediate object*, differentiated from the *dynamic object* that exists only in pure potentiality and cannot be represented by any sign. **(chapter 3.1)**

In the present study, this concept is used within an interactionist perspective, understanding representamen and interpretant as action and reaction. This allows the development of the immediate mathematical object as expressed in these signs to be reconstructed in a *semiotic reconstructive analysis (SRA)*. **(chapter 5.2.2)**

- **Multimodal sign** The idea of a multimodal sign takes into account that gestures are always interpreted in their relation to speech and may also have a relation to co-occurring inscription. In a multimodal sign, the three modalities gesture, inscription, and speech are considered to form signs intertwined in their interplay, shaping together the immediate mathematical object of the entire (multimodal) utterance. **(chapter 4.4)**

- **Semiotic composition** The semiotic composition describes simultaneously used semiotic resources constituting the representamen of the multimodal sign. **(chapter 4.4)**

Speech The term 'speech' refers to the verbal expression following syntactic and semantic rules that enable communication. The analysis of speech acts is carried out on three levels – locationary, illocutionary, and perlocutionary level, following Austin (1962).

Within-functions As within-functions of gestures are understood those functions of gestures **(chapter 5.1)** that take part in shaping the immediate mathematical object of a multimodal sign. The within-function of a gesture is formed by two aspects, concerning its relationship to speech and its relationship to inscription: On the one side, gestures may have a *specifying function* **(chapters 7.1.1 & 7.1.2)** respecting them enriching the verbal expression by providing additional information about the immediate mathematical object. On the other side, their relationship to inscription can be described by the referential level the gesture is performed on **(chapter 7.1.3)**.

12.2 Glossary of Important Gestures Referred to in This Study

Axis of reflection

Iconic-symbolic gesture established in the gesture space in PA7 (Rosa and Lisa elaborating the parabola task). The gesture is introduced as axis of reflection in a process of packing an information bundle concerning the 'shape of the curve' **(chapter 7.2.1, example 2.1.b)**, its use co-timed to the verbal expression 'perpendicular bisector' causes confusion **(section 7.2.2.1, example 2.1c)**.

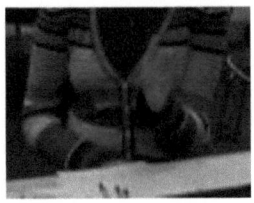

Grasping hand

A metaphorical gesture established in the context of substitution in CF7 (Rosa and Lisa elaborating the task dealing with the continued fraction). The gesture is associated to 'grasping' the element and 'putting' it somewhere, e.g. below the fraction bar of the successor. It can already be recognized in its Growth Point and takes important part in the packing of the information bundle 'recursive substitution' **(Chapter 8.2)**.

Shape of the curve

An iconic-symbolic associated gesture established *by each of the three pairs of students for the elaborations of the parabola task* (PA5, PA6, PA7), starting from a dynamic embedding of the shape within the folding sheet but also performed in some form (one branch or both) in the gesture space.

12.2 Glossary of Important Gestures Referred to in This Study 273

Example of gestures associated with 'shape of the curve', taken from chapter 7.2.1:

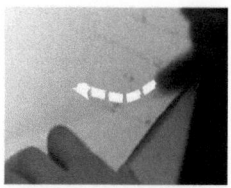

performed as embedded in the folding sheet (referential level 2)

performed within the gesture space (referential level 3), one branch

performed in the gesture space (referential level 3), two branches, symmetric

Part IV
Appendix

The appendix shall provide further information that facilitates the reading and understanding of the presented results. For that reason, the blank tasks are presented as well as an analysis of each task providing possible ways to solve it and presenting difficulties that may occur. While the analysis of the task already served to explore the interpretation frame to understand the students learning processes, it is also provided to give the reader better access to the mathematics behind the excerpts and examples presented. Furthermore, evidence for the within-functions and for the epistemic functions of gestures is listed, and the condensed process diagrams used for the comparisons presented in chapter 9 are given.

A Tasks

On the following pages are presented the three original tasks as given to the students. To make them comprehensible also for those readers that do not speak German, English translations are added in text bubbles.

a) Geometric-Algebraic Task Dealing With the Parabola as Geometrical Locus (PA)

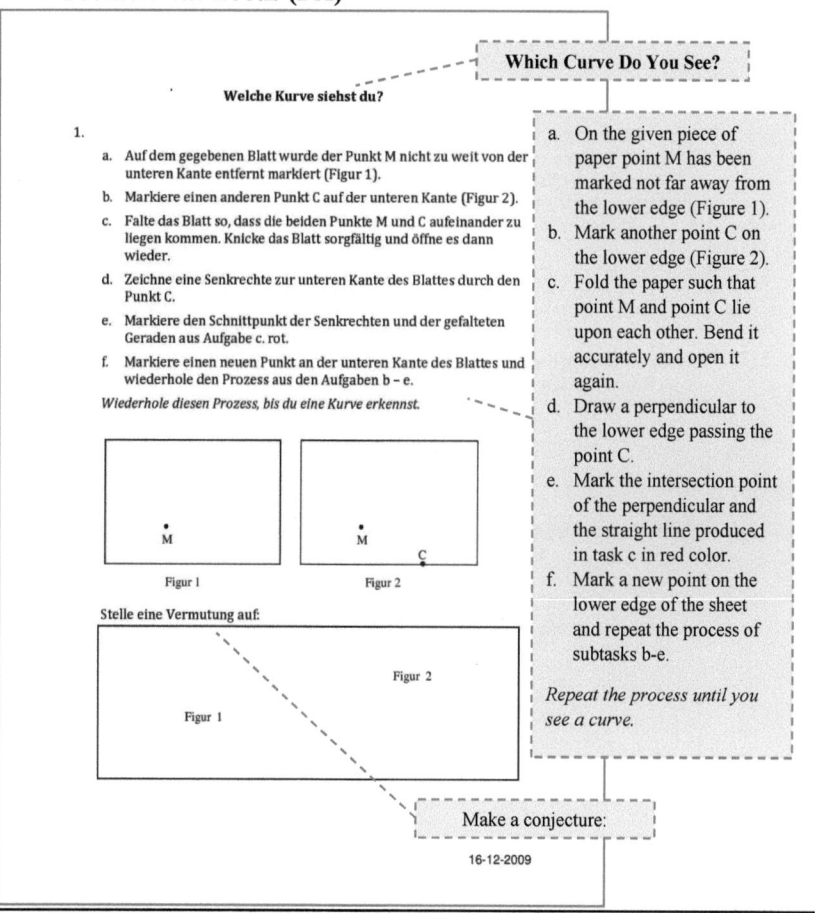

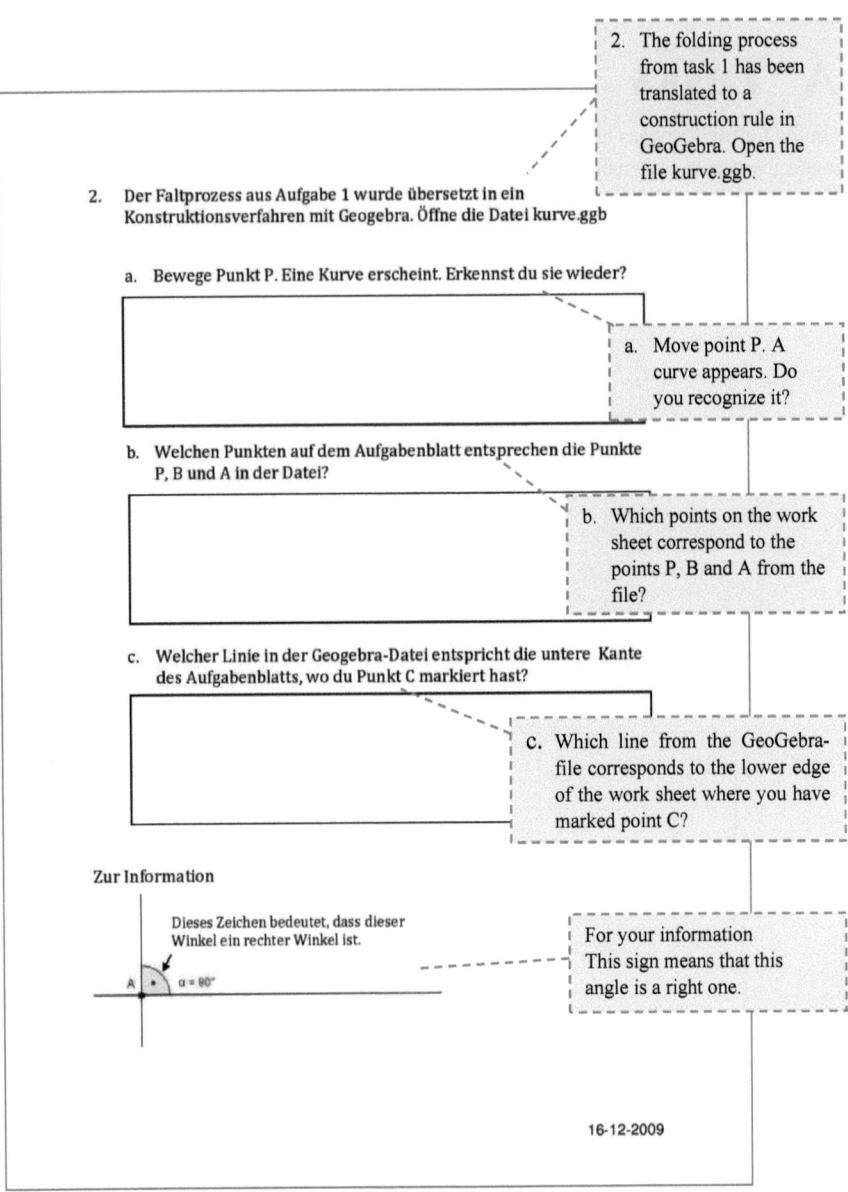

3. Betrachte einen beliebigen Punkt A auf der Kurve.
    a. Markiere den Abstand zwischen A und B sowie den Abstand zwischen A und der Gerade g. Stelle eine Vermutung auf.

    b. Begründe deine Vermutung.

3. Look at an arbitrary point A on the curve.
    a. Mark the distance between A and B as well as the distance between A and the straight line g. Make a conjecture.
    b. Justify your conjecture.

Remark: At this point, printouts as seen on the right side, representing a possible situation of the GeoGebra-diagram have been given to the students.

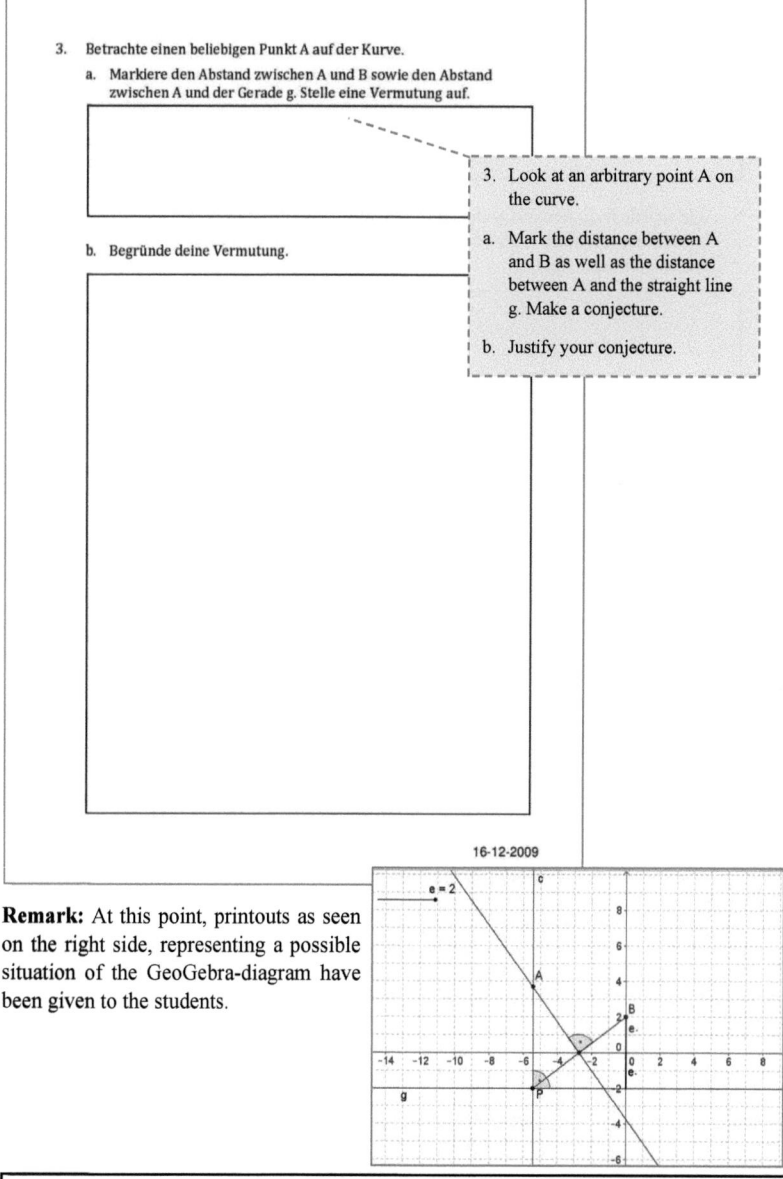

This task has been developed in the project "Effective knowledge construction in interest-dense situations" (funded by the German-Israeli-Foundation for Scientific Research and Development, Grant No. 946-357.4/2006). It is a revised version of a task designed by Nava Gilboa with specific contribution of Julia Cramer.

4.
   a. Welche Kurve hast du in Aufgabe 1 und 2a erkannt?

> a. Which curve did you recognize in tasks 1 and 2a?

   b. Wie würdest du jemand anderen davon überzeugen, dass es sich tatsächlich um diese Kurve handelt? Nutze alles, was du bisher herausgefunden hast.

> b. How would you convince somebody that it is actually this curve? Use everything you found out until now.

   c. Wie könntest du deine Ergebnisse zusammenfassend formulieren?

> c. How could you summarize your results?

16-12-2009

This task has been developed in the project "Effective knowledge construction in interest-dense situations" (funded by the German-Israeli-Foundation for Scientific Research and Development, Grant No. 946-357.4/2006). It is a revised version of a task designed by Nava Gilboa with specific contribution of Julia Cramer.

5. Ein Kreis ist eine Menge von Punkten, die alle dieselbe Entfernung von einem festen Punkt haben (dem Mittelpunkt).

Beschreibe in einer ähnlichen Weise die Parabel als Menge von Punkten.

5. A circle is a set of points that all have the same distance from a fixed point (the center).

Describe the parabola in a similar way as a set of points.

This task has been developed in the project "Effective knowledge construction in interest-dense situations" (funded by the German-Israeli-Foundation for Scientific Research and Development, Grant No. 946-357.4/2006). It is a revised version of a task designed by Nava Gilboa with specific contribution of Julia Cramer.

b) Arithmetic-Analytic task Dealing With a Sequence Defined by a Continued Fraction and the Limit of This Sequence (CF)

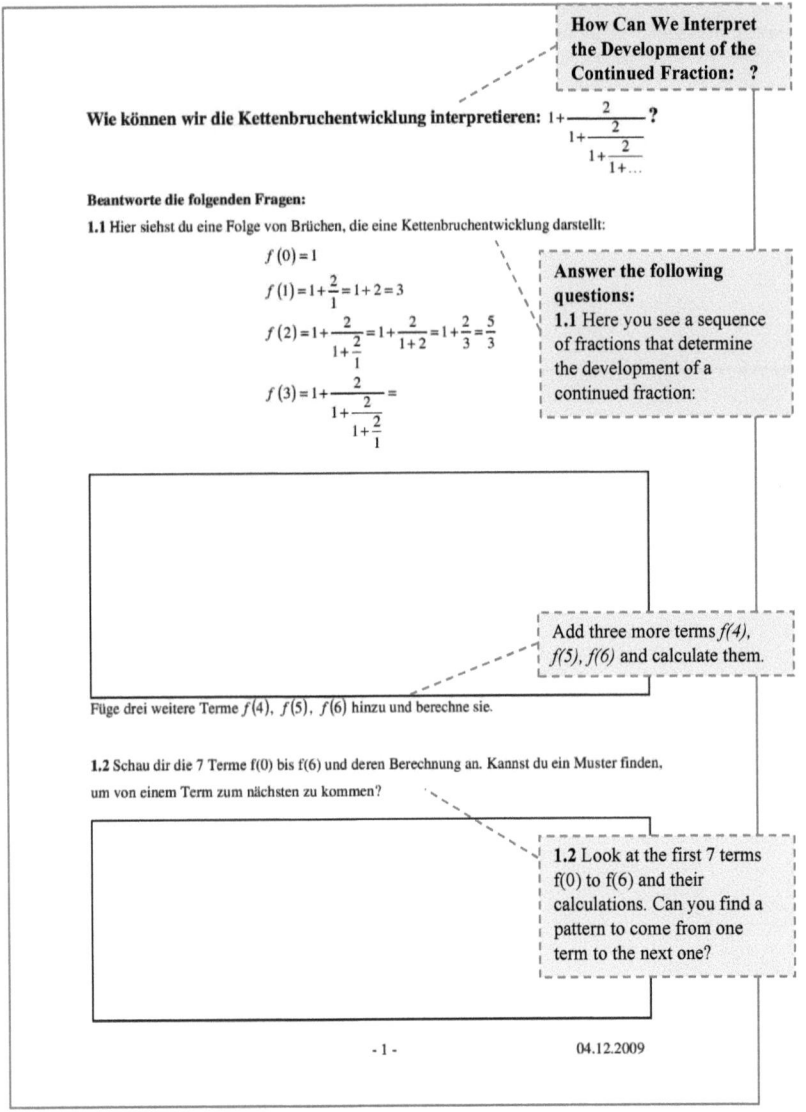

How Can We Interpret the Development of the Continued Fraction: ?

Wie können wir die Kettenbruchentwicklung interpretieren: $1+\cfrac{2}{1+\cfrac{2}{1+\cfrac{2}{1+\ldots}}}$?

Beantworte die folgenden Fragen:

1.1 Hier siehst du eine Folge von Brüchen, die eine Kettenbruchentwicklung darstellt:

$f(0) = 1$

$f(1) = 1 + \frac{2}{1} = 1 + 2 = 3$

$f(2) = 1 + \cfrac{2}{1+\frac{2}{1}} = 1 + \frac{2}{1+2} = 1 + \frac{2}{3} = \frac{5}{3}$

$f(3) = 1 + \cfrac{2}{1+\cfrac{2}{1+\frac{2}{1}}} =$

Answer the following questions:

1.1 Here you see a sequence of fractions that determine the development of a continued fraction:

Füge drei weitere Terme $f(4)$, $f(5)$, $f(6)$ hinzu und berechne sie.

Add three more terms $f(4)$, $f(5)$, $f(6)$ and calculate them.

1.2 Schau dir die 7 Terme f(0) bis f(6) und deren Berechnung an. Kannst du ein Muster finden, um von einem Term zum nächsten zu kommen?

1.2 Look at the first 7 terms f(0) to f(6) and their calculations. Can you find a pattern to come from one term to the next one?

The task has been designed in the project "Effective knowledge construction in interest-dense situations" (funded by the German-Israeli-Foundation for Scientific Research and Development, Grant No. 946-357.4/2006) with specific contribution of Nava Gilboa.

1.3 Erkläre das Muster – wieso funktioniert es?

1.3 Explain the pattern – why does it work?

2.1 Berechne weitere Terme der Folge, bis du 20 Terme der Folge hast. Nutze dabei das Muster, das du gefunden hast. Fülle die folgende Tabelle aus. Nutze einen Taschenrechner, um die Kettenbrüche der Folge als Dezimalzahlen darzustellen. Schreibe alle Nachkommastellen ab, die der Taschenrechner anzeigt.

2.1 Calculate further terms of the sequence until you have 20 terms. Use the pattern that you have found. Complete the following table. Use a calculator to represent the continued fractions of the sequence as decimal numbers. Write down all decimal places displayed by the calculator.

	Bruchzahl	Dezimalzahl
$f(0)$	1	1
$f(1)$	3	3
$f(2)$	$\frac{5}{3}$	1.666666667
$f(3)$	$\frac{11}{5}$	
$f(4)$		

- 2 -

A Tasks

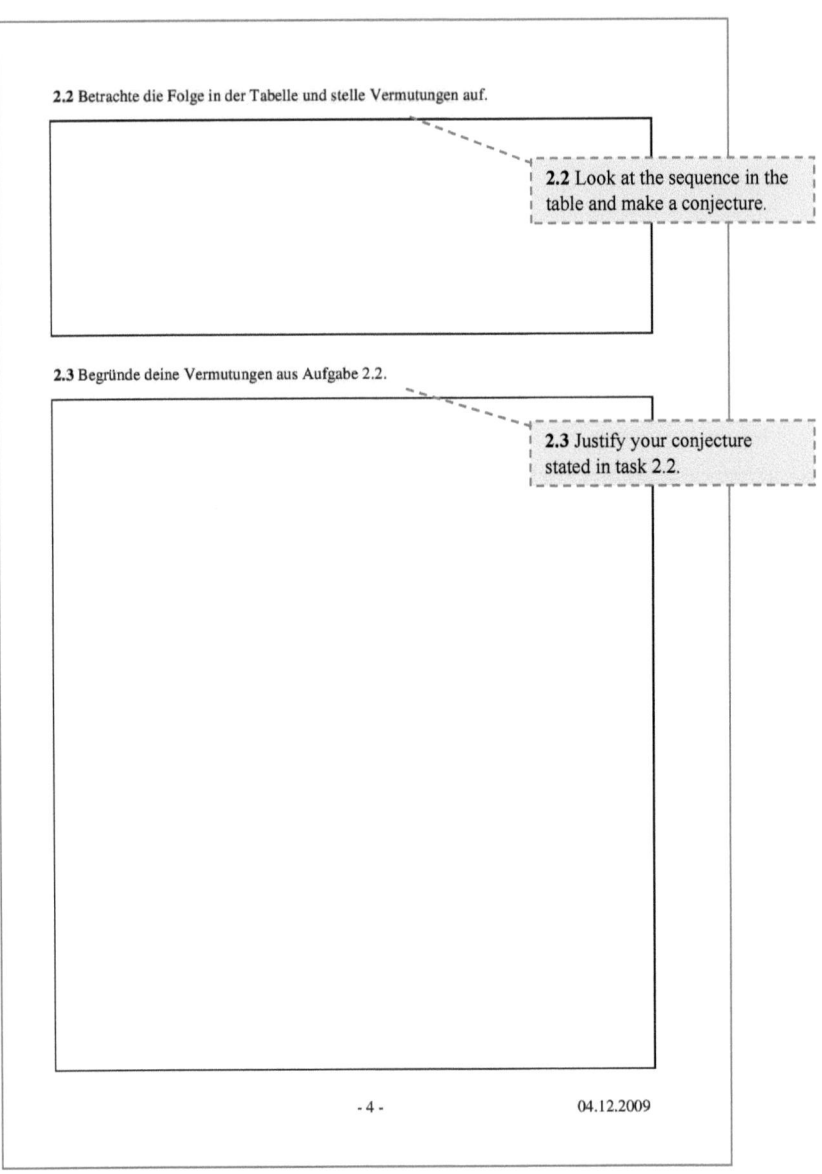

2.2 Betrachte die Folge in der Tabelle und stelle Vermutungen auf.

2.2 Look at the sequence in the table and make a conjecture.

2.3 Begründe deine Vermutungen aus Aufgabe 2.2.

2.3 Justify your conjecture stated in task 2.2.

- 4 -   04.12.2009

The task has been designed in the project "Effective knowledge construction in interest-dense situations" (funded by the German-Israeli-Foundation for Scientific Research and Development, Grant No. 946-357.4/2006) with specific contribution of Nava Gilboa.

2.4 Wieso begründet dies deine Vermutungen aus Aufgabe 2.2?

> **2.4** Why does this justify your conjecture stated in subtask 2.2?

Zusatzaufgabe

3. Eine Übung zur Reflexion:

Wie kann man diesen Ausdruck interpretieren: $\sqrt{1+\sqrt{1+\sqrt{1+\sqrt{1+\sqrt{1+...}}}}}$?

> **Additional task:**
> **3.** An exercise for reflection:
> How can one interpret this expression:... ?

Vielen Dank! ☺

04.12.2009

The task has been designed in the project "Effective knowledge construction in interest-dense situations" (funded by the German-Israeli-Foundation for Scientific Research and Development, Grant No. 946-357.4/2006) with specific contribution of Nava Gilboa.

c) Task on Logical Reasoning Using the Principle of Mathematical Induction (IN)

Des Königs Berater

Stell dir vor, du möchtest einer der Berater des Königs werden. Hierfür musst du eine Logikprüfung bestehen. Alle Berater, die diese Aufnahmeprüfung selbst bereits bestanden haben und deren Regeln kennen, sind zu dieser Prüfung eingeladen. Auf diesem Treffen bittet der König alle Berater und den Prüfling, in einem Kreis Platz zu nehmen. Er setzt jedem im Kreis einen Hut auf und sagt: Mindestens einer von euch trägt einen Hut, der mit einem `X´ markiert ist. Jeder hat die Aufgabe herauszufinden, ob sein/ihr Hut mit einem `X´ markiert ist oder nicht. Im Fünfminutentakt wird ein Gong erklingen. Jeder von euch, der nach einem oder mehreren Gongschlägen zu der Schlussfolgerung gekommen ist, dass sein/ihr Hut markiert ist, hebt beim nächsten Gongschlag die Hand." Durch die kreisförmige Sitzordnung sieht jeder Berater und auch der Prüfling die Hüte aller anderen, aber nicht seinen eigenen. Niemand darf sprechen.

Deine Aufgabe ist es, darüber nachzudenken, wie du diese Prüfung bestehen kannst.

The King's Consultants

Imagine you want to become one of the king's advisors. Therefore, you have to pass a logical examination. All the advisors that have already passed this examination and know the rules are invited to this examination. At the meeting, the king asks all consultants and the examinee to sit in a circle. He puts a hat on everyone's head and says: "At least one of you has a hat that is marked with an `X´. Everybody has to figure out whether the own hat is marked with an `X´. Every five minutes, a bell will ring. Each one of you that come to the conclusion that the own hat is marked shall raise his hand with the next ring of the bell." Due to the fact that the consultants and the examinee sit in a circle, everyone sees the hats of all the others but not their own. No one is allowed to speak.

It is your task to think about how you can pass this examination.

Du bist im Kreis und siehst n markierte Hüte. Beschreibe deine Strategie. Erkläre, wieso deine Strategie funktioniert.

You are sitting in the circle and see n marked hats. Describe your strategy. Explain why your strategy works.

This task has been designed in the context of the project "Effective knowledge construction in interest-dense situations" (funded by the German-Israeli-Foundation for Scientific Research and Development, Grant No. 946-357.4/2006) by Raz Harel.

B Analysis of the Tasks

In the following, the mathematical background of the three tasks set to the students is unfold. For this, different ways to solve the tasks mathematically are given, as well as possible pitfalls that may emerge. Analyzing the tasks in that regard has two main functions:

1. A detailed analysis of the mathematical potential of the task *sensitizes the interpreter* for the reconstruction of the students' epistemic process. The knowledge students construct can be wide-ranging and they may follow ways that have not been intended when the task was constructed. Although an analysis of the task can never predict every possible solution effectively carried out by the students, it unfolds possibilities that can be taken into consideration when analyzing the epistemic processes.
2. Second, the analysis serves the reader. The considered tasks are quite complex and providing possible ways to solve it shall help the reader to understand the interpretations and results presented in this study.

a) Geometric-Algebraic Task Dealing With the Parabola as Geometrical Locus (PA)

The goal of this task is to develop an understanding of the parabola as geometrical locus. This requires detaching from the parabola as function represented in a coordinate system in order to draw on its definition as a set of points having equal distance from the focal point and the directrix.

Didactical Remarks: The entire task is influenced by the creation of the folding representation carried out in subtask 1 in the beginning of the working process. The folding serves several purposes throughout the task:

1) **Exploring:** Folding the paper according to the instruction creates a product on which the first requested goal of "recognizing a curve" (subtask 1.1) can be explored. Furthermore, the instruction leaves a certain degree of idiosyncrasy within the process of folding in so far, that the students can choose where to construct the points and how many points shall be constructed. This helps to immerse oneself into this folding process and the product created by it.

2) **Applying** the *geometrical implementation of the folding activity*: Folding one point on top of another one creates a folding line. Geometrically, the folding line constitutes the axis of reflection on which the two points are reflected. The folding activity can thus be implemented geometrically as an act of reflecting. In this task, this analogy needs to be seen to justify that the distance between points A and B equals the distance between points A and P in the printout (subtask 3). This requires to be aware of the correspondence between the entities in both representations, explicated in subtask 2.
3) **Generalizing** the equality of the distances in order to define the parabola: The folding can be applied to substantiate that $\overline{AP} = \overline{AB}$ in the specific folding representation. However, the points are constructed by repeatedly following the same construction. The equality of the lengths of the segments thus holds for **every** point A and the corresponding point P on the directrix as grounded in the folding process and provide a defining property for the points constituting a parabola.

Solution of the Task:

Subtask 1: The repeated execution of the instruction given in subtask 1 leads to the production of an envelope curve of a parabola. Furthermore, the points on the parabola corresponding to the tangent lines that constitute the envelope curve are marked within the representation. The curve requested to be seen can thus be identified in the set of tangent lines or in the set of points constructed, both times suggesting the shape of the curve through idealization. The conjecture to be made shall thus identify the curve to be a parabola. However, it is not explicitly requested to state a conjecture regarding the type of curve that has been identified. Other possible conjectures that can be expected are listed in the following:

- The folding lines are tangents on the curve.
- The curve is symmetric (Or stated more precisely, symmetric to a perpendicular to the lower edge running through point M).
- The minimum point of the curve lies directly below point M.
- The aperture of the curve depends on the distance of the point M to the lower edge (Or more specifically: The aperture increases with decreasing distances of point M to the lower edge and decreases with increasing distance. It can also be hypothesized whether the relation between aperture and distance of point M to the lower edge may be inversely proportional.)
- Changing the instruction to choose points on the upper instead of on the lower edge flips the curve upside down.

"Wrong" observations can concern the identification of the type of the curve, for example as exponential function. This may be grounded in constructing not enough points to predict the shape of the curve. If points are only constructed on one side, the symmetry may not become apparent when not concluding that the principle of folding is the same on both sides, causing the symmetric structure.

Subtask 2: Subtask 2 serves to link the entities as represented in the folding diagram to a dynamic representation of a parabola within a coordinate system, constructed in a GeoGebra environment. In the dynamic diagram, a point P can be moved on a straight line $g = g(x) = -2$. With this, point A moves too and this movement traces a curve. A point B is fixed on $(0,2)$. The entities in the GeoGebra environment correspond to those in the folding sheet representation as follows:

Table B.1: Correspondence between entities represented in the GeoGebra environment and the folding sheet

GeoGebra	Folding Sheet
point P	Point chosen on the lower edge of the sheet (named C in the instructions)
point B	Point M (fixed point in the middle of the sheet; point on which chosen point on lower edge is folded upon)
point A	Point of intersection of the folding line and the perpendicular through the corresponding point on the lower edge (marked with a red cross)
straight line g	Lower edge of the folding sheet; edge where points for construction have been chosen

Furthermore, adjusting a scroll bar in the dynamic GeoGebra-environment allows the value e, determining the vertical position of point B, to be varied in discrete steps between 0 to 2. This value of e corresponds to the half distance of the fixed point from the lower border in the folding sheet.

Moving point P on the directrix and with this also point A on the parabola suggests the correspondence between these two points and allows the students to explore different cases concerning the position of point P. Varying the value for e and tracing the curve for these different values makes it possible to investigate empirically how the position of the focal point influences the graphical representation of the function.

Subtask 3: A printout of one concrete situation that can be adjusted in the GeoGebra environment makes it possible to mark the distances as requested in subtask 3a. This is represented in Fig. B.*1*.[1]

---

[1] In the following, it is referred to the points, segments etc. according to their notations used in the printout diagram.

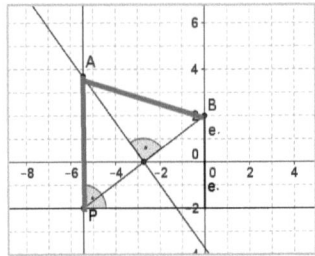

Fig. B.1: Cutout of the printout diagram given to the students, segments added in green

In this regard, the conjecture suggested to be stated concerns the observation that the distance between points A and B equals the distance between point A and the straight line g. The distance between a point and a straight line is defined as the minimal distance between them, that is the distance from the point to the foot of the perpendicular to the straight line through the point.[2] While this observation can be made and tested empirically, for example by measuring the distances in the concrete example given in the printout, justifying the conjecture requires more general considerations. As has already been explained in the didactical remark above, this conjecture can be justified by the geometrical implementation of the folding activity as act of reflecting. That is, the folding line (or tangent line as it can be identified in the GeoGebra representation) provides a mapping $R \colon \mathbb{R}^2 \to \mathbb{R}^2$ with

(1) If $v \in \mathbb{R}^2$ is a point on the folding line, then $R(v) = v$.
(2) If $v \in \mathbb{R}^2$ is a point **not** on the folding line, then
   (a) $[vR(v)] \perp$ folding line
   (b) The distance between point $R(v)$ and folding line equals the distance between point v and the folding line.

The axis of reflection goes through point A by construction so that $R(A) = A$. Denote with C the intersection point of the axis of reflection with $[PB]$ as represented in Fig. B.2.

---

[2] This latter definition is not necessarily known to the students and may cause a problem on which the interviewer may react.

Parabola as Geometrical Locus

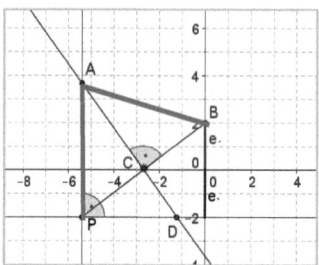

Fig. B.2: Cutout of the printout diagram given to the students, points C and D, and segments $[AP]$ and $[AB]$ added

Since point C lays on the folding line, it is $R(C) = C$. Two of the fundamental properties of a reflection mapping are that (i) straight lines are mapped on straight lines and that (ii) the lengths of a segment equals the length of the segment it is mapped on. As a reflection mapping, R maps straight lines on straight lines such that distances stay the same after mapping. With this follows that $\overline{AP} = \overline{R(AP)}$, and with A mapped on $R(A)$ and P mapped on $R(P) = B$ it is

$$\overline{AP} = \overline{R(AP)} = \overline{R(A)R(P)} = \overline{R(A)B} = \overline{AB}.$$

Further possible observations that can be made concern the congruency of the triangles $\triangle ACB$ and $\triangle APC$. This congruency follows immediately from the triangles being reflected on the tangent line: It is $\overline{R(AP)} = \overline{AB}$, $[AC] = [R(AC)]$, and with $R(P) = B$, both points have the same distance from the axis of reflection, hence $\overline{CB} = \overline{PC}$. Corresponding in the lengths of the three sides, both triangles are congruent. With $[CP]$ perpendicular on the folding/tangent line, the triangle $\triangle CPD$ with D the intersection point of the tangent and the directrix (see Fig. B.2) and triangle $\triangle CAP$ are similar: Both triangles have a right angle in C. Since they 'share' the side $[PC]$ that divides the right angle in point P such that with $\alpha := \angle CPA$ and $\beta := \angle DPC$, it is $|\alpha| + |\beta| = 90°$. With sum of angles in a triangle equals 180°, it follows for the third angles $|\angle PAC| = 90° - |\alpha| = |\beta| = |\angle DPC|$ and $|\angle CDP| = 90° - |\beta| = |\alpha| = |\angle CPA|$. With all angles corresponding, the two triangles are similar. Furthermore, the triangle $\triangle APD$ has a right angle in P, shares the angle in point A with $\triangle CAP$ and the angle in point D with $\triangle CPD$. Because of this, $\triangle APD$ is also similar to $\triangle CPD$ and to $\triangle CAP$.

Subtask 4: Task 4a explicitly claims for identifying the type of the curve as a parabola. Convincing somebody on this conjecture can be done by showing that the function equation of the curve has the form $f(x) = ax^2 + bx + c$.

An empirical way to find the function equation makes use of the concrete values as given in the GeoGebra environment:

- Three points on the curve, noted by means of their x- and y-coordinates, are sufficient to set up a system of linear equations to determine a function equation of second degree. However, this approach already uses the condition that the function equation has the form that shall be proven.
- Gathering points can lead to seeing a structural relation between x- and y-value from which can be concluded the function equation.

Since the function equation shall be stated generally as being independent of the value e, both approaches need to be conducted for several values of e. The resulting specific function equations can be compared in order to find a general structure in them and in dependence of e. Possible specific function equations to find are the following, suggested by the values that can be easily adjusted by the scroll bar:

Table B.2: Function equations for different values of e

Value of e	Function Equation
1	$f_1(x) = \frac{1}{4}x^2$
2	$f_2(x) = \frac{1}{8}x^2$
0.5	$f_{0.5}(x) = \frac{1}{2}x^2$
0.25	$f_{0.25}(x) = x^2$

In these equations, it can be seen the structure in the factor such that the general function equation follows as

$$f_e(x) = \frac{1}{4e}x^2$$

However, the empirical approach leads to a general function equation that verifies the conjecture of the curve being a parabola but it does not provide a scientific proof that this function equation is actually valid for all $e \in \mathbb{R}_{>0}$. For this, general properties of the curve need to be considered independently from the specific value of e. The function equation can be determined as general relation between x- and y-value for an arbitrary point $A(x,y)$ on the parabola. Three possible solutions are presented in the following, all of them make use of relations between different entities as represented in the printout diagram, but all are grounded in recent conjectures made in subtask 2:

Conjecture: The equation $f(x) = \frac{1}{4e}x^2$ describes the parabola for all $e \in \mathbb{R}_{>0}$.

- **Proof Using the Pythagorean Theorem:**

    The application of the Pythagorean theorem requires the consideration of an additional straight line in the printout diagram, the perpendicular to the y-axis running through point $A(x, y)$. Let F be the point in which this new auxiliary line intersects the y-axis (Fig B.3).

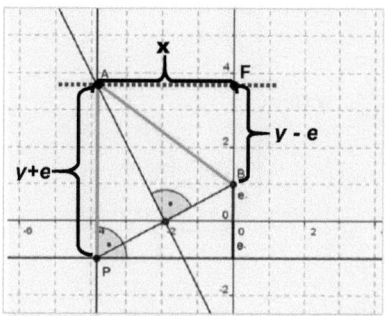

Fig. B.3: Printout diagram with straight line through points A and F added

Proof:
It is
$\overline{AF} = |x|$, $\overline{BF} = |y - e|$ and with $\overline{AB} = \overline{AP}$ (shown in subtask 2) also,
$\overline{AB} = |y + e|$. Furthermore it is $e \in \mathbb{R}_{>0}$.
Since $[AF]$ is orthogonal on the y-axis, the angle in point F is a right one and according to the Pythagorean Theorem it is

$$\overline{AF}^2 + \overline{BF}^2 = \overline{AB}^2$$

and thus

$$x^2 + (y - e)^2 = (y + e)^2$$

$$\Leftrightarrow \quad x^2 + y^2 - 2ye + e^2 = y^2 + 2ye + e^2$$

$$\Leftrightarrow \quad 4ye = x^2$$

$$\Leftrightarrow \quad y = \frac{x^2}{4e} = \frac{1}{4e}x^2$$

Any point A on the curve is thus determined by the quadratic function equation

$$f(x) = y = \frac{1}{4e}x^2.$$

q.e.d.

- **Proof Using the Triangle Altitude Theorem**

This proof requires the reference to other entities than the proof presented before.

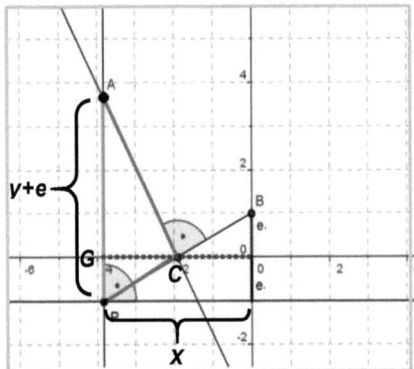

Fig. B.4: Printout diagram modified to apply the triangle altitude theorem

It can be seen that the length of the altitude through point C in the triangle ΔAPC is $\left|\frac{x}{2}\right|$: This follows immediately from the triangles ΔPCG and ΔBCO (with O origin) having sides of equal length $\overline{GP} = \overline{OB} = e$ and $\overline{PC} = \overline{CB}$ and their angles in C being alternate angles. With that also the third side needs to be equal: $\overline{CG} = \overline{CO}$. Since $\overline{GC} + \overline{CO} = x$, it is $\overline{GC} = \overline{CO} = \left|\frac{x}{2}\right|$.

The triangle altitude theorem states that:

> Given a right triangle with altitude h, dividing the hypotenuse in two segments p and q. Then it is $h^2 = p \cdot q$.

That is

$$\left(\frac{x}{2}\right)^2 = y \cdot e$$

$$\Leftrightarrow \quad \frac{x^2}{4} = y \cdot e$$

$$\Leftrightarrow \quad \frac{x^2}{4e} = y = \frac{1}{4e}x^2$$

Any point A on the curve is thus determined by the quadratic function equation

$$f(x) = y = \frac{1}{4e}x^2.$$

q.e.d.

- **Proof Using the Similarity of Triangles $\triangle AGC$ and $\triangle CGP$**

In the same way as has been done for subtask 3 for the triangles $\triangle APD$ and $\triangle PDC$, the similarity for the triangles $\triangle AGC$ and $\triangle CGP$ can be proven as lemma to the proof of the validity of the quadratic function equation.

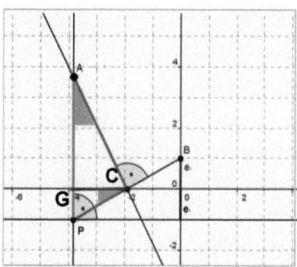

Fig. B.5: Printout diagram modified to make visible the similarity of the triangles $\triangle GAC$ and $\triangle GCP$

Both triangles, $\triangle AGC$ and $\triangle CGP$, have a right angle in G (see Fig. B.5) such that the respective other two angles in each triangle sum up to 90°. Since $[AC]$ is perpendicular to $[BP]$, it is $|\angle ACP| = 90°$ according to construction. This angle $\angle ACP$ is divided by $[GC]$ in two angles, $\angle ACG$ and $\angle GCP$, each one belonging to one of the triangles $\triangle AGC$ and $\triangle CGP$. Therefore, it is $|\angle GCP| = 90° - |\angle ACG| = |\angle GAC|$. Having two angles of equal value, also the third angle need to correspond and the triangles are similar.

For similar triangles holds that corresponding sides relate in the same way, such that
$$\frac{\overline{GC}}{\overline{GP}} = \frac{\overline{AG}}{\overline{GC}}$$

and

$$\frac{\left(\left|\frac{x}{2}\right|\right)}{e} = \frac{y}{\left(\left|\frac{x}{2}\right|\right)}$$

$$\Leftrightarrow \quad \left(\frac{x}{2}\right)^2 = y \cdot e$$

$$\Leftrightarrow \quad \frac{x^2}{4} = y \cdot e$$

$$\Leftrightarrow \quad y = \frac{x^2}{4e}$$

Any point A on the curve is thus determined by the quadratic function equation

$$f(x) = y = \frac{1}{4e}x^2.$$

q.e.d.

Subtask 5: In the last subtask it is requested to detach from the parabola as a function and directed towards seeing the parabola as a geometrical locus. The correct answer for this subtask is given by the following definition:

A parabola is a set of points that all have same distance from the focal point and the directrix.

However, the notions of "focal point" and "directrix" are not familiar such that the students are expected to carry out the definition by using the concrete notations of the point as given in the printout/GeoGebra-environment or the folding sheet. The reference to the definition of the circle - " A circle is a set of points all having same distance from a fixed point, the center" – and the request to define the parabola-definition in a **similar** way suggests starting with "a parabola is a set of points". Furthermore, the concrete formulation prompts searching for distances.

Using the notation of the printout diagram, in subtask 3 it has already been found that for any given point A on the parabola, it is $\overline{AB} = \overline{AP}$. This given point corresponds to a specific point that can be constructed in the folding diagram, hence it is constructed as point on the parabola. The translations needed to be made in order to formulate the requested definition of the parabola are manifold:

(i) The given point A in the printout diagram corresponds to a possible constructed point in the folding diagram. Furthermore, the correspondence to the other entities in both diagrams needs to be seen. This is prepared in subtask 2.

(ii) The folding activity corresponds to a reflection. This justifies the equality of the lengths of the segments $\overline{AB} = \overline{AP}$ (subtask 3).

(iii) The concrete case presented stands for any point $A(x, y)$ on the parabola and it is $\overline{AB} = \overline{AP} = y + e$ (subtask 4).

The definition of the circle requires one comparison to find a generality for all points: the comparison of the distance to the center. The case is different for the parabola: On the one hand, comparing distances for one specific but arbitrary point it can be seen that it has same distance to the focal point and the directrix. On the other hand, a comparison between the points leads to the observation that this distance is different for any two points (that are not reflected on the axis through the focal point). However, taking into account that this distance always equals $y + e$ for any point $A(x, y)$ on the parabola, this difference is only a numerical, not a relational one. Thus, comparing the relations and seeing the generality in these thus provides the core aspirations and difficulties of this task.

b) Arithmetic-Analytic Task Dealing With a Sequence Defined by a Continued Fraction and the Limit of This Sequence (CF)

The elaboration of the arithmetic-analytic task is directed towards determining a recursive rule for a sequence defined by the development of a continued fraction and finding out about its limit. To justify this limit, different observations need to be made on the relations between different elements of the sequence, emanating from the calculation of concrete elements.

Subtask 1: The task deals with the continued fraction (CF)

$$1 + \cfrac{2}{1+\cfrac{2}{1+\cfrac{2}{\dots}}},$$

The sequence $(f(x))_{x \in \mathbb{N}}$ defined by this CF starts with $x = 0$ and has given starting values

$$f(0) = 1$$

$$f(1) = 1 + \frac{2}{1} = 1 + 2 = 3$$

$$f(2) = 1 + \cfrac{2}{1+\cfrac{2}{1}} = 1 + \frac{2}{1+2} = 1 + \frac{2}{3} = \frac{5}{3}$$

$$f(3) = 1 + \cfrac{2}{1+\cfrac{2}{1+\cfrac{2}{1}}} =$$

All but the last given fraction $f(3)$ are simplified and calculated up to the form of a fraction term. The equal sign in front of the last fraction prompts doing the same for $f(3)$, although it is not explicitly requested. The notation of "$f(x)$" suggests the reference to the sequence understood as function in x such that the representation of the first four elements allows making observations and conjectures on the structure of the sequence:

    (i) x corresponds to the number of "+" in the CF.
    (ii) x corresponds to the number of "2"'s in the CF.
    (iii) x corresponds to the number of fraction bars in the CF.
    (iv) One element always becomes the denominator of the whole CF in the next element.
    (v) $f(1)$ always substitutes the least denominator from one element to the next one.

(vi) It is always added "$\frac{2}{1}$" to the least denominator from one element to the next one.

(vii) It is always complemented "$1 + \frac{2}{}$" in the top of a CF from one element to the next one.

The first three observations concern explicit rules to determine an element $f(x)$ of the sequence, the last four observations concern recursive rules. While the former three do not provide an aid for the requested calculation of the elements $f(4), f(5)$ and $f(6)$, the latter four may be helpful for this. It however needs to be considered that the application of rules according to conjectures (v) and (vi) is restricted to the representation of the element as CF. Using these approaches with the element represented as fraction terms leads to misleading results, for example

! $f(4) = \frac{11}{1+\frac{2}{1}}$ according to (v) or

! $f(4) = \frac{11}{5+\frac{2}{1}}$ according to (vi).

To avoid "false friends" in these rules, it is thus necessary to state them as

(v) $f(1)$ always substitutes the **one in the** least denominator from one element to the next one.

(vi) It is always added "$\frac{2}{1}$" to the **one in the** least denominator from one element to the next one.

Seen from another perspective, two of the recursive rules concern substitution strategies, two of them complementing strategies. In mathematical symbolic notation, this is summarized in Fig. B.6 below.

Only the strategies working with the entire element, hence those following observations (iv) and (vii), can be represented as formula:

$$f(x+1) = 1 + \frac{2}{f(x)} \qquad (1)$$

with $x \in \mathbb{N}$ and $f(0) = 1$

Other strategies for a recursive rule that can be made after the calculation of the first seven elements concern relations between numerator and denominator of two consecutive elements represented as fraction term:

There are $k, l \in \mathbb{N}, \gcd(k, l) = 1$

$$f(x) = \frac{k}{l}, \text{ and it is } f(x+1) = \frac{2k \pm 1}{k} \qquad (2)$$

and the algebraic sign in the numerator changes alternatingly. Although this rule can already be seen to constitute the following element, it does more likely become apparent in subtask 2.

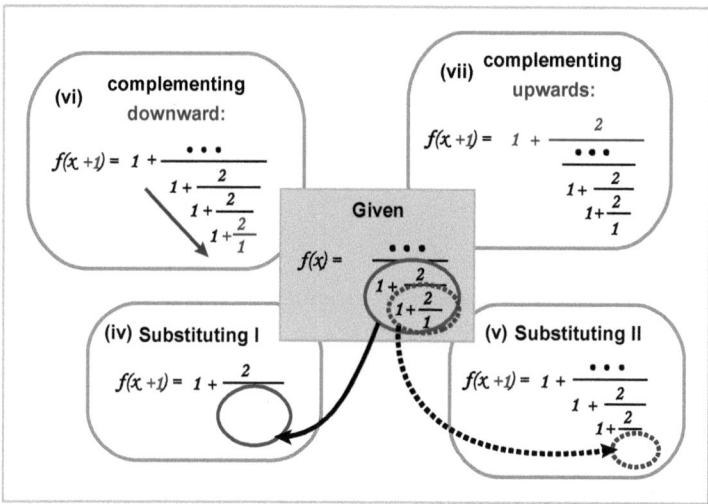

Fig. B.6: Strategies to determine one element f(x+1) from the previous one, f(x)

Subtask 2 starts with calculating the first 20 elements of the sequence and filling them in a table, represented both as fraction term and as decimal number. Emanating from these representations, conjectures shall be made on the sequence. Conjecture (2) as has already been mentioned as possible observation to be made with respect to a recursive strategy for subtask 1.2, becomes more evident by a larger amount of concrete examples to consider. Other observations that can be made concern the decimal numbers and their decimal places:

- Beginning with $f(2)$, the integer part always alternates between 1 and 2. It is $f(2n) = 1. ...$ and $f(2n + 1) = 2. ....$ for every $n \in \mathbb{N}_{>0}$.
- Beginning with $f(4)$ the decimal places for elements of the form $f(2n)$ start with a 9, those of the form $f(2n + 1)$ with a 0.
- The amount of nine's, respectively zero's in the decimal places becomes larger with increasing number of the element of the sequence. More precisely the number of nine'/zero's following the decimal point increases more and more.

Mathematically, this leads to the hypothesis that $\lim_{x \to \infty} (f(x))_{x \in \mathbb{N}} = 2$.

Proof of $\lim\limits_{x \to \infty} (f(x))_{x \in \mathbb{N}} = 2$: Approach 1

One way to prove that $\lim\limits_{x \to \infty} (f(x))_{x \in \mathbb{N}} = 2$ uses that, if two disjoint subsequences $(f_1(x))_{x \in \mathbb{N}}$ and $(f_2(x))_{x \in \mathbb{N}}$ converge to a value a and $f_1(x) \cup f_2(x) = f(x)$, also the sequence $(f(x))_{x \in \mathbb{N}}$ converges to a.

The sequence $f(x)$ can be split up in two disjoint subsequences $f_1(x) \cup f_2(x) = f(x)$,

$$f_1(x) := f(2x) \qquad (3)$$

and $\qquad f_2(x) := f(2x + 1)$

with $x \in \mathbb{N}$.

To prove $\lim\limits_{x \to \infty} (f(x))_{x \in \mathbb{N}} = 2$, it needs to be shown that

$$\lim\limits_{x \to \infty} (f_1(x))_{x \in \mathbb{N}} = 2,$$

$$\lim\limits_{x \to \infty} (f_2(x))_{x \in \mathbb{N}} = 2.$$

A lemma used for this is grounded in a further observation that can be made for the fraction terms:

For any arbitrary element $f(x)$ of the sequence $(f(x))_{x \in \mathbb{N}}$, there are $k, l \in \mathbb{N}$, $\gcd(k, l) = 1$ such that $f(x) = \frac{k}{l}$ and it holds that $f(x) = \frac{k}{l}$ has distance $\frac{1}{l}$ to 2.

More precisely: For any arbitrary element $f_1(x)$ of the sequence $(f_1(x))_{x \in \mathbb{N}}$ there are $a, b \in \mathbb{N}$, $\gcd(a, b) = 1$ such that $f_1(x) = \frac{a}{b}$ is $\frac{1}{b}$ less than 2, that is $f_1(x) = 2 - \frac{1}{b}$. (With analogue conditions it is $f_2(x) = \frac{c}{d} = 2 + \frac{1}{d}$ with $c, d \in \mathbb{N}$, $\gcd(c, d) = 1$ for every element of $(f_2(x))_{x \in \mathbb{N}}$.)

The prove is conducted similarly for both subsequences such that without loss of generality, it suffices to present the proof for $(f_1(x))_{x \in \mathbb{N}}$.

Lemma 1.1:

Let $(f_1(x))_{x \in \mathbb{N}}$ be defined as in (3). Then for any $n \in \mathbb{N}$ there are $a_n, b_n \in \mathbb{N}$ with $\gcd(a_n, b_n) = 1$ such that $f_1(n) = \frac{a_n}{b_n} = 2 - \frac{1}{b_n}$.

Proof:
Applying the recursive formula (1) two times leads to

$$f(x + 2) = 1 + \frac{2}{f(x+1)} = 1 + \frac{2}{1 + \frac{2}{f(x)}}$$

and with that

Continued Fraction

$$f(2n+2) = f_1(n+1) = 1 + \frac{2}{1+\frac{2}{f_1(n)}} \text{ for any } n \in \mathbb{N}, f_1(0) = 1.$$

Using mathematical induction the lemma can be proven:

(BS) Basis step: For $n = 0$, it is $f(0) = f_1(0) = 1 = \frac{a_0}{b_0} = \frac{1}{1} = 2 - \frac{1}{1}$. Since $gcd(1,1) = 1$, the induction hypothesis is proven to be valid for $n = 0$.

(IS) Inductive step: Assuming the validity of the induction hypothesis for some $n \in \mathbb{N}$: There are $a_n, b_n \in \mathbb{N}$, $gcd(a_n, b_n) = 1$, such that $f_1(n) = \frac{a_n}{b_n} = 2 - \frac{1}{b_n}$.
Then

$$f_1(n+1) = 1 + \frac{2}{1+\frac{2}{f_1(n)}} = 1 + \frac{2}{1+\frac{2}{2-\frac{1}{b_n}}}.$$

This can be transformed:

$$f_1(n+1) = 1 + \frac{2}{1+\frac{2}{2-\frac{1}{b_n}}}$$

$$= 1 + \frac{2}{1+\frac{2}{\frac{2b_n-1}{b_n}}}$$

$$= 1 + \frac{2}{1+\frac{2b_n}{2b_n-1}}$$

$$= 1 + \frac{2}{\frac{2b_n-1}{2b_n-1}+\frac{2b_n}{2b_n-1}}$$

$$= 1 + \frac{2}{\frac{4b_n-1}{2b_n-1}}$$

$$= 1 + \frac{4b_n-2}{4b_n-1}$$

$$= \frac{4b_n-1}{4b_n-1} + \frac{4b_n-2}{4b_n-1}$$

$$= \frac{8b_n-3}{4b_n-1} \qquad (= \frac{8b_n-2}{4b_n-1} - \frac{1}{4b_n-1}) \qquad (4)$$

$$= 2 - \frac{1}{4b_n-1}$$

Letting $a_{n+1} = 8b_n - 3$ and $b_{n+1} = 4b_n - 1$ and noticing that $gcd(a_{n+1}, b_{n+1}) = 1$, the lemma is proven.

The latter equality (4) can be gotten from two different approaches:

1. Defining $p(n) := 8b_n - 3 = a_{n+1}$ and $q(n) := 4b_n - 1 = b_{n+1}$ two linear poynomials. With polynomial long division it is

$$\frac{p(n)}{q(n)} = \frac{8b_n - 3}{4b_n - 1} = 2 - \frac{1}{4b_n - 1}$$

It is thus $gcd(8b_n - 3, 4b_n - 1) = 1$.

2. Alternatively, it is
$$f_1(n+1) = \frac{8b_n - 3}{4b_n - 1}$$
$$= \frac{2(4b_n - 2) - 1}{4b_n - 1}$$
$$= 2 - \frac{1}{4b_n - 1}.$$
q.e.d.

It is however more likely to expect that $8b_n - 3$ and $4b_n - 1$ being coprime is substantiated following the idea that $2(4b_n - 1) = 8b_n - 2$ is the direct neighbor of $8b_n - 3$. Since they have a difference of only 1, they cannot have a common divisor other than 1. It follows that $gcd(8b_n - 3, 4b_n - 1) = 1$.

Using this lemma it can be proven that the sequence $(f_1(x))_{x \in \mathbb{N}}$ converges to 2:

Theorem 1.2: With $(f_1(x))_{x \in \mathbb{N}}$ as defined in (3), it is $\lim_{x \to \infty} (f_1(x))_{x \in \mathbb{N}} = 2$.

Proof:
According to the monotone convergence theorem, a monotonic bounded sequence converges to its supremum.

Following Lemma 1.1, there are $a_n, b_n \in \mathbb{N}$ coprime such that $f_1(n) = \frac{a_n}{b_n}$. With Lemma 1.1, equation (4) and because $\frac{1}{b_n} > \frac{1}{4b_n - 1}$ for any $n \in \mathbb{N}$ it is furthermore

$$f_1(n) = 2 - \frac{1}{b_n} < 2 - \frac{1}{4b_n - 1} = f_1(n+1). \qquad (5)$$

The sequence is thus monotonically increasing. Moreover, with $(b_n)_{n \in \mathbb{N}}$ the sequence given by the denominator, it is $\lim_{n \to \infty} (b_n)_{n \in \mathbb{N}} = \infty$, consequently $\lim_{n \to \infty} \left(\frac{1}{b_n}\right)_{n \in \mathbb{N}} = 0$ and

$$\lim_{x \to \infty} (f_1(x))_{x \in \mathbb{N}} = \lim_{n \to \infty} \left(2 - \frac{1}{b_n}\right)_{n \in \mathbb{N}} = 2.$$
q.e.d.

In the same way, it is shown that there are $c_n, d_n \in \mathbb{N}$ with $gcd(c_n, d_n) = 1$ such that $f_2(n) = \frac{c_n}{d_n} = 2 + \frac{1}{d_n}$ and that $f_2(n+1) = 2 + \frac{1}{4d_n + 1}$. It is also $\lim_{x \to \infty} (f_2(x))_{x \in \mathbb{N}} = 2$ following analogue conclusions. In sum, this proves that both disjoint subsequences of $(f(x))_{x \in \mathbb{N}}$ converge to 2 and with that also $\lim_{x \to \infty} (f(x))_{x \in \mathbb{N}} = 2$.
q.e.d.

Without presuming that the students make use of a convergence criterion, they may use equation (5) to justify that the distance to two becomes smaller with each step. That it never reaches two can then be justified by the existence of the remainder that approaches, but never equals zero.

The splitting of the sequence in two disjoint subsequences does not necessarily need to be carried out: A similar approach works directly with the entire sequence $(f(x))_{x \in \mathbb{N}}$ but distinguishes the cases of x even and x uneven. The proof given above can be slightly modified for this approach.[3]

An alternative, more sophisticated proof uses the Banach fixed-point theorem, applied on a contraction through which any element of the sequence is mapped on the next one:[4]

Proof of $\lim_{x \to \infty} (f(x))_{x \in \mathbb{N}} = 2$: Approach 2

This proof uses the idea that 2 is the fixed-point of a contraction φ, defined to map an arbitrary element $f(n)$ of the sequence $(f(x))_{x \in \mathbb{N}}$ on its successor $f(n+1)$. Based on the recursive rule (1) and $f(1) = 3, f(2) = \frac{5}{3}$, it is $M = [\frac{5}{3}, 3]$ and

$$\varphi: M \to M \text{ with } s \mapsto 1 + \frac{2}{s}.$$

The **Banach fixed-point theorem** states the following:

Let (M, d) be a non-empty complete metric space and $\varphi: M \to M$ a contraction mapping in M. Then φ admits a unique fixpoint a in M. Furthermore, the sequence $f(x+1) = \varphi(f(x))$ converges to the fixpoint a for every initial value in M.

To show that $\lim_{n \to \infty} f(x) = 2$, it needs to be proven that (a) the conditions of the Banach fixed-point theorem are fulfilled and that (b) the fixpoint equals 2.

a) It is $(M, d) = ([\frac{5}{3}, 3], |\cdot|)$ is a metric space in $(\mathbb{R}, |\cdot|)$. $M = [\frac{5}{3}, 3] \neq \emptyset$ is complete as closed interval in the complete space $\mathbb{R}$. To prove φ to be a contraction, it needs to be shown that there exists a Lipschitz constant $\lambda \in [0,1)$ with

$$|\varphi(s_1) - \varphi(s_2)| \leq \lambda |s_1 - s_2|$$

for all $s_1, s_2 \in M$.

Without loss of generality let $s_1 < s_2$ such that $[s_1, s_2]$ a closed interval in M. Using the mean value theorem (φ is continuous in $[s_1, s_2]$ and differentiable in (s_1, s_2)), there is a $\zeta \epsilon (s_1, s_2)$ such that

---

[3] Priwitzer (2010, pp. 17-18), Janßen (2010, pp. 16-18) and Behrens (2013, pp. 23-24) carry out the proof that way.
[4] This approach is carried out similarly to the proof presented in Priwitzer (2010, pp. 17-18) and Janßen (2010, pp. 16-18).

$$\frac{|\varphi(s_1)-\varphi(s_2)|}{|s_1-s_2|} = \left|\frac{\varphi(s_1)-\varphi(s_2)}{s_1-s_2}\right| = |\varphi'(\zeta)|$$

$$\leq \sup\{\varphi'(s): s\in M\}$$

$$= \sup\left\{\left|-\frac{2}{s^2}\right|: s\in M\right\}$$

$$= \frac{2}{\left(\frac{5}{3}\right)^2} = \frac{18}{25} =: \lambda$$

Hence for every $s_1, s_2 \in M$ it is $|\varphi(s_1) - \varphi(s_2)| \leq \frac{18}{25}|s_1 - s_2|$ and φ is a contraction mapping. Consequently, there is a point $a \in M$ with $\varphi(a) = a$.

b) It is known that $\varphi(a) = 1 + \frac{2}{a} = a$ and thus

$$a + 2 = a^2$$
$$\Leftrightarrow \quad a^2 - a + 2 = 0$$
$$\Leftrightarrow \quad a = 2 \vee a = -1$$

The two solutions of the equation $a + 2 = a^2$ are thus $a_1 := 2$ and $a_2 := -1$. Observing that $a_1 \in M \not\ni a_2$ it is $a = a_1 = 2$ the fixpoint of $\varphi(s)$ and the limit of $f(x)$.

Since $f(1) = 3 \in M$ is a possible initial value, the convergence of $(f(x))_{x \in \mathbb{N}}$ to 2 follows immediately.

q.e.d.

Rather than a formal proof, the students are more likely expected to explore the convergence towards 2 based on empirical observations and justify it generally by assuming that the distance to 2 becomes smaller with each step.

The additional subtask has not been posed to the students. Because of this, it is refrained from an analysis of this reflection-task.

c) Task on Logical Reasoning Using the Principle of Mathematical Induction (IN)

The task on logical reasoning uses the principle of mathematical induction and for that, it is shortened by IN. The task deals with a word-problem and does not provide any further representations of the situation apart from the text given. The presentation of the situation framing the task, that is the setting of the examination (see original task given in attachment A), is followed by the actual question to investigate:

> "You are in the circle and see n marked hats. What is your strategy? Explain why your strategy works."

The description of the situation provides the condition that there is at least one marked hat.

1. Exploring the Situation to Find a Strategy:

Following this, the situation can be explored by abstracting from the total number of consultants and starting with the case of seeing $n = 0$ marked hat.

- Case 1: If $n = 0$, then I wear a marked hat because of the condition that there is at least one marked hat. That means, I have to raise my hand to signal at the first ring of the bell.

Assuming now the case of seeing $n = 1$ marked hat:

- Case 2: If $n = 1$, then I cannot decide at the first ring of the bell: While the condition is fulfilled with the one hat I see, mine could be marked as well.
    - Case 2.1: There is only one marked hat in the circle. Then the consultant who is wearing the marked hat that I do see does not see any other marked hat. As a conclusion, he is in case 1 and raises his hand at the first ring of the bell.
    - Case 2.2: There are two marked hats in the circle. Then the other consultant is in the same situation as I am, seeing my marked hat. That is, he will not raise his hand at the first ring of the bell, since my hat could be the only one marked. Therefore, none of us raises his hand at the first ring and we both know that the other one also sees one marked hat. This other marked hat thus must be our hat. Consequently, both of us conclude to wear a marked hat from the reaction of the others at the first ring of the bell and raise our hands at the second ring.

In the case of $n = 2$, that is of seeing two marked hats, the situation is as follows:

- Case 3: If $n = 2$, then I cannot decide at the first ring of the bell. Furthermore, the two consultants wearing a marked hat cannot decide at the first ring either, since both of them see at least one marked hat. But in case I also wear a marked hat, then they see two marked hats. There are thus two cases to consider:

- Case 3.1: There are only two marked hats in the circle. Then case 2.2 occurs and both consultants raise their hands at the second ring of the bell.
- Case 3.2: There are three marked hats in the circle. Then all three of us are in the same situation, seeing two marked hats and waiting for the others to react at the second ring. Therefore, nobody signals at the second ring and each of us knows that there must be a third marked hat on the own head. Consequently, all three of us conclude to wear a marked hat from the reaction of the others at the second ring of the bell and raise our hands at the third ring.

From these examples the following strategy can be concluded:

When seeing n marked hats, one has to wait for the nth ring of the bell to draw a conclusion on one's own status from the reaction of the consultants wearing a marked hat: In case they raise their hands, the own hat is not marked. In case they do not raise their hands, they are in the same situation as I am and see n marked hats. Consequently, I wear a marked hat and need to raise my hand at the $(n+1)^{th}$ ring of the bell.

The conjecture to prove is thus: If I see n marked hats, then I know about my own status at the nth ring of the bell. In case I conclude to wear a marked hat, I raise my hand at the $(n+1)^{th}$ ring.

2. Proof

The conjecture is formulated in dependence of $n \in \mathbb{N}$. That is, it can be proven by applying the principle of mathematical induction. Following this principle, the predicate $E(n)$ is true for all $n \in \mathbb{N}$ if and only if

(1) $E(0)$ is true

and

(2) For every $n \in \mathbb{N}$ we have that if $E(n)$ is true than $E(n+1)$ is true.

(1) is called the *basis step* and (2) **the** *inductive step*. The conjecture that $E(n)$ is true is called the *induction hypothesis*.

(1) The case of $n = 0$ has already been explicated above, in order to explore the situation.

(2) Given the situation of seeing $n+1$ marked hats and assuming the induction hypothesis to be true. Then there are two cases to consider:

- If I do not wear a marked hat, each of the consultants I see wearing a marked hat sees one hat less than I do, hence only n marked hats. Since the induction hypothesis is considered true for seeing n marked hats, they know about their status at the nth ring of the bell and raise their hand at

the $(n + 1)^{th}$ ring of the bell. Concluding from this, I know that I do not wear a marked hat.
- If I do wear a marked hat, each of the consultants I see wearing a marked hat is in the same situation as I am; all of them seeing $n + 1$ marked hats. Since the inductions hypothesis is considered to be true, all of us conclude that everybody sees more than n marked hats from the observation that nobody raised his hand at the $(n + 1)^{th}$ ring of the bell. That is, we all know about wearing a marked hat after the $(n + 1)^{th}$ and raise our hands at the $(n + 2)^{th}$ ring of the bell.

This proves the predicate right.

A formal prove of the applicability of the strategy thus embraces the use of the principle of mathematical induction, and a distinction of cases.

3. Additional Task

In case the students solve the task quickly, it can be decided to pose an additional task to them. For this, the students are claimed to consider the situation without the condition of knowing that there is at least one marked hat. They are asked to find out whether the strategy they found does still work.

The knowledge of the existence of at least one marked hat allows to conclude about the own status in the case of seeing no marked hat, hence for the case of n=0. Modifying the situation by not providing this condition concerns the *basis step* that cannot be proven anymore. Following the principle of mathematical induction, the strategy does not work anymore.

C Condensed Process Diagrams of the EDEs

PA5

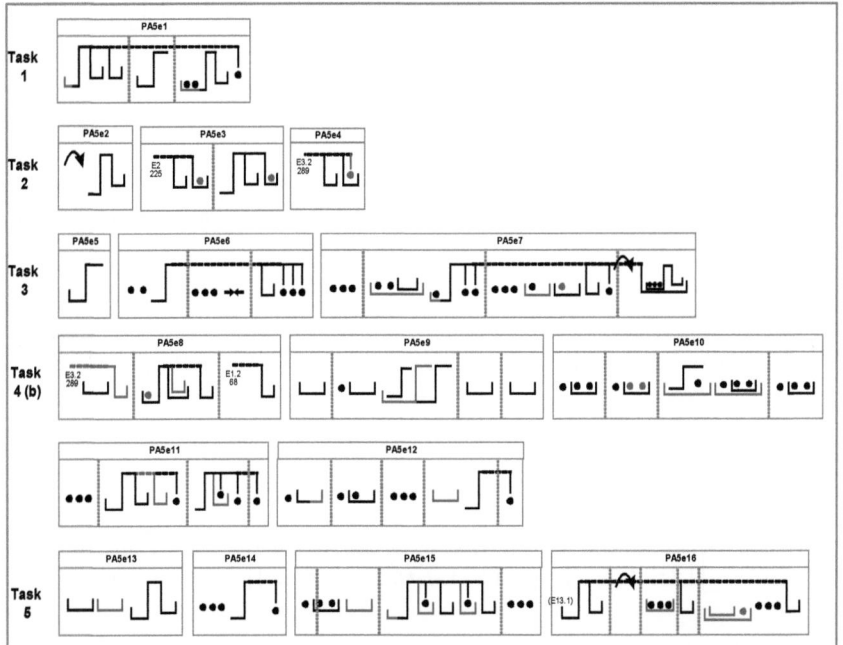

Fig. C.1: Condensed process diagram of the EDEs of PA7 (sorted with respect to subtasks)

PA7

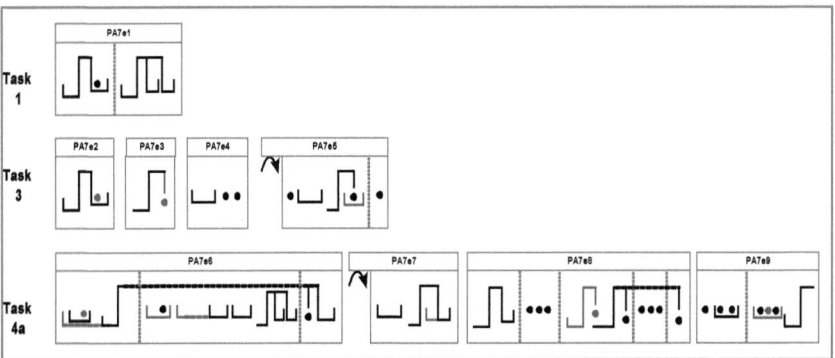

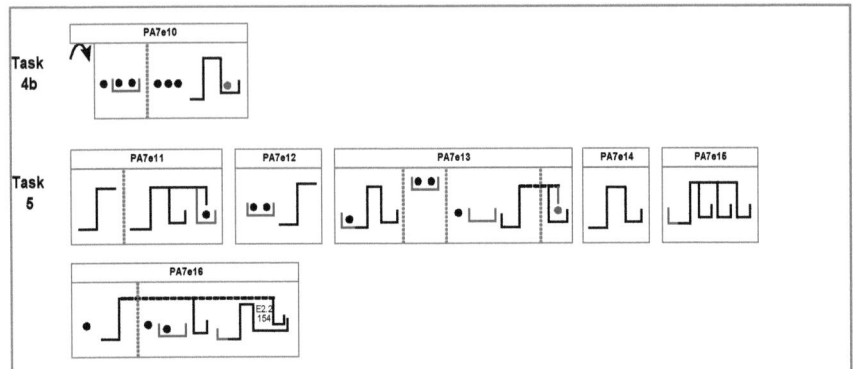

Fig. C.2: Condensed process diagrams of the EDEs of PA7 (sorted with respect to subtasks)

PA6

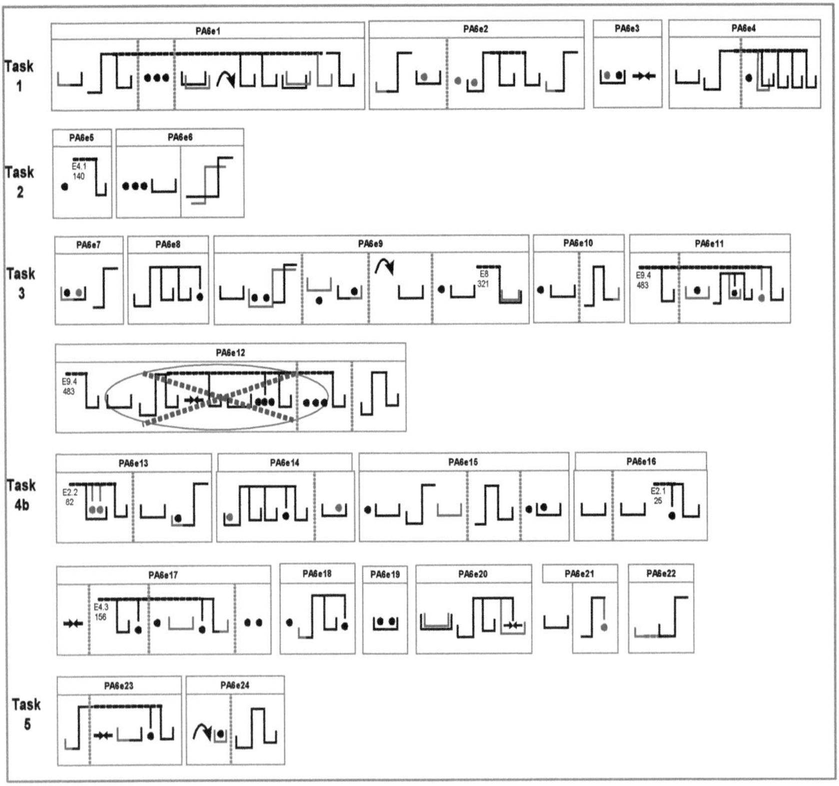

Fig. C.3: Condensed process diagrams of the EDEs of PA6 (sorted with respect to subtasks)

CF7

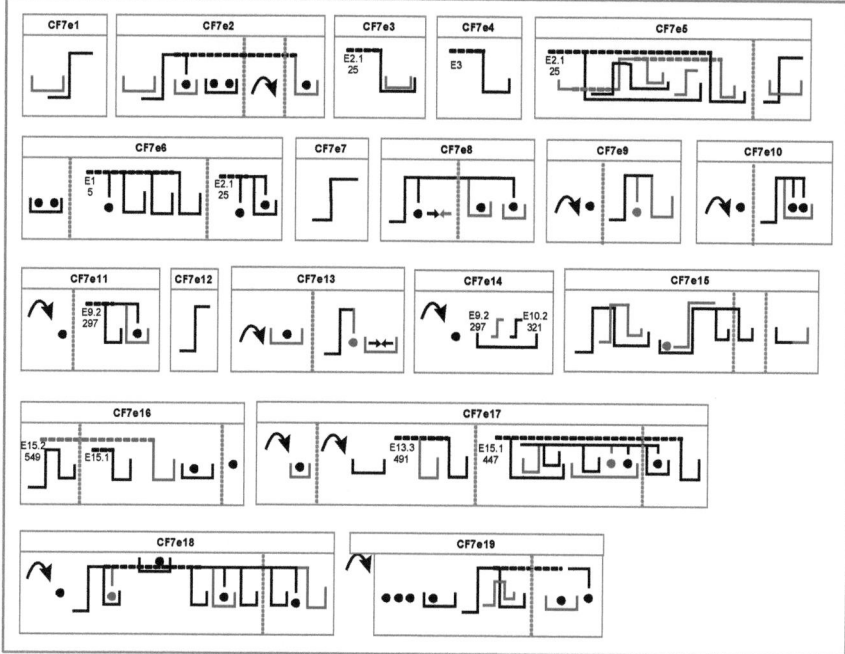

Fig. C.4: Condensed process diagrams of the EDEs of CF7

IN7

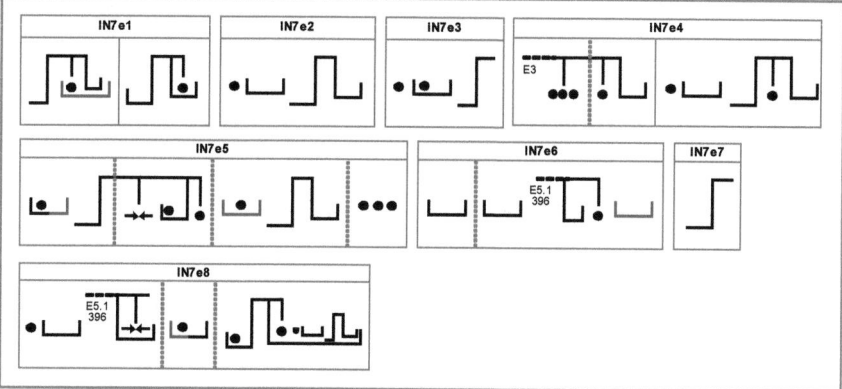

Fig. C.5: Condensed process diagrams of the EDEs of CF7

D The Inscriptions Produced by the Students

On the following pages, the inscriptions produced by the students while working out the task are documented. Additionally, the single sheets are tagged with "work sheet" "printout" and "note sheet", as well as enumerations.

a) PA5

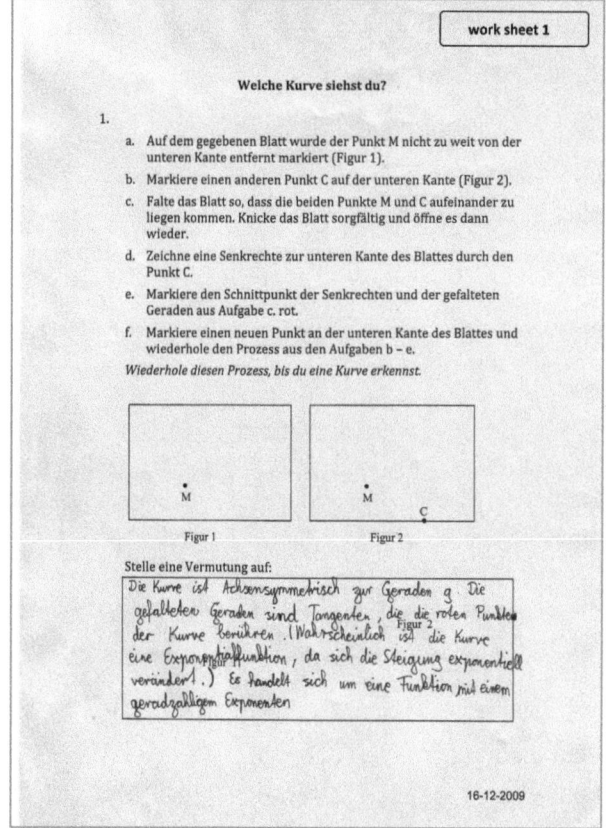

The data was gathered by Angelika Bikner-Ahsbahs, Julia Cramer, Thomas Janßen, and Jacob Priwitzer in the project „Effective knowledge construction in interest-dense situations", funded by the German Israeli foundation for Scientific Research and Development (Grant No. 946-357.4/2006).

314   The Students' Inscriptions

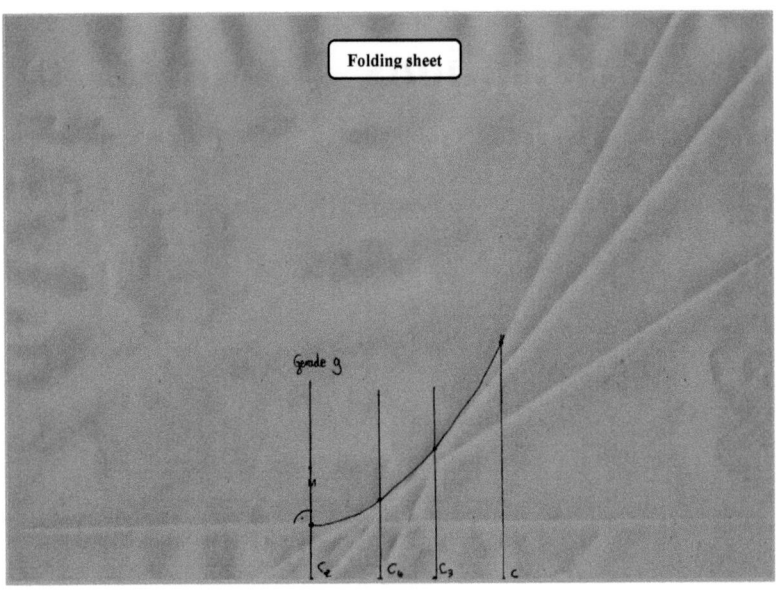

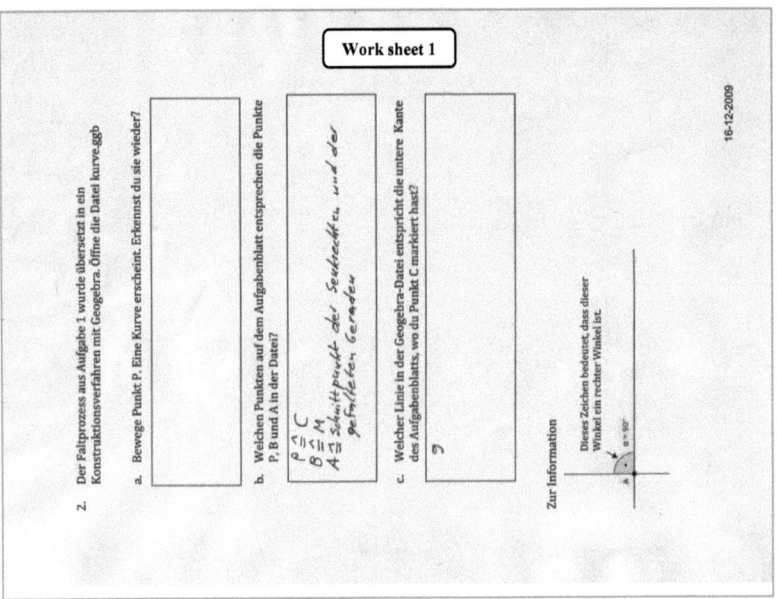

The data was gathered by Angelika Bikner-Ahsbahs, Julia Cramer, Thomas Janßen, and Jakob Priwitzer in the project „Effective knowledge construction in interest-dense situations", funded by the German Israeli foundation for Scientific Research and Development (Grant No. 946-357.4/2006).

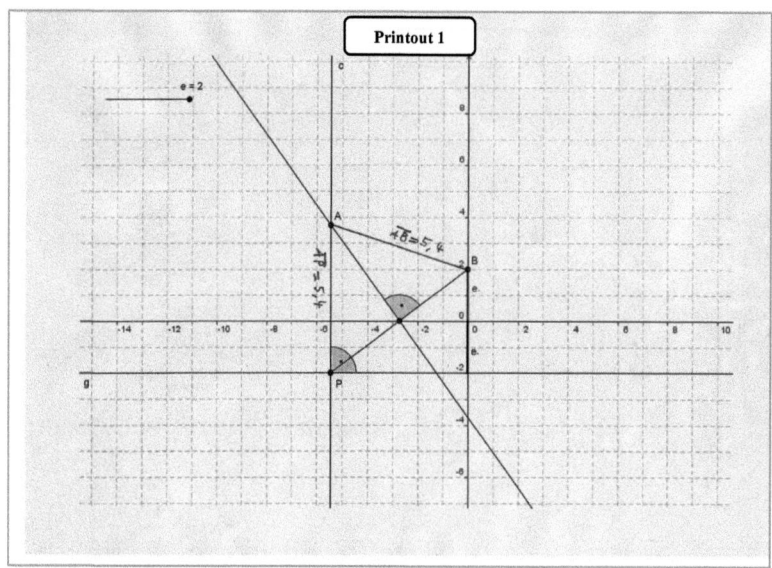

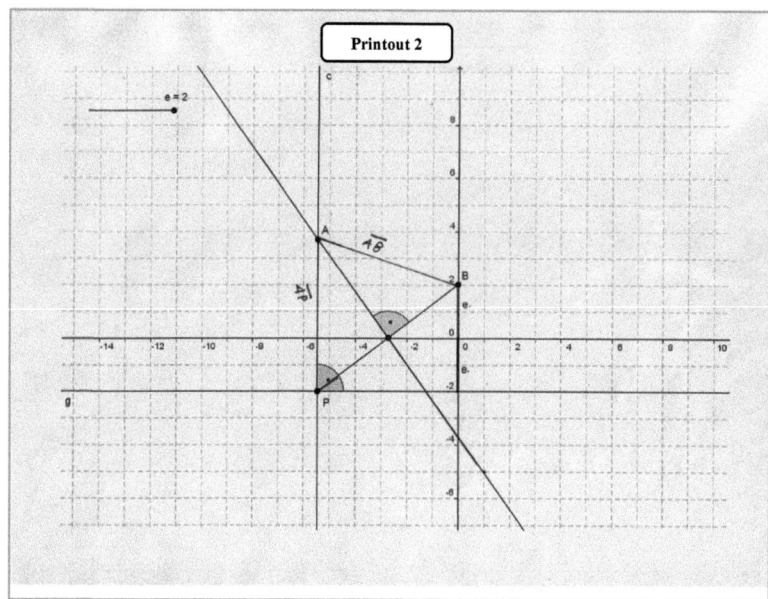

The data was gathered by Angelika Bikner-Ahsbahs, Julia Cramer, Thomas Janßen, and Jakob Priwitzer in the project „Effective knowledge construction in interest-dense situations", funded by the German Israeli foundation for Scientific Research and Development (Grant No. 946-357.4/2006).

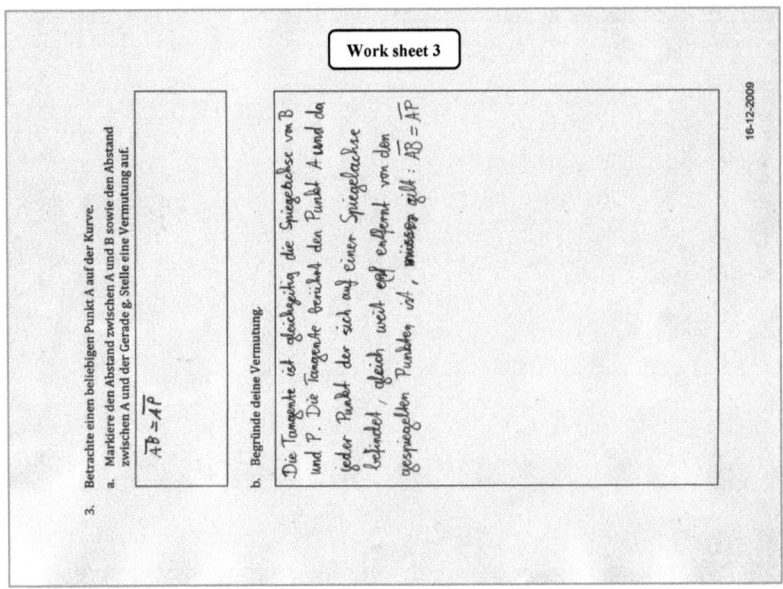

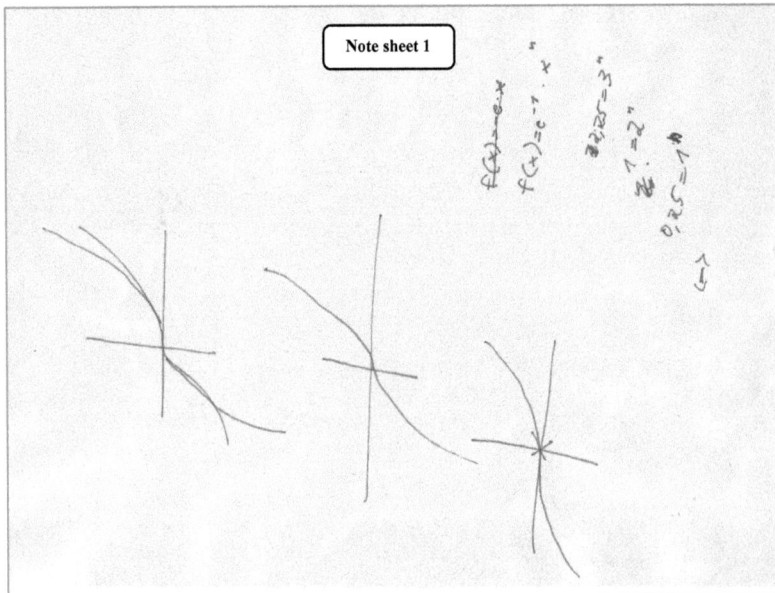

The data was gathered by Angelika Bikner-Ahsbahs, Julia Cramer, Thomas Janßen, and Jakob Priwitzer in the project „Effective knowledge construction in interest-dense situations", funded by the German Israeli foundation for Scientific Research and Development (Grant No. 946-357.4/2006).

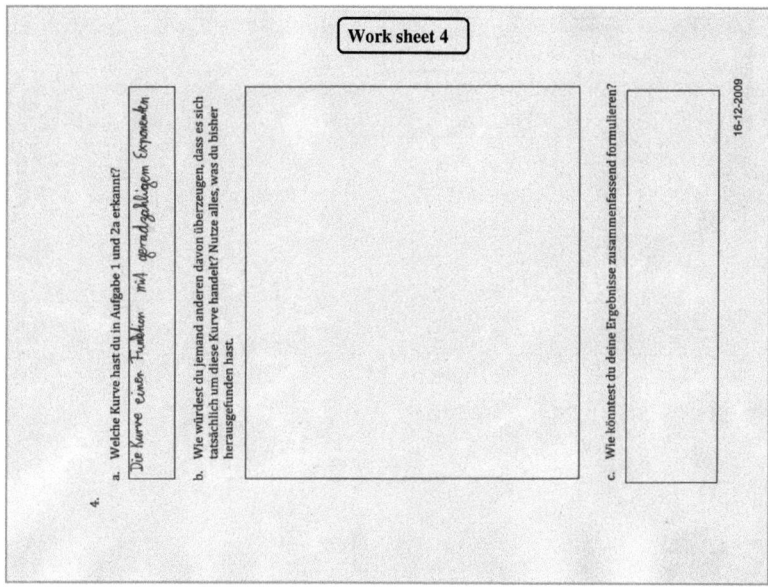

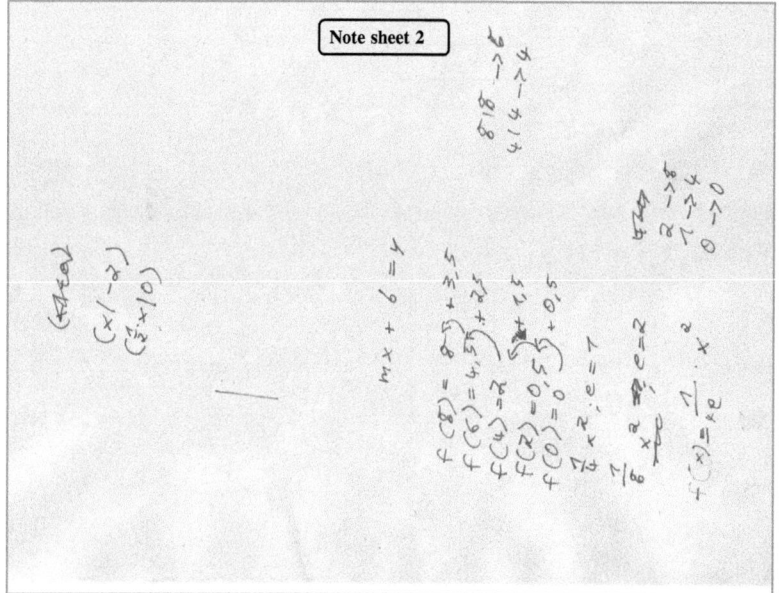

The data was gathered by Angelika Bikner-Ahsbahs, Julia Cramer, Thomas Janßen, and Jakob Priwitzer in the project „Effective knowledge construction in interest-dense situations", funded by the German Israeli foundation for Scientific Research and Development (Grant No. 946-357.4/2006).

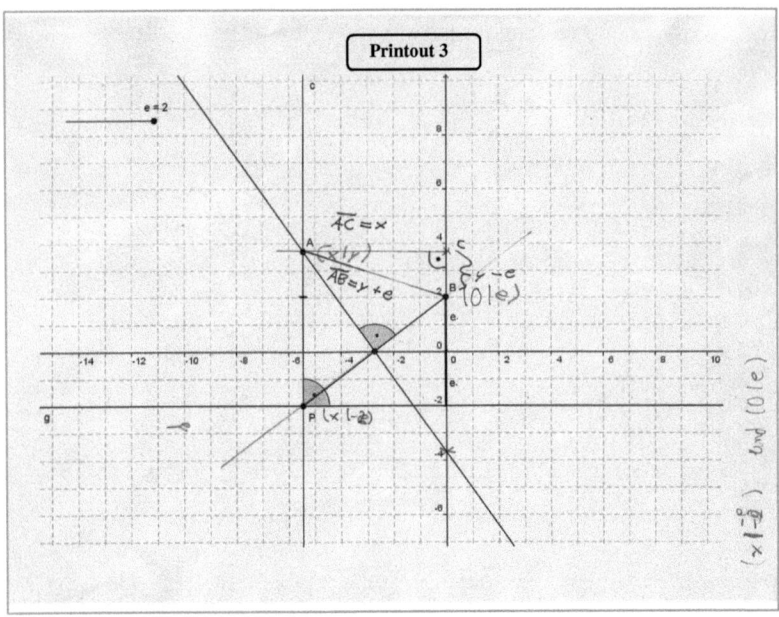

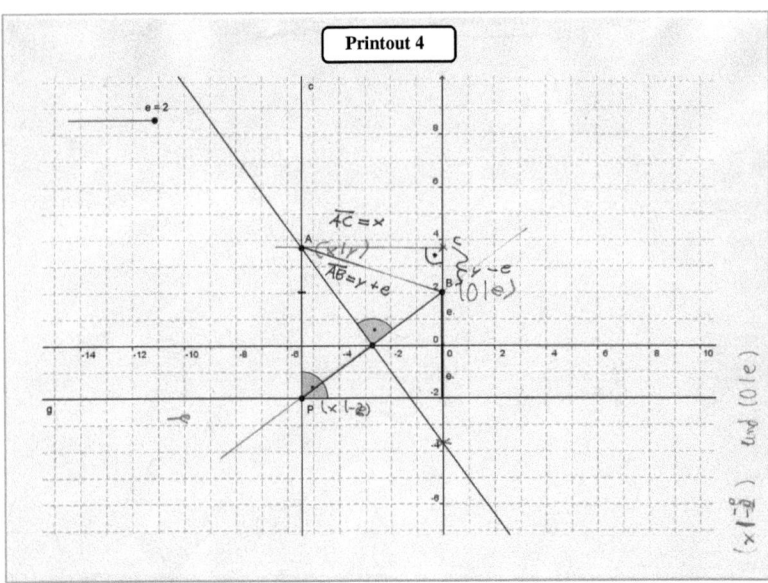

The data was gathered by Angelika Bikner-Ahsbahs, Julia Cramer, Thomas Janßen, and Jakob Priwitzer in the project „Effective knowledge construction in interest-dense situations", funded by the German Israeli foundation for Scientific Research and Development (Grant No. 946-357.4/2006).

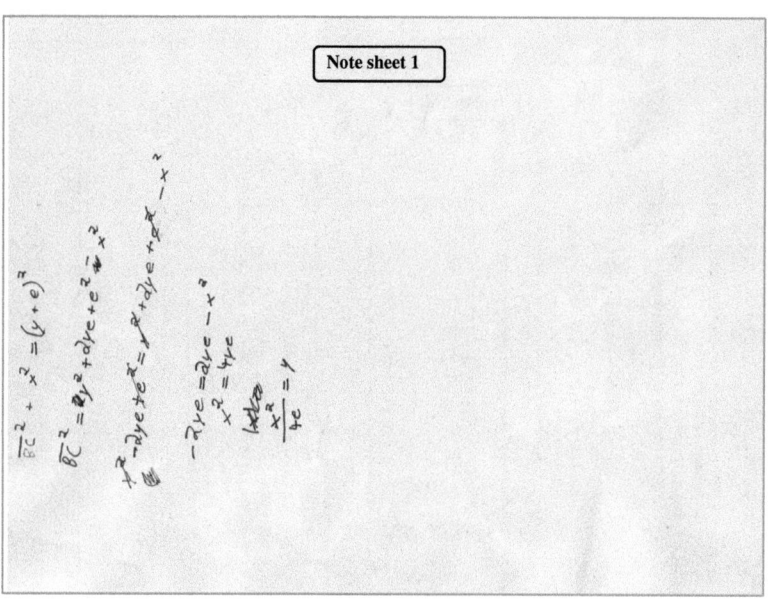

Note sheet 1

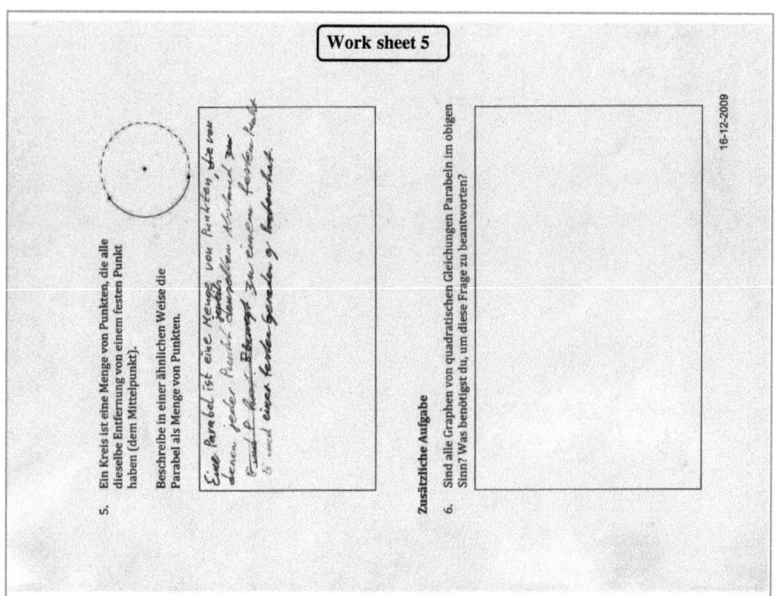

Work sheet 5

b) PA7

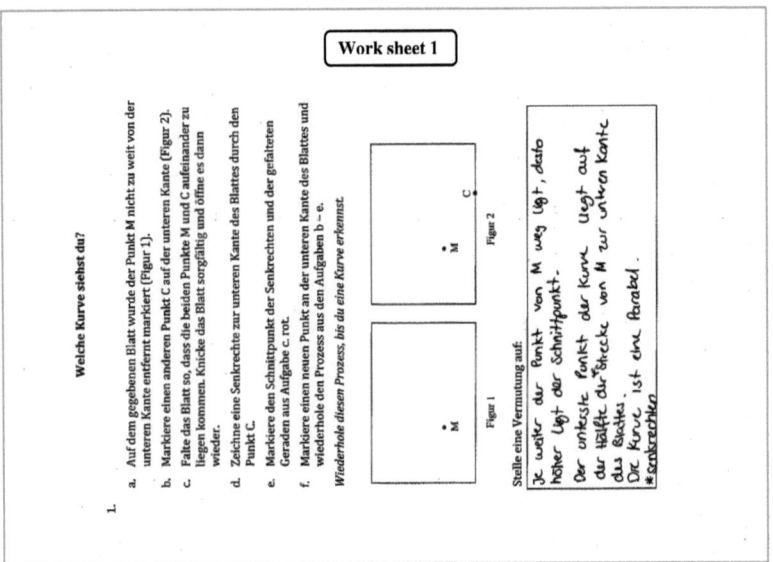

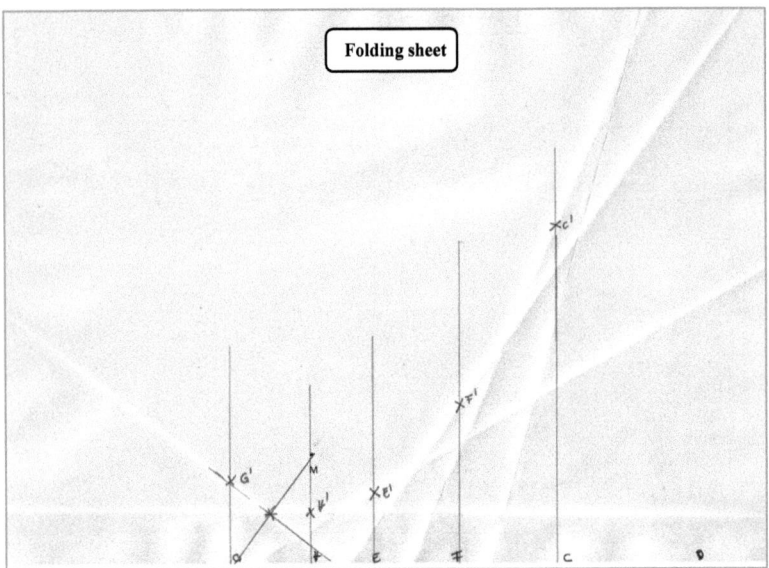

The data was gathered by Angelika Bikner-Ahsbahs, Julia Cramer, Thomas Janßen, and Jakob Priwitzer in the project „Effective knowledge construction in interest-dense situations", funded by the German Israeli foundation for Scientific Research and Development (Grant No. 946-357.4/2006).

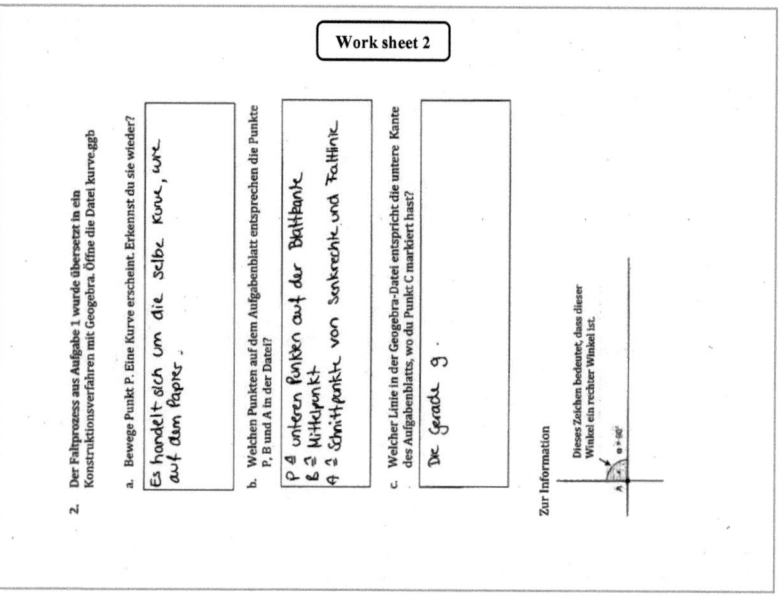

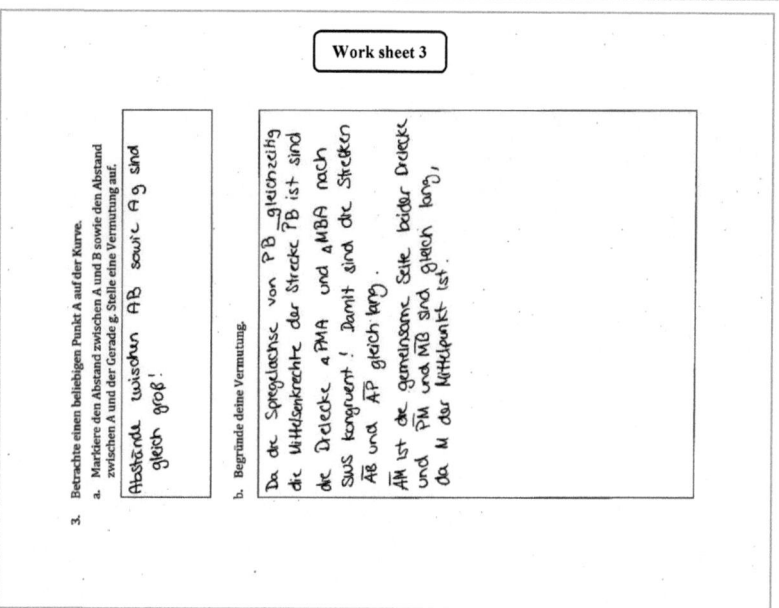

The Students' Inscriptions

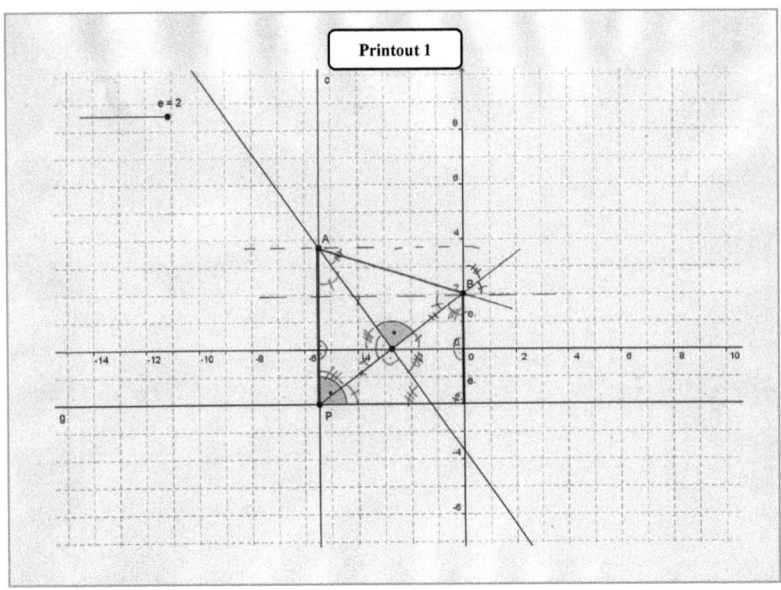

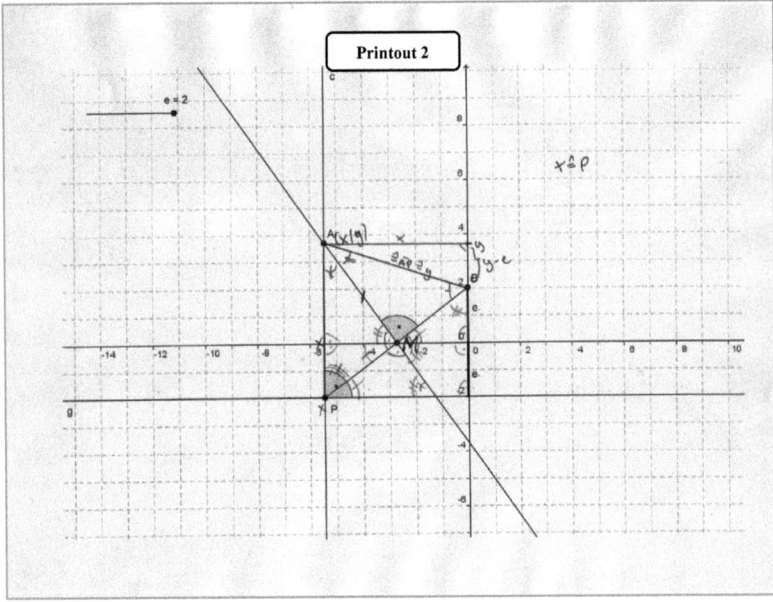

The data was gathered by Angelika Bikner-Ahsbahs, Julia Cramer, Thomas Janßen, and Jakob Priwitzer in the project „Effective knowledge construction in interest-dense situations", funded by the German Israeli foundation for Scientific Research and Development (Grant No. 946-357.4/2006).

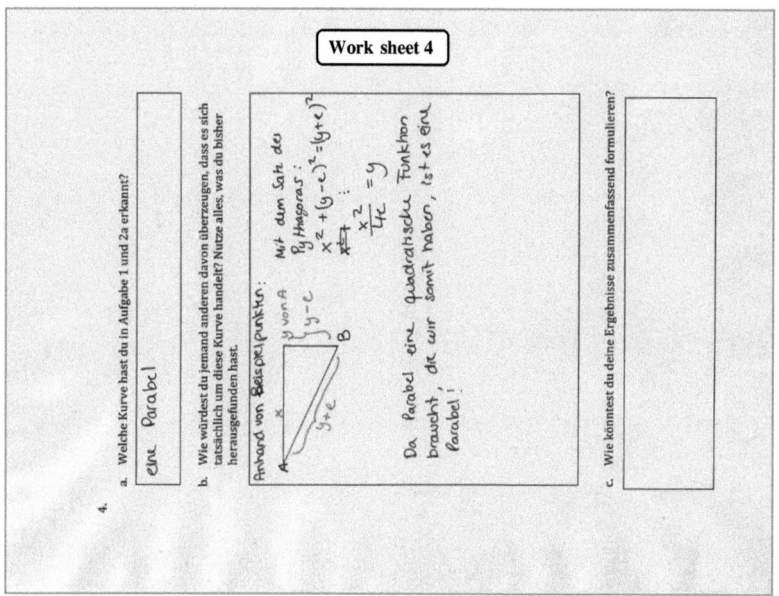

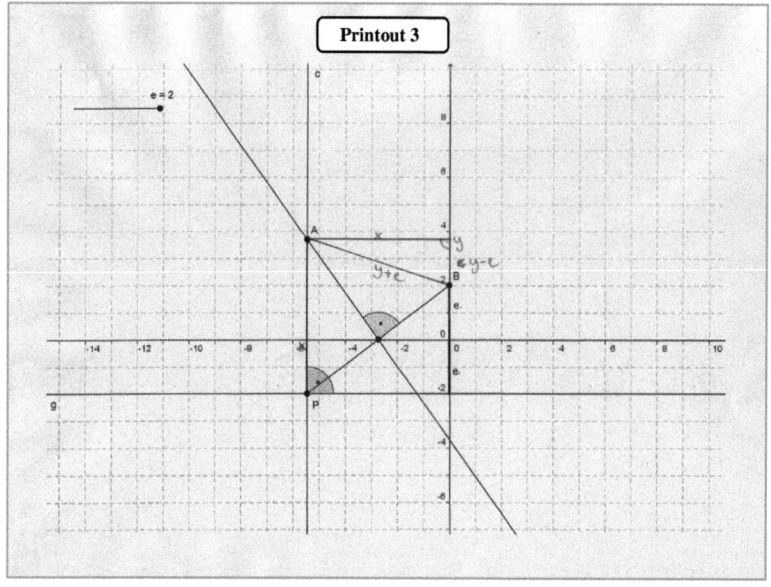

Note sheet 1

$e=1$ $6;9$ $8;8$ $2;2$

      $4;4$ $e=2$ $4;2$ $e=0,5$ $4,8$

      $2;1$

      $3; 2,25$ $2,05$

     ↙ $6;4,5$

             $3; 1,12$

$1 \cdot \left(\frac{x}{2}\right)^2 = f(x)$ $8:4=2$ $2 \cdot 2^2 = 8$ $\left(\frac{8}{4}\right)^2 \cdot 2$

                 $4:4=1$ $2 \cdot 1=2$

                 $2:4=0,5$ $2 \cdot 0,5^2 = 0,5$

           $2\left(\frac{x}{4}\right)^2 = f(x)$

$e \cdot \left(\frac{x}{2e}\right)^2 = f(x)$

Note sheet 2

$x^2 + (y-e)^2 = (y+e)^2$

$x^2 + y^2 - 2ye + e^2 = y^2 + 2ye + e^2 \mid -e^2 \; -y^2$

$x^2 - 2ye = 2ye \mid +2ye$

$x^2 = 4ye \mid :4e$

$\frac{x^2}{4e} = y$ $e \cdot \frac{x^2}{4e^2}$

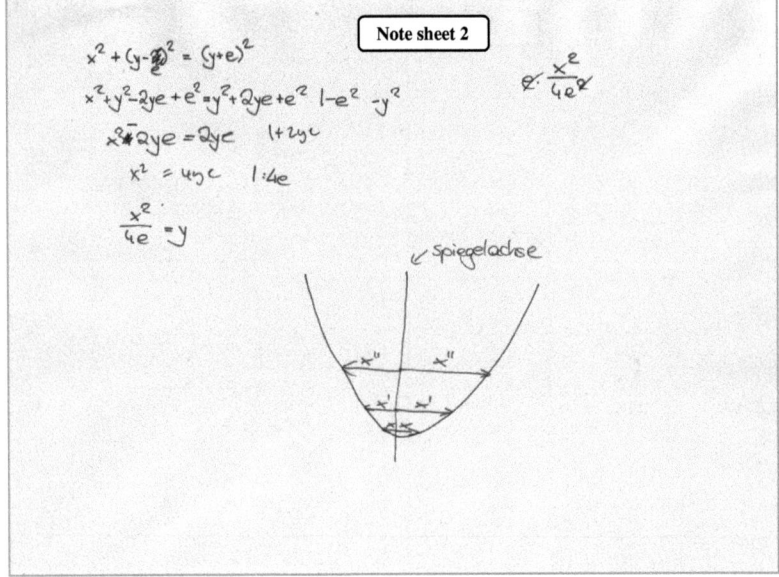

← Spiegelachse

The data was gathered by Angelika Bikner-Ahsbahs, Julia Cramer, Thomas Janßen, and Jakob Priwitzer in the project „Effective knowledge construction in interest-dense situations", funded by the German Israeli foundation for Scientific Research and Development (Grant No. 946-357.4/2006).

Work sheet 5

5. Ein Kreis ist eine Menge von Punkten, die alle dieselbe Entfernung von einem festen Punkt haben (dem Mittelpunkt).

   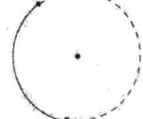

   Beschreibe in einer ähnlichen Weise die Parabel als Menge von Punkten.

   ~~Eine Parabel ist eine Menge von Punkten, die jeweils dieselbe Entfernung vom festen Punkt B und~~

   Eine Parabel ist eine Menge von Punkten, die jeweils denselben Abstand zum festen Punkt B und der Geraden g haben.

Zusätzliche Aufgabe

6. Sind alle Graphen von quadratischen Gleichungen Parabeln im obigen Sinn? Was benötigst du, um diese Frage zu beantworten?

c) PA6

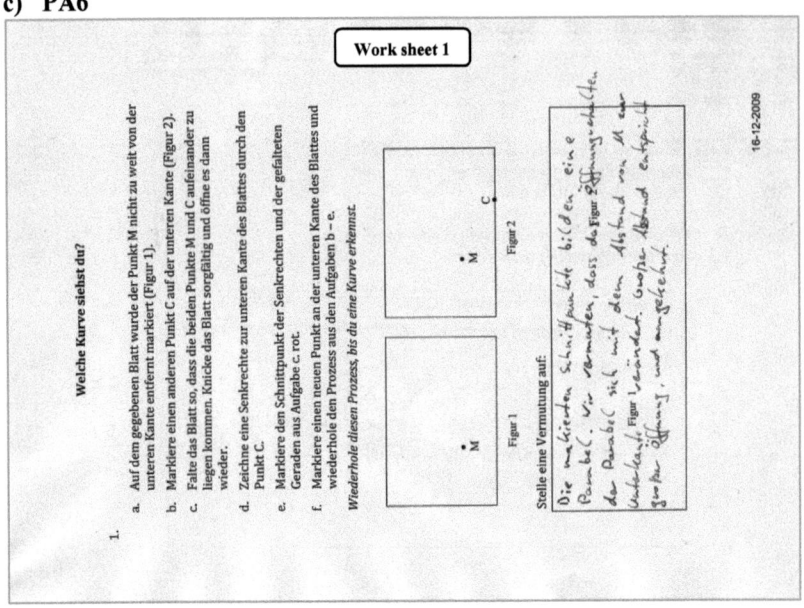

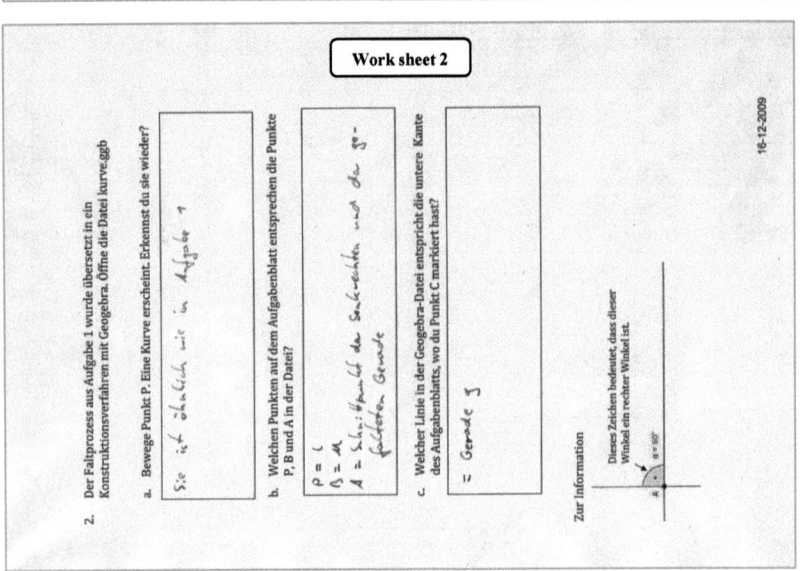

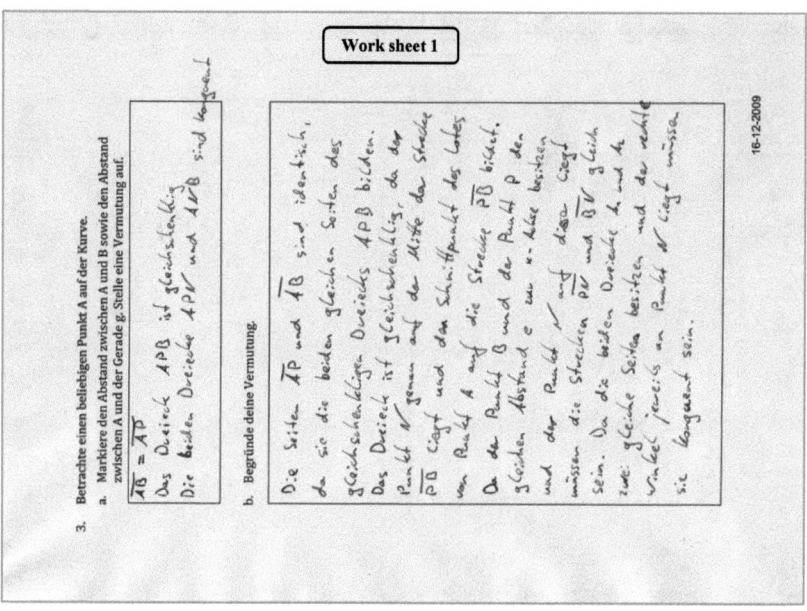

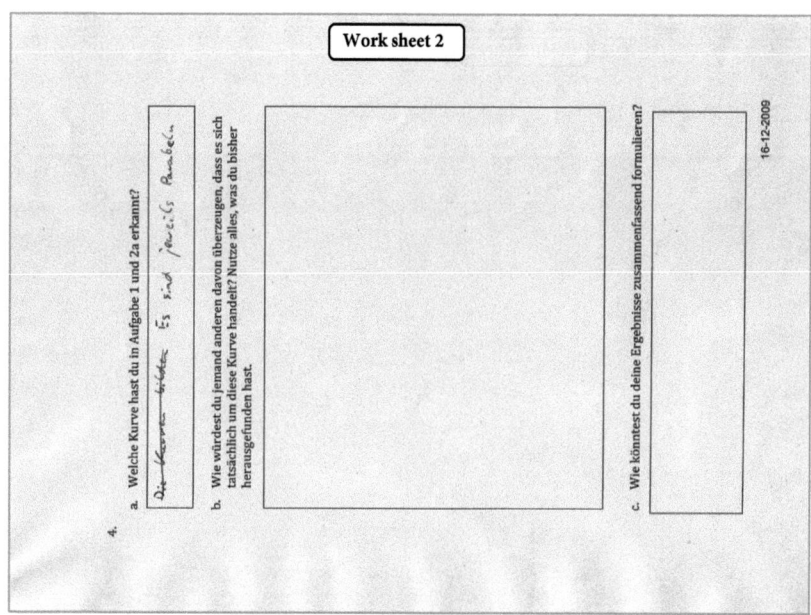

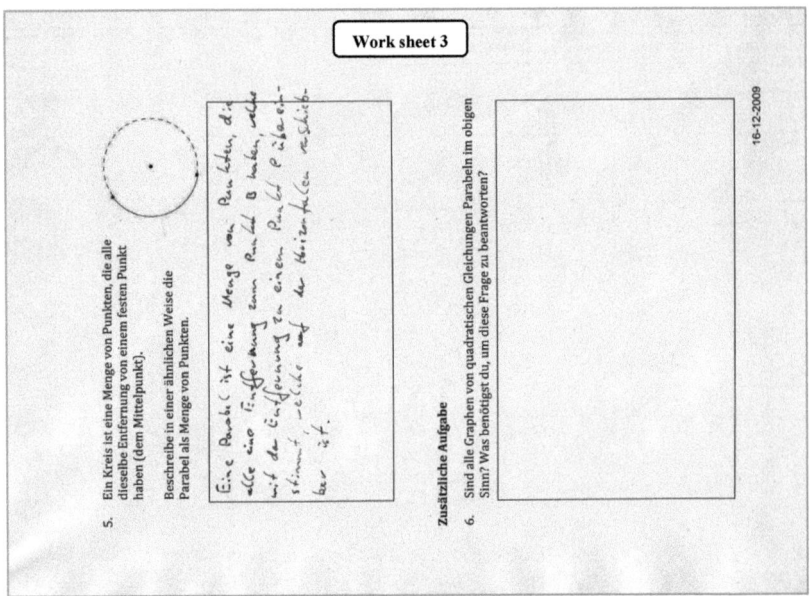

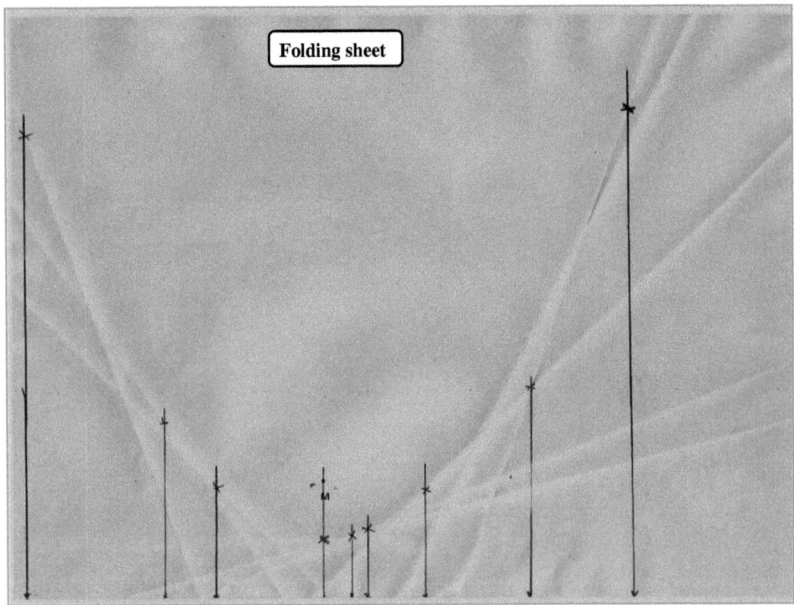

The data was gathered by Angelika Bikner-Ahsbahs, Julia Cramer, Thomas Janßen, and Jakob Priwitzer.

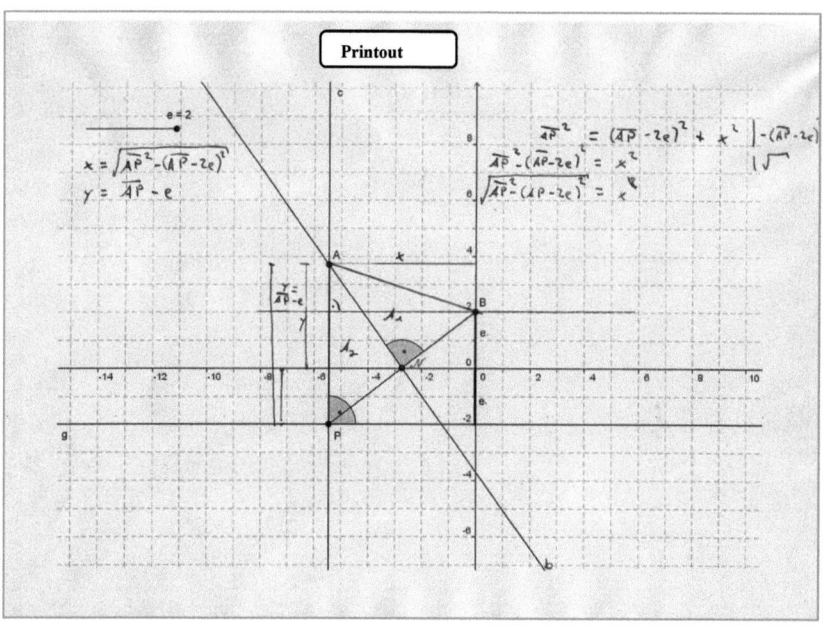

d) CF7

Work sheet 1

Wie können wir die Kettenbruchentwicklung interpretieren: $1+\dfrac{2}{1+\dfrac{2}{1+\dfrac{2}{1+\ldots}}}$?

Beantworte die folgenden Fragen:

1.1 Hier siehst du eine Folge von Brüchen, die eine Kettenbruchentwicklung darstellt:

$f(0) = 1$

$f(1) = 1 + \dfrac{2}{1} = 1 + 2 = 3$

$f(2) = 1 + \dfrac{2}{1+\dfrac{2}{1}} = 1 + \dfrac{2}{1+2} = 1 + \dfrac{2}{3} = \dfrac{5}{3}$

$f(3) = 1 + \dfrac{2}{1+\dfrac{2}{1+\dfrac{2}{1}}} = \dfrac{11}{5}$

$f(4) = \dfrac{21}{11}$

$f(5) = \dfrac{43}{21}$

$f(6) = \dfrac{85}{43}$

Füge drei weitere Terme $f(4)$, $f(5)$, $f(6)$ hinzu und berechne sie.

1.2 Schau dir die 7 Terme f(0) bis f(6) und deren Berechnung an. Kannst du ein Muster finden, um von einem Term zum nächsten zu kommen?

$1 + \dfrac{2}{f(x-1)} = f(x)$

For $f(0)$ gilt: $f(0) = 1$

$1 \leq x < \infty$

nur ganze Zahlen

-1- 04.12.2009

The data was gathered by Angelika Bikner-Ahsbahs, Julia Cramer, Thomas Janßen, and Jakob Priwitzer in the project „Effective knowledge construction in interest-dense situations", funded by the German Israeli foundation for Scientific Research and Development (Grant No. 946-357.4/2006).

Work sheet 2

1.3 Erkläre das Muster – wieso funktioniert es?

Da sich das Muster immer wieder in sich selbst wiederfindet handelt es sich um eine Rekursion. Der darauffolgende Term baut sich aus $1 + \frac{2}{\text{dem vorherigen Term}}$ auf.

2.1 Berechne weitere Terme der Folge, bis du 20 Terme der Folge hast. Nutze dabei das Muster, das du gefunden hast. Fülle die folgende Tabelle aus. Nutze einen Taschenrechner, um die Kettenbrüche der Folge als Dezimalzahlen darzustellen. Schreibe alle Nachkommastellen ab, die der Taschenrechner anzeigt.

	Bruchzahl	Dezimalzahl	
f(0)	1	1	$2-1$
f(1)	3	3	$2+1$
f(2)	$\frac{5}{3}$	1.666666667	$2-\frac{1}{3}$
f(3)	$\frac{11}{5}$	2,2	$2+\frac{1}{5}$
f(4)	$\frac{21}{11}$	1.909090909	$2-\frac{1}{11}$
f(5)	$\frac{43}{21}$	2.047619048	$2+\frac{1}{21}$
f(6)	$\frac{85}{43}$	1.976744186	$2-\frac{1}{43}$

Work sheet 3

f(7)	$\frac{171}{85}$	2.011764706
f(8)	$\frac{341}{171}$	1.994152047
f(9)	$\frac{683}{341}$	2.002932551
f(10)	$\frac{1365}{683}$	1.998535871
f(11)	$\frac{2731}{1365}$	2,000732601
f(12)	$\frac{5461}{2731}$	1,999633834
f(13)	$\frac{10923}{5461}$	2,000183117
f(14)	$\frac{21845}{10923}$	1,999908415
f(15)	$\frac{43691}{21845}$	2,000045777
f(16)	$\frac{87381}{43691}$	1,999977112
f(17)	$\frac{174763}{87381}$	2,000011444
f(18)	$\frac{349525}{174763}$	1,999994278
f(19)	$\frac{699051}{349525}$	2,000002861

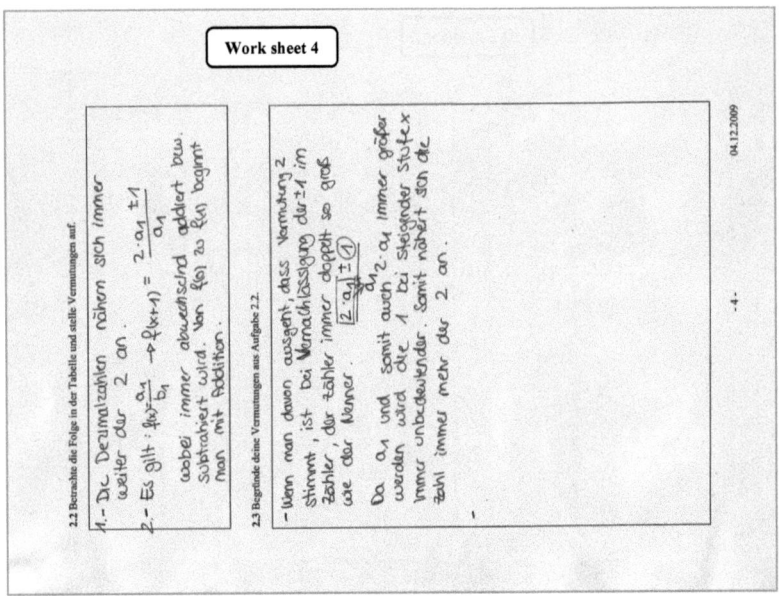

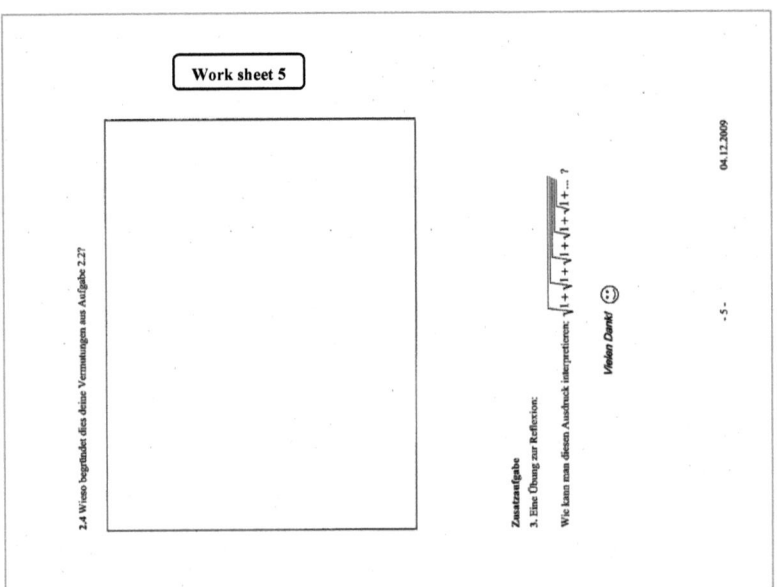

The data was gathered by Angelika Bikner-Ahsbahs, Julia Cramer, Thomas Janßen, and Jakob Priwitzer in the project „Effective knowledge construction in interest-dense situations", funded by the German Israeli foundation for Scientific Research and Development (Grant No. 946-357.4/2006).

(1.1) — Note sheet 1

$$f(4) = 1 + \cfrac{2}{1+\cfrac{2}{1+\cfrac{2}{1+\cfrac{2}{\cancel{1}}}}} = 1 + \cfrac{2}{11/5} = 1 + \cfrac{10}{11} = \boxed{\tfrac{21}{11}}$$

$$f(3) = 1 + \tfrac{2}{5/3} = 1 + \tfrac{6}{5} = \boxed{\tfrac{11}{5}} = 2{,}2$$

$$f(5) = 1 + \tfrac{2}{\tfrac{21}{11}} = 1 + \tfrac{2 \cdot 11}{21} = 1 + \tfrac{22}{21} = \boxed{\tfrac{43}{21}}$$

$f(5) = \cancel{1 + \tfrac{2}{\ldots}} \quad 1 + \tfrac{2}{21/11} = \tfrac{22}{21}$

$f(6) = 1 + \tfrac{2}{22/21} = 1 + \tfrac{42}{\ldots}$

$$f(6) = 1 + \tfrac{2}{\tfrac{43}{21}} = 1 + \tfrac{42}{43} = \boxed{\tfrac{85}{43}}$$

(1.3) — Note sheet 2

$1 + \tfrac{2}{1} \qquad 1 + \cfrac{2}{1+\tfrac{2}{1}} \qquad \boxed{1+\tfrac{2}{1}} \searrow \quad 1 + \cfrac{2}{\boxed{1+\tfrac{2}{1}}}$

$1 + \tfrac{2}{1} \qquad \boxed{1 + \cfrac{2}{1+\tfrac{2}{1}}}$

(2.1)

$\tfrac{12}{3} = 4 \qquad 1 + \cfrac{2}{\tfrac{2731}{1365}} = 1 + \cfrac{2 \cdot 1365}{\cancel{2731}/1365} = 1 + \cancel{\tfrac{2731}{\ldots}}$

$$1 + \tfrac{2 \cdot 1365}{2731} = \tfrac{2730}{2731} + 1 = \tfrac{5461}{2731}$$

The data was gathered by Angelika Bikner-Ahsbahs, Julia Cramer, Thomas Janßen, and Jakob Priwitzer in the project „Effective knowledge construction in interest-dense situations", funded by the German Israeli foundation for Scientific Research and Development (Grant No. 946-357.4/2006).

(2.2)

$$1 + \frac{2}{f(x-1)}$$

Note sheet 3

$$\frac{a_1}{b_2} \rightarrow \frac{2a_1 \pm 1}{a_1} = \frac{a_2}{b_2}$$

$$\Rightarrow \frac{2(2a_1+1)}{2a_1+1} = \frac{4a_1+2}{2a_1+1}$$

(2.3)

$\dfrac{2 \cdot a_1 (\pm 1)}{a_1}$ → wird immer unbedeutender im Vergleich, da $a_1 + 2a_1$ immer

$\cancel{\dfrac{4a_1}{2a_1+1} + \dfrac{2}{4a_1}}$

$1 + \dfrac{\frac{2}{x}}{y}$ $f(3) = 1 + \dfrac{2}{\frac{5}{3}} = 1 + \dfrac{2 \cdot 3}{5}$

$= 1 + \dfrac{2 \cdot y}{x}$ $= \dfrac{1 \cdot 5}{5} + \dfrac{6}{5}$

Note sheet 4

$\dfrac{a}{b} = \dfrac{2a \pm 1}{a} = \dfrac{2a}{a} \pm \dfrac{1}{a} = 2 \pm \dfrac{1}{a}$

$\rightarrow 1 + \dfrac{2}{2 + \frac{1}{a}} \rightarrow 1 + \cancel{\dfrac{2}{2} + \dfrac{2}{1/a}}$

$\rightarrow \cancel{1 + 1 + \dfrac{2a}{1}}$

$\rightarrow 1 + \dfrac{2}{\frac{2 \cdot a}{a} + \frac{1}{a}}$

$1 + \dfrac{3}{4}$ $\rightarrow 1 + \dfrac{2}{\frac{2a+1}{a}} \rightarrow \dfrac{1}{1} + \dfrac{2a+1}{2a+1} \cancel{\dfrac{}{}} = e$

$\dfrac{2a+1+2a}{2a+1} \rightarrow \dfrac{4a+1}{2a+1}$

The data was gathered by Angelika Bikner-Ahsbahs, Julia Cramer, Thomas Janßen, and Jakob Priwitzer in the project „Effective knowledge construction in interest-dense situations", funded by the German Israeli foundation for Scientific Research and Development (Grant No. 946-357.4/2006).

$$1 + \frac{2a}{2a+1} \rightarrow 2 + \frac{2a-(2a+1)}{2a+1} \rightarrow 2 + \frac{2a-2a-1}{2a+1} \rightarrow 2 + \frac{-1}{2a+1}$$

Note sheet 5

$$\rightarrow 2 - \frac{1}{2a+1}$$

$$2 + \frac{1}{a} \rightarrow \boxed{2 - \frac{1}{(2a+1)}} \stackrel{!}{=} 2 - \frac{1}{b}$$

$f(1)\ 2 + \frac{1}{1} = \text{③}\quad a=1$

$f(2)\ 2 - \frac{1}{2\cdot 1+1} = \boxed{2-\frac{1}{3}} = \frac{2\cdot 3}{3} - \frac{1}{3} = \boxed{\frac{5}{3}}\quad a=1\ b=3$

$2 + \frac{1}{\cancel{3}} = 2 + \cancel{\tfrac{1}{3}} = \cancel{\tfrac{10}{3}} + \tfrac{3}{3} = $

$2 + \tfrac{1}{3} = $

$\quad\quad\quad 2\cdot(2\cdot 1+1)+1$
$\quad\quad\quad 4+2+1 = 7$

Note sheet 6

$$2 - \tfrac{1}{b} \rightarrow 1 + \frac{2}{2-\tfrac{1}{b}} \rightarrow 1 + \frac{2}{\tfrac{2b}{b} - \tfrac{1}{b}} \rightarrow 1 + \frac{2}{\tfrac{2b-1}{b}} \rightarrow 1 + \frac{2b}{2b-1}$$

$$2 + \frac{2b-(2b-1)}{2b-1} \rightarrow \boxed{2 + \frac{1}{2b-1}}\quad 3.$$

$f(2) = 2 - \frac{1}{[3] \rightarrow b}$

$f(3) = 2 + \frac{1}{(2\cdot 3)-1} = \frac{10}{5} + \frac{1}{5} = \underline{\underline{\frac{11}{5}}}$

The data was gathered by Angelika Bikner-Ahsbahs, Julia Cramer, Thomas Janßen, and Jakob Priwitzer in the project „Effective knowledge construction in interest-dense situations", funded by the German Israeli foundation for Scientific Research and Development (Grant No. 946-357.4/2006).

e) IN7

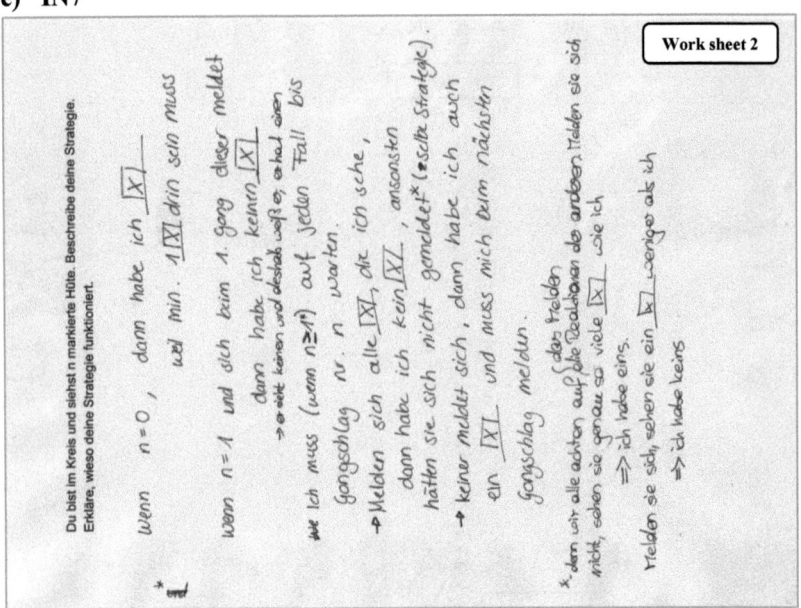

Note sheet 1

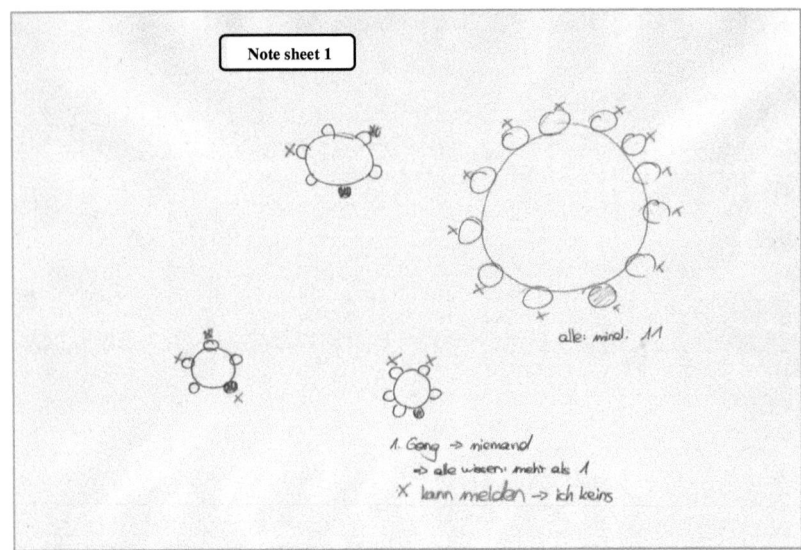

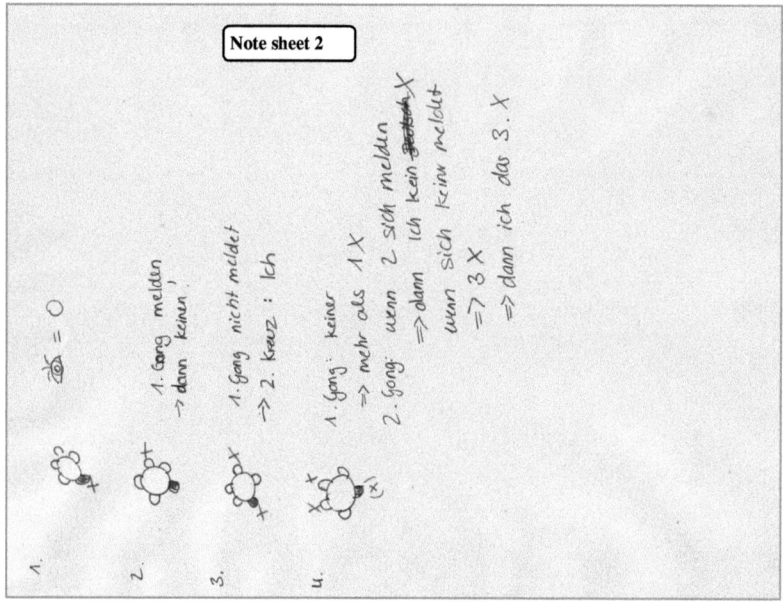

The data was gathered by Angelika Bikner-Ahsbahs, Julia Cramer, Thomas Janßen, and Jakob Priwitzer in the project „Effective knowledge construction in interest-dense situations", funded by the German Israeli foundation for Scientific Research and Development (Grant No. 946-357.4/2006).

E Example of a Multimodal SRA

The following figure presents the multimodal sequential reconstructive analysis of the first part of the scene PA5e1.1, lines 49-54. The interpretation space for the verbal utterances and the interpretation of the gestures are given in chapter 5.2.3.

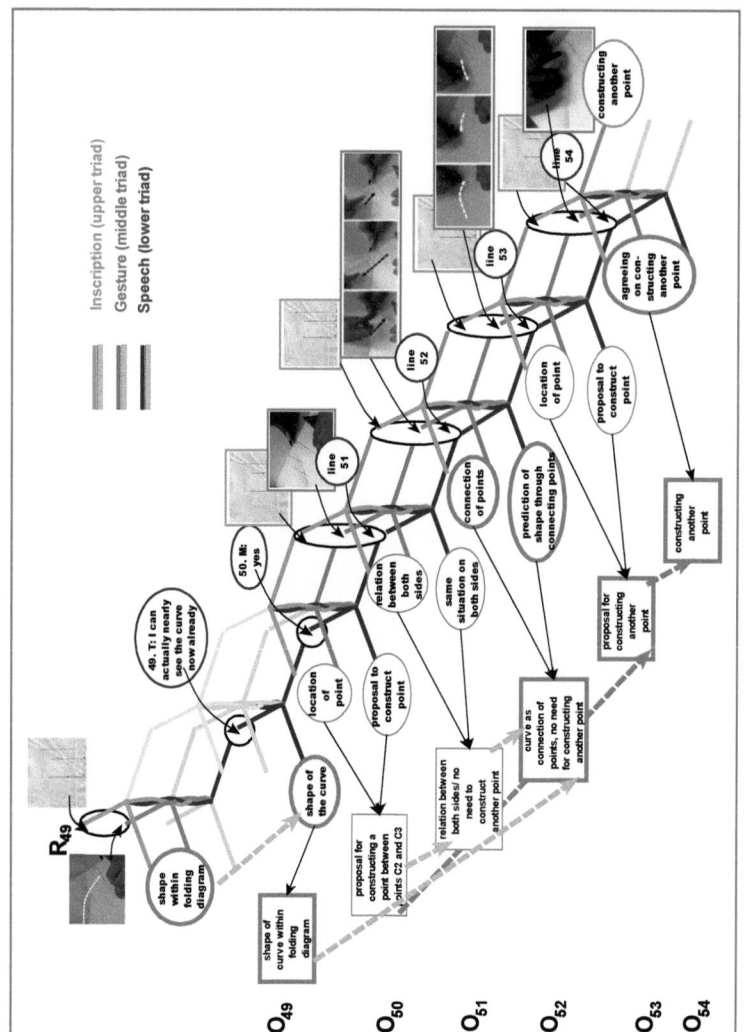

Fig. E.1: *Multimodal SRA for the Interaction in PA5e1.1, lines 49-54*

Table of Figures

Fig. 3.1:   A Peircean understanding of signs as a triadic relation between representamen (R), object (O), and interpretant (I) 20

Fig. 3.2:   Kendon's Continuum 27

Fig. 3.3:   Gesture space when sitting: In front of the body, framed by shoulders and the table.
a) preparation phase of a gesture, b) stroke of a gesture 27

Fig. 3.4:   Exemplified gesture unit, containing two gesture phrases 28

Fig. 3.5:   Overview on relations between gestures examined in this study 33

Fig. 4.1:   Multimodal sign shaped by the synchronous relationship between the three semiotic sets speech, gesture, and inscription. The 'semiotic composition' forms the representamen of the multimodal sign. 50

Fig. 5.1:   'Larger whole' in which the functions of gestures are investigated 54

Fig. 5.2:   The Idea of the first step of the semiotic sequence analysis to reconstruct meaning as developing in social interaction (following Bikner-Ahsbahs 2005, pp. 197-198, p. 208) 60

Fig. 5.3:   Tim and Mike's folding sheet after constructing three points (M is the point of reference for performing the folding) 62

Fig. 5.4:   Semiotic reconstructive analysis of the first part of the **first** scene in the elaboration of the parabola-task by the students Mike (M) and Tim (T): A first suggestion on the shape of the curve that becomes visible in the folding product is stated and shall be tested by constructing an additional point. 65

Fig. 5.5:   Semiotic reconstructive analysis of the first part of the **second** scene in the elaboration of the parabola-task by the students Mike and Tim:

	The additional point confirms the first hypothesis and the suggested curve becomes fixed within the folding sheet. 66
Fig. 5.6:	A three-component-model for analyzing the meaning of gestures within the semiotic bundle 68
Fig. 5.7:	Tim's gesture in anticipation to his utterance in line 49 69
Fig. 5.8:	Components influencing the interpretation space for disclosing the students' interpretation of the gesture 49 69
Fig. 5.9:	Gesture accompanying the utterance in line 51 70
Fig. 5.10:	Components influencing the interpretation space for disclosing the students' interpretation of the gesture 51 70
Fig. 5. 11:	Gesture co-timed to "exactly the same" as part of Mike's utterance in line 52 71
Fig. 5.12:	Components influencing the interpretation space for disclosing the students' interpretation of the gesture 52 71
Fig. 5.13:	Tim's gesture starting while Mike still speaks (52/53) 72
Fig. 5.14:	Components influencing the interpretation space for disclosing the students' interpretation of the gesture 52/53 72
Fig. 5.15:	Gesture accompanying Mike's proposal in line 54 73
Fig. 5.16:	Components influencing the interpretation space for disclosing the students' interpretation of the gesture 54 73
Fig. 5.17:	Tim's gesture, performed from left to right: "shall I now try to draw this curve approximately' " (55) 74
Fig.5.18:	Components influencing the interpretation space for disclosing the students' interpretation of the gesture 55 75
Fig. 5.19:	Gesture sketched from right to left in line 58 75
Fig. 5.20:	Components influencing the interpretation space for disclosing the students' interpretation of the gesture 58 75

Table of Figures

Fig. 5.21: Schematic representation of the SRA (left hand side) in comparison to the schematic representation of its refined, multimodal version (right hand side) 77

Fig. 5.22: Multimodal SRA of the first utterance of scene PA5e1.1 78

Fig. 6.1: Initial situation given as example for following the folding instructions 92

Fig. 6.2: Possible outcome of the construction of the folding sheet following the instructions 92

Fig. 6.3: Possible situation from the GeoGebra environment, fixed as printout.
Left side: Without trace;
Right side: Two traces produced for two different values of e 93

Fig. 6.4: Horizontal and vertical axis of analysis 98

Fig. 6.5: Example of an excerpt of a condensed process diagram 100

Fig. 6.6: Example of determining an 'epistemic-dense episode' (PA7e2) 102

Fig. 6.7: Condensed diagrams of epistemic-dense episode; episode CF7e10, scenes 10.1 and 10.2 106

Fig. 6.8: Example of a table of an EDE that served for the simultaneous representation of the outcomes of different steps of the analysis. Column 5, as well as the numbers noted in column 3 concern the results that are described in chapter 7. The green arrow next to utterance 51 as noted in column 2 denotes the holding of a gesture. The arrows in columns 4 and 5 denote possible relations between the codings as they arose in the course of the analysis 107

Fig. 7.1: Folding sheet of Mike and Tim as visible in the beginning of PA5e5-49 (represented with higher contrast to provide better visibility to the folding lines; tagging of points enlarged) 114

Fig. 7.2: Folding diagram in the beginning of scene 2 in PA5e1, 66-68
(a) the students' original diagram represented with tagging of point M enlarged

	(b) perspective used in the quoting of the gestures in the transcript (inscribed curve highlighted to provide better visibility)	116
Fig. 7.3:	Idealized representation of the case "five consultants, two have a marked hat"	118
Fig. 7.4:	Printout with auxiliary line fixed perpendicular to the y-axis through point A, in the beginning of PA7-849 (a) Students' original represented with higher contrast and tagging of points A, B, and P enlarged. Auxiliary line is red, perpendicular to the y-axis through point A. (b) perspective used in the quoting of the gestures in the transcript (color added to optimize visibility: red auxiliary line; white, dotted line segment [AB])	122
Fig. 7.5:	Relation between specifying (dark) and non-specifying (light) illustrators (in %)	126
Fig. 7.6:	Amount of specifying-gestures specifying a certain number of aspects (in %)	126
Fig. 7.7:	Guideline for the coding of specifying-functions of gestures	128
Fig. 7.8:	The situation after having produced the traces in the GeoGebra-diagram represented in a part of the split-screen. Left side: left front perspective. Right side: screen	129
Fig. 7.9:	Cutout of GeoGebra-diagram as visible on the screen. The arrows indicate the segments with length e	129
Fig. 7.10:	Rosa's inscription representing the case of five consultants when everybody wears a marked hat (inscription on the right side: "4th ring: nobody")	133
Fig. 7.11:	Shift of reference from the concrete inscription into the gesture space	139
Fig. 7.12:	Folding diagram as elaborated in line 65	143
Fig. 7.13:	Folding sheet as constructed just before the curve is first suggested to be seen in line 58 (tag for point M enlarged)	146

Fig. 7.14:	Integration of the within-functions as influencing the shaping of the immediate object	158
Fig. 8.1:	Printout diagram in the beginning of scene PA5e7.3, line 494 (tags of points enlarged)	163
Fig. 8.2:	List of x- and y-coordinates of points on the parabola, gathered for different values for e (PA7e8-667)	170
Fig. 8.3:	GeoGebra-diagram as visible on the screen in the beginning of PA6e7-268	174
Fig. 8.4:	Cutout of the table in subtask 2.1 as filled out by the students in CF7e9-296	177
Fig. 8.5:	Location and movement of the pointing gestures in lines 299 and 301 (a) **Pointing gesture in line 299:** "one always has this one] [beneath the fraction bar.]" (b) **Pointing gesture in line 301:** "[that one then] [beneath the fraction bar.]]"	178
Fig. 8.6:	Folding sheet (a) and GeoGebra situation (b) as present in the beginning of excerpt 3.2a, line 399 (a) points on the lower edge tagged with (from left to right): G, H, E, F with corresponding constructed points G', H', E', F' marked with a red cross	182
Fig. 8.7:	Interviewer's fixed transformation of a general element $f(x) = \frac{a}{b}$ (note sheet 4)	186
Fig. 8.8:	Lisa's and Rosa's notation of the pattern concerning numerator and denominator as fixed in subtask 2.2 (CF)	186
Fig. 8.9:	Visualization of two consecutive elements of the sequence, where $2 + \frac{1}{a}$ is the successor of a general fraction $f(x) = \frac{a}{b}$ (note sheet 5)	187
Fig. 8.10:	Table on work sheet 2 as filled out by the students in the beginning of CF7e18-1063	187
Fig. 8.11:	Detail of the table on work sheet 2	189

Fig. 8.12:	First four elements of the sequence $(f(x))_{x \in \mathbb{N}}$ defined by the given continued fraction, fixed on work sheet 1	192
Fig. 8.13:	Indications in CF7e1-5: (a) "[[the one]]", (b) "[that stands down there]"	193
Fig. 8.14:	GeoGebra diagram in the beginning of PA5e3 (trace thickened to optimize visibility)	193
Fig. 8.15:	Indications in CF7e2-25: (a) "[this thing there]" (25b), (b)"[that]" (25c)	201
Fig. 8.16:	Pointing (red arrow) and tracing (light blue, dotted arrow) (a) "[one plus two divided by one plus two" (b) "two thirds]"	201
Fig. 8.17:	Lisa's and Rosa's solution for subtask 1.2 of the CF-task. The values considered for x are noted on the right hand side as "$1 \leq x < \infty$" and below, more specifically, "only integer numbers"	208
Fig. 8.18:	Hands shaped in line 183	212
Fig. 8.19:	Indication simultaneous to "[because [this] thing]" (185)	213
Fig. 8.20:	Shape and position of hands in the second part of line 185, similar to the shape in line 183	213
Fig. 8.21:	Movement of the right hand co-timed to "[becomes only one ,one more]" (185/187)	213
Fig. 8.22:	Virtual indication of "[this one]]" (187)	214
Fig. 8.23:	Indication co-timed to "one two up there]" (189)	214
Fig. 8.24:	(a-e) Gesture co-timed to "[[and this two up there] [we always divide by that before-]]" (189)	215
Fig. 8.25:	a) "[[,and then that becomes]" b) "[,also always- one more]]" (189)	216
Fig. 8.26:	Metaphorical representation of the tree-metaphor ("[[every tree] is in [itself a tree]]", 193)	217

Fig. 8.27:	Position of the hands in the beginning of the gesture co-timed to "[is again a thousand times-] ,in itself again.]" (193/195)	217
Fig. 8.28:	Movement of the right hand co-timed to "[is again a thousand times-] ,in itself again.]" (193/195)	218
Fig. 8.29:	Condensed process diagram of Scene CF7e5.1	223
Fig. 9.1:	*Horizontal Comparison:* Percentage of gestures that specify one (1), two (2), three (3), or four (4) aspects	229
Fig. 9.2:	*Horizontal Comparison:* Amount of specifying-gestures specifying a certain aspect on a certain referential level (percentages given in the table below)	229
Fig. 9.3:	*Horizontal comparison:* Amount of specifying-gestures on each of the three referential levels (in %)	230
Fig. 9.4:	*Vertical comparison:* Amount of specifying-gestures on each of the three levels (in %)	232
Fig. 9.5:	*Vertical comparison:* Percentage of gestures that specify one (1), two (2), three (3), or four (4) aspects	233
Fig. 9.6:	*Vertical comparison:* Amount of specifying-gestures specifying a certain aspect on a certain referential level (percentage given in the table below)	233
Fig. 9.7:	Lisa and Rosa's notation of the pattern concerning numerator and denominator as fixed for subtask 2.2 (CF7)	235
Fig. 9.8:	Visualization of two consecutive elements of the sequence, where $2 + \frac{1}{a}$ is the successor of a general fraction $f(x) = \frac{a}{b}$	236
Fig. 9.9:	Components influencing the interpretation space for disclosing the students' interpretation of the gesture CF7-908	236
Fig. 9.10:	Components influencing the interpretation space for disclosing the students' interpretation of the gesture CF7-914	237
Fig. 9.11:	Gesture-connecting action leading into structure seeing	240

Fig. 9.12: Condensed process diagram of the epistemic-dense episodes (PA7) 240

Fig. 9.13: Condensed process diagram of the epistemic-dense episodes (CF7).....241

Fig. 9.14: Condensed process diagram of the epistemic-dense episodes (IN7) 242

Fig. B.1: Cutout of the printout diagram given to the students, segments added in green 290

Fig. B.2: Cutout of the printout diagram given to the students, points C and D, and segments $[AP]$ and $[AB]$ added..291

Fig. B.3: Printout diagram with straight line through points A and F added........293

Fig. B.4: Printout diagram modified to apply the triangle altitude theorem 294

Fig. B.5: Printout diagram modified to make visible the similarity of the triangles $\triangle CAG$ and $\triangle GCP$...295

Fig. B.6: Strategies to determine one element f(x+1) from the previous one, f(x)..299

Fig. C.1: Condensed process diagram of the EDEs of PA7 (sorted with respect to subtasks)..309

Fig. C.2: Condensed process diagrams of the EDEs of PA7 (sorted with respect to subtasks)..310

Fig. C.3: Condensed process diagrams of the EDEs of PA6 (sorted with respect to subtasks)..310

Fig. C.4: Condensed process diagrams of the EDEs of CF7................................311

Fig. C.5: Condensed process diagrams of the EDEs of CF7................................311

Fig. E.1: Multimodal SRA for the Interaction in PA5e1.1, lines 49-54 339

Table Directory

Table 6.1:	Transcription key	96
Table 6.2:	Signs representing the epistemic actions accomplished	99
Table 6.3:	Signs representing other actions accomplished with respect to the working process	100
Table 6.4:	Pictograms marking the use of non-verbal and non-gestural semiotic resources	103
Table 7.1:	Development of the associated sign referring to 'shape of the curve' in PA5	143
Table 7.2:	Use of the associated sign to elaborate the immediate mathematical object	145
Table 7.3:	Development of the associated sign referring to 'shape of the curve' and linking it to the associated sign 'axis of reflection' in PA7	149
Table 7.4:	Associated sign referring to 'increasing numbers' in CF7	156
Table 8.1:	Comparison of extracting-gestures in both excerpts	179
Table 8.2:	Development and use of the associated gesture 'grasping hand'	219
Table 9.1:	Overview on the epistemic-dense episodes identified within the three PA-data sets	228
Table 9.2:	Overview on the frequencies of gestures in the epistemic-dense episodes of the PA-data sets	228
Table 9.3:	Overview on the epistemic-dense episodes identified within the three data sets concerning Rosa and Lisa's elaborations of the three tasks	231

Table 9.4: Overview on the specifying-gestures in Rosa and Lisa's epistemic-dense episodes 232

Table B.1: Correspondence between entities represented in the GeoGebra environment and the folding sheet 289

Table B.2: Function equations for different values of e 292

Springer Spektrum

springer-spektrum.de

Springer Spektrum Research
Forschung, die sich sehen lässt

Ausgezeichnete Wissenschaft

Werden Sie AutorIn!

Sie möchten die Ergebnisse Ihrer Forschung in Buchform veröffentlichen?

Seien Sie es sich wert. Publizieren Sie Ihre Forschungsergebnisse bei Springer Spektrum, dem führenden Verlag für klassische und digitale Lehr- und Fachmedien im Bereich Naturwissenschaft I Mathematik im deutschsprachigen Raum.

Unser Programm Springer Spektrum Research steht für exzellente Abschlussarbeiten sowie ausgezeichnete Dissertationen und Habilitationsschriften rund um die Themen Astronomie, Biologie, Chemie, Geowissenschaften, Mathematik und Physik.

Renommierte HerausgeberInnen namhafter Schriftenreihen bürgen für die Qualität unserer Publikationen. Profitieren Sie von der Reputation eines ausgezeichneten Verlagsprogramms und nutzen Sie die Vertriebsleistungen einer internationalen Verlagsgruppe für Wissenschafts- und Fachliteratur.

Ihre Vorteile:

Lektorat:
- Auswahl und Begutachtung der Manuskripte
- Beratung in Fragen der Textgestaltung
- Sorgfältige Durchsicht vor Drucklegung
- Beratung bei Titelformulierung und Umschlagtexten

Marketing:
- Modernes und markantes Layout
- E-Mail Newsletter, Flyer, Kataloge, Rezensionsversand, Präsenz des Verlags auf Tagungen
- Digital Visibility, hohe Zugriffszahlen und E-Book Verfügbarkeit weltweit

Herstellung und Vertrieb:
- Kurze Produktionszyklen
- Integration Ihres Werkes in SpringerLink
- Datenaufbereitung für alle digitalen Vertriebswege von Springer Science+Business Media

Sie möchten mehr über Ihre Publikation bei Springer Spektrum Research wissen? Kontaktieren Sie uns.

Marta Schmidt
Springer Spektrum | Springer Fachmedien
Wiesbaden GmbH
Lektorin Research
Tel. +49 (0)611.7878-237
marta.schmidt@springer.com

Springer Spektrum I Springer Fachmedien Wiesbaden GmbH

MIX
Papier aus verantwortungsvollen Quellen
Paper from responsible sources
FSC® C105338
www.fsc.org

If you have any concerns about our products,
you can contact us on
ProductSafety@springernature.com

In case Publisher is established outside the EU,
the EU authorized representative is:
Springer Nature Customer Service Center GmbH
Europaplatz 3, 69115 Heidelberg, Germany

Printed by Libri Plureos GmbH
in Hamburg, Germany